THE FOOD CRISIS IN PREHISTORY

THE FOOD CRISIS IN PREHISTORY

Overpopulation and the Origins of Agriculture

MARK NATHAN COHEN

New Haven and London, Yale University Press

Designed by Sally Sullivan and set in IBM Press Roman type. Printed in the United States of America by The Alpine Press, South Braintree, Massachusetts.

Library of Congress Cataloging in Publication Data

Cohen, Mark Nathan.

The food crisis in prehistory.
Bibliography: p.
Includes index.
1. Man, Prehistoric–Food. 2. Man, Prehistoric–Population. 3. Agriculture–Origin. 4. Food supply–History. I. Title.
GN799.F6C64   338.1'9'3   76-41858
ISBN 0-300-02016-3 (cloth); 0-300-02351-0 (paper).

12   11   10   9   8   7   6   5   4   3

CONTENTS

PREFACE

Before proceeding, let me say a few words about the use of archaeological data in this book. The arguments presented here are based on one major premise: I believe that the events leading up to the emergence of agriculture in various regions of the world demonstrate remarkable parallelism, and I believe that this parallelism not only permits but demands that some common underlying force or factor be found operating in all world regions, not necessarily to the exclusion of local variables, but in conjunction with those variables. The major thrust of this book is to demonstrate the similarity of events in different world regions while at the same time demonstrating that these events are plausibly linked with population pressure.

This goal forces me to synthesize the archaeological data at a very generalized level—a level that is foreign to recent archaeological work and one that tends to be frowned upon. It is currently fashionable in our science to place heavy emphasis on the uniqueness of local events and to see systematic feedback between cultural and natural variables only on the local level. At present, the broadest syntheses which even attempt to provide a professional documentation of sources are continental in scope. These are works such as Willey's (1966, 1971) summaries of North and South American prehistory. Broader outlines such as Clark's (1969) *World Prehistory* or Fagan's (1974) *Men of the Earth* tend to be more impressionistic in outline and more selective, as well as more popularly oriented, in their bibliographies. Moreover, these broad synthetic treatments of archaeological data are intended primarily as textbooks. Broad syntheses aimed at making a theoretical point to a professional or sophisticated student audience are very rare. My aim, however, is to demonstrate that the data take on different appearance (and, I believe, are more accurately interpreted) when viewed at a greater distance. The resulting level of synthesis and abstraction will be uncomfortable to many scholars, but I believe that it is justified not only by the scope of the hypothesis presented but also by the very nature of archaeological data. Our data are *samples* of the residues left by prehistoric human behavior and as such are subject to a number of sources of bias and error. Archaeological samples are extremely small in proportion to the universe from which they come. Moreover, they are subject to a good deal of random and nonrandom disturbance. Recognition of these problems tends to make archaeologists conservative. Although caution is warranted, the *kind* of conservatism which we choose is often counterproductive. We tend to turn our attention away from general patterns to local sequences. But, as statistical samples get poorer, it is detail that is lost first, not overall patterns. The poorer the samples, the more we should direct ourselves to discernible broad generalizations that emerge despite sampling error rather than to local detail. I believe that a relatively high level of abstraction and generalization is a strong, rather than a weak, use of archaeological materials.

Such a broad focus is, of course, a good protection only against random error or local biases in samples or their interpretation: if we are dealing with systematic biases in the data, then broadened perspectives provide no safety. There are in fact a number of ways in which the broad conclusions described below are affected by *systematically* biased preservation. In those cases careful consideration will have to be given to determine whether a certain pattern could result from poor preservation alone or whether it is in fact a valid historical pattern. The choice is not always clear.

In order to obtain the type of synthesis that I am seeking, I have broken another of the rules of good scholarship by making heavy use of secondary or even tertiary sources rather than relying solely on primary materials. One reason for this is purely practical. If an enquiry is to be couched in worldwide terms, as I maintain that this one must be, then practical limits of time and expertise must temper my dealings with the primary literature. I could not possibly synthesize the prehistory of various world regions as efficiently or accurately as the experts in those areas. If I am to work with broad regional patterns I must deal with the literature which describes those patterns. This use of secondary sources, however, also acts as an important check on my own imagination, and it is an important guarantee to the reader. It is one thing for me to see similar patterns everywhere and it is quite another to demonstrate that experts in various world regions see and describe similar patterns. These experts of course have their own biases, but as an aggregate they are presumably neutral on the particular questions raised in this book.

There is still the possibility that the patterns which I have collected and which are documented in the variety of regional studies cited are similar because all of the scholars in question share certain biases. I have tried to counter this by getting a reasonable sample of scholars from major world regions. Nonetheless, even though the individuals referred to have differing positions within the mainstream of anthropology, it may well be that we have created certain patterns out of the data because that whole mainstream is biased in certain directions. Perhaps archaeologists find what they expect to find; and what they expect to find is what they have seen reported from other regions. Kraybill (in press) has suggested in effect that the interpretation of stone tool functions is systematically biased against the recognition of tools for vegetable processing because of the masculine/hunting orientation of most scholars in the field. And various colleagues, commenting on my book in the manuscript stage, have questioned whether some of the pattern which I observe in the literature might not result from just such broad-based biases in reporting and interpretation. In a number of instances below, I have attempted to meet such questions head on, but in other, more subtle, but pervasive ways, such biases probably remain. Some of these undoubtedly are my own; a few may well have broad subscription among professionals in the field. Perhaps the crystallizing of these prejudices will help us to reconsider some of our interpretive trends.

One final point should be made. The intent of this book is to present a hypothesis in orderly form; to demonstrate that the hypothesis fits reasonably well our accumulated observations both from ethnography and archaeology; and to suggest that the hypothesis explains a number of otherwise anomalous observations. My intent is to show that this one hypothesis *could* be used to account for certain observed patterns. What follows is not intended to be a comprehensive, or even a balanced, review of world prehistory. It is intended to be a defense of a thesis, and as such it is intentionally argumentative. It has been my observation that simple hypotheses boldly defended are often the best teaching tools and the best spurs to research.

Acknowledgements

I would like to express my thanks to Michael Harner, Morton Fried, and Edward Lanning, who started me thinking about certain issues; to Richard Robbins, Gordon Pollard, and Shirley Gorenstein, who provided criticism, advice, and moral support; and to my family, friends, and neighbors, who simply had to go on living in the vicinity while I was writing. Most important, I would like to thank Thomas Lynch for his careful and constructive criticism of this manuscript, and Fekri Hassan, who, although he probably won't like the book, has nonetheless helped to improve it considerably by his criticism of some of my arguments.

Preface to the Third Printing

This printing contains one substantive change from the text of previous printings. In returning to my original sources I found that I had misrepresented some of the facts, though not the spirit, of two publications concerning the economic behavior of the natives of Arnhem Land, Australia (McCarthy and McArthur 1960; McArthur 1960.) For this printing I have rewritten the first paragraph of page 31, which deals with these sources, in an attempt to treat them with greater accuracy.

1 THE PROBLEM OF AGRICULTURAL ORIGINS

"Cultural man has been on earth for some 2,000,000 years; for over 99% of this period he has lived as a hunter-gatherer. Only in the last 10,000 years has man begun to domesticate plants and animals. . . . *Homo sapiens* assumed an essentially modern form at least 50,000 years before he managed to do anything about improving his means of production. . . . To date, the hunting way of life has been the most successful and persistent adaptation man has ever achieved" (Lee and DeVore 1968:3).

This statement by Richard Lee and Irven DeVore introducing a symposium on hunting and gathering populations raises a serious question about the nature of culture change, and in particular about the episode of economic and social change commonly referred to as the "Neolithic Revolution," which witnessed the transition from hunting and gathering to agriculture as man's major mode of subsistence. Animal populations and human populations as well are characteristically extremely conservative in maintaining adaptive postures and structures which have proved successful. A long and well-studied paleontological record suggests that in the biological world change does not occur for its own sake (Romer 1933) but rather results from altered selective pressures which necessitate adjustive modifications on the part of the population involved. If, as the archaeological record indicates, hunting and gathering was such a successful mode of adaptation over such a long period of time, and if most human populations are as conservative as anthropologists have observed them to be, we are faced with answering the question why this form of adaptation was ever abandoned.

The problem of explaining the origins of agriculture has been a major focus of anthropological enquiry throughout the history of the discipline. Through much of this history, however, our interpretations were constricted by the limited development of some of the supporting sciences, by limited anthropological use of these sciences, and by a number of theoretical biases deeply rooted in the history of our own

1

discipline. These biases can, I believe, be traced primarily to two sources.

The first of these major influences was Thomas Malthus, the early nineteenth-century economist who, in a series of essays on population, developed a model of population growth which has come to dominate most thinking in the social and biological sciences. His model stressed the essentially dependent nature of population as a variable responding to limits set for it by the available food supply, while tending to ignore the possible importance of population growth as a stimulus to technological change and increased food production.

The second major influence, which is harder to trace to its origins but which can be seen clearly developing in the works of the nineteenth-century evolutionary anthropologists, is the tendency to classify human cultures into sharply delimited sets representing evolutionary stages, and furthermore to define these sets or stages primarily on the basis of technology: the type of tools manufactured or, later, the methods employed in the food quest. Thus Lewis Henry Morgan (1877), in the comprehensive statement on the nature of cultural evolution which has become the model for most subsequent work, suggests that the process of culture change was primarily one of accumulating technological capacities and that the level of technological development achieved was the primary determinant of the level of the culture system as a whole. Moreover, his history of human culture describes a succession of well-defined and relatively static stages separated from one another by fairly well-defined, abrupt transitions.

The combination of these two influences led anthropology toward a one-sided view of the relationship between technology and population growth. Technology has been considered an independent variable; technological changes, often visualized as occurring in rapid, revolutionary spurts, have been seen to modify the carrying capacity of the environment; population, in this view, simply adjusts to the new limits in a Malthusian sense without having any significant effect on technological change. Thus, the history of population growth has been assumed to have a stepped pattern in which periods of stable equilibrium alternated with periods of rapid population growth.

Perhaps the most explicit modern statement of these two themes and the resulting model of population growth is found in the work of V. Gordon Childe (1951), which relates the development of European civilization to a series of technological revolutions (for example, the "Neolithic" or "food-producing" revolution) involving relatively short-term massive reorganization of technology which resulted in periods of rapid population growth and reorganization of social institutions.

This background has tended to skew theories and research on agricultural origins in certain directions. One approach of long standing was to view the origins of agriculture as part of the natural process of cultural evolution. There was a tendency to perceive agriculture as a new conceptual level dependent primarily on the achievement of sufficient knowledge and sophistication. A good deal of attention was paid to tracing the history of the evolutionary patterns and identifying the time, place, and probable circumstances in which agriculture first emerged. But little attention was paid to *why* agriculture was adopted as an economic strategy in preference to hunting and gathering. Agriculture was seen as offering such significant and obvious economic advantages to human populations that once the appropriate level of knowledge was achieved acceptance of the new economy would be axiomatic.

A second common approach was to assume that agriculture emerged in response to some stress or disequilibrium which upset the traditional hunting and gathering subsistence patterns. Because of the prevailing Malthusian assumptions, however, it was thought that this stress would have to originate outside the human realm, in some sort of environmental change or aberration. Because of the obvious temporal correlation, climatic events associated with the end of the Pleistocene typically loomed large in such explanations. However, because of the limited development of the associated sciences, until recently such events were dimly and often erroneously perceived.

In recent years there have been a number of major practical and theoretical advances in our study of agricultural origins. One of the most important practical advances has been the use of pollen analysis in the diagnosis of climate and other environmental variables and the study of economic change (Faegri and Iverson 1964). But similar advances have occurred and are occurring in the treatment of macro-organic remains and of inorganic aspects of the environment (Butzer 1971). In the realm of theory, the breakthroughs that might be considered most important have occurred outside archaeology. The work of Richard Lee and others (see chapter 2) has helped to break up old stereotypes of hunting and gathering groups while providing an accurate picture of hunting and gathering economies against which the emergence of agriculture can be measured. In the field of archaeology itself, the work of Robert Braidwood and his associates (various publications) can be said to represent the beginning of the modern era of research. Braidwood and his colleagues achieved great sophistication in defining the ecology of early agriculture and the consequences of the new economy. But, more important for present purposes, they also introduced the modern style of interdisciplinary studies focusing multifaceted scientific research on the problem. Although his work was not directed primarily toward early agriculture, J. G. D. Clark (1952, 1954) also played a major part in refining the concept of economic interpretation in archaeology.

More recently, the work of Coe and Flannery (1964, 1967) and of MacNeish (1958, Byers [ed.] 1967) has led us to view early agricultural development in a framework of regional geography and evolving human settlement patterns rather than concentrating on locating individual sites. The "agricultural revolution" is now seen as involving progressive changes in the distribution of human populations over a landscape of varied microenvironments. At the same time, a series of recent contributions, particularly those of Flannery (1965, 1968, 1969) and of Binford (1968), have forced us to deal with agricultural origins as a process rather than an event. We have been led to view the emergence of agriculture as a modification of systematic behavior patterns linking man and his resources and to evaluate the factors which interact to modify these systems. It is no longer sufficient to show where or when agriculture was discovered; it is essential to explain how and why human behavior patterns were modified. Most significant, both Flannery (1968) and Binford (1968) have reintroduced the notion of stress or disequilibrium as a causative factor and applied this concept within a well-defined framework of interacting variables.

In what is probably the best theoretical analysis of emerging agriculture yet offered, Flannery (1968) demonstrated how a complex but stable system of interaction between man and his resources might be modified in the direction of agriculture in response to the availability of mutant forms of wild maize. After describing the various resources available for human exploitation and analyzing the behavior patterns involved in their use, Flannery argued that mutant maize forms would encourage man to invest more of his time in the production and harvest of maize at the expense of other resources. By placing the beginning of agriculture in a context of competitive economic strategies, Flannery provided a realistic model of the behavioral patterns and cultural choices involved in this new technology. Because the various behavior patterns are described as systematically related, we are able to reconstruct the potential impact of various external forces and the potential importance of various cultural choices.

However, if our knowledge of the change process has been greatly advanced in recent years, the explanations put forward to account for stress and provide the motivation for change are still, I think, largely inadequate. The failure is due primarily to our failure to view the problem of agricultural origins in broad enough temporal and geographic perspective. Probably in response to the excessive zeal of early diffusionists who explained the arrival of agriculture in one region largely on the basis of the dissemination of ideas from another, modern archaeologists have stubbornly insisted on studying the causes of agriculture within the limits of local archaeological sequences and local variables. *The most striking fact about early agriculture, however, is precisely that it is such a universal event.* Slightly more than 10,000 years ago, virtually all men lived on wild foods. By 2,000 years ago the overwhelming majority of people lived by farming. In the four million year history of *Homo sapiens,* the spread of agriculture was accomplished in about 8,000 years. As Charles Reed (in press-b), pointed out at a recent symposium on agricultural origins, the problem is not just to account for the beginnings of agriculture, but to account for the fact that so many human populations made this economic transition in so short a time. This new perspective suggests a subtle shift in the focus of our enquiries. Until recently prehistorians have made an artificial distinction between the "invention" or first appearance of agriculture in a region and its spread as an economic system. Students of agricultural origins have tended to focus only on "primary hearths," such as the Middle East with its unique problems. Other scholars have analyzed the spread of agriculture and noted why, for example, parts of North America, but not all of the continent, accepted the new economy. There has been a failure to recognize that we are in fact talking about one common problem: why human populations chose agriculture as a strategy over hunting and gathering and why so many of the world's people chose it during one brief time span, regardless of whether they "invented" or "learned" the concept. What we need, then, is a theory which can account for the widespread similarity of culture changes occurring in a variety of environments and which can account for the approximately synchronous nature of these changes. The theory should account not only for the "invention" of agriculture but also for its acceptance and the widespread economic transformation of human society which resulted. It is the latter rather than the invention per se which is the important historical event. Ideally, the theory should also explain why agriculture emerged in the particular time span it did.

Flannery's (1968) analysis errs, I believe, in its emphasis on mutation as the factor promoting disequilibrium and adjustment. Mutations are random but repetitive aspects of the environment. Mutant maize would have been available on any number of occasions prior to the time when man first expanded his use of this resource. Clearly, we must account for the fact that the human population at a certain time responded significantly to a mutation which had previously been ignored. It is the human response, not the environment, which has changed. A further problem with Flannery's emphasis on the importance of mutations is that he fails to account for the fact that people in various parts of the world were apparently simultaneously undergoing processes very similar to the ones he described. Some of these were dealing with other maize populations with mutation patterns of their own. Others were dealing with totally different plant species.

Binford (1968) argues that the basic stimulus to agriculture was population "pressure" or the development of population density in excess of what could be supported in any region by hunting and gathering. To this extent, of course, I believe that he is essentially correct. Binford, however, worked within some of the theoretical constraints of traditional anthropology, and he did not believe that overpopulation or population pressure could develop as a matter of course in any region. He was forced to construct a complicated and, I believe, untenable hypothesis to account for the buildup of population pressure and to view this pressure as emerging only in selected locations. Briefly, Binford argued that rising post-Pleistocene sea levels created a favorable coastal habitat for sedentary human populations, which then grew rapidly, shunting off daughter populations toward the interior where they intruded on the territories of inland hunting and gathering groups. These areas of intrusion were the zones of overpopulation in which domestication first occurred. The concept of specialized zones of population pressure is an interesting one which may have considerable significance for microanalysis of emergent agriculture within particular regions. Flannery (1969) has even suggested that something like Binford's model would work in the Middle East. However, I think that this specific focus significantly underplays the extent to which the emergence of agriculture and the population buildup preceding it were general phenomena.

More important objections can be raised to Binford's use of sea-level changes as the factor initiating the process. There is a certain attractiveness to his hypothesis because rising sea level is one environmental change which has a worldwide distribution. On the other hand, coastlines are irregular. Rising sea level would alter the position, configuration, and ecology of the coast, but it would affect each segment of coast in a way peculiar to its own particular configuration. The argument that sea-level change would have a general tendency to improve the coastal habitat or promote parallel sedentary coastal adaptations throughout the world is untenable. Moreover, fluctuating sea level has been a characteristic of coastal environments throughout human evolution; human populations have presumably encountered a number of such changes. It is not clear why only the post-Pleistocene rise in sea level would have triggered this response.

In addition, it should be pointed out that Binford's hypothesis lacks general support in the archaeological record. For one thing, in many of the world's early agricultural centers there is no evidence of the kind of inland migration patterns that he suggests (see, for example, Meyers 1971). More important, there is no reason to believe that the kind of settled, intensely exploitative populations that he describes were wholly or even primarily confined to seacoasts. In fact, as I hope to show, the post-Pleistocene coastal exploitation seems to be only one facet of a much more general phenomenon. Finally, the hypothesis places a heavy emphasis on the post-Pleistocene timing of the events leading to agriculture, while in fact, the significant patterns (including coastal exploitation) were emerging well before the period to which Binford refers.

Similar problems are encountered in attempting to deal with any of a variety of climatic explanations for agriculture. Childe (1951), for example, argued that agriculture occurred in the Middle East when man and his potential domesticates were forced into intimate and mutually dependent relations during a drought which was presumed to have occurred at the end of the Pleistocene. Modern research has cast doubt on the postulated drought and Childe's theory is generally discredited (G. Wright 1971; Binford 1968; Braidwood and Howe 1960; Braidwood 1967), but in any case the hypothesis could be of only local significance. Similarly H. E. Wright (in press) has suggested that agriculture might have begun in response to the appearance of large dense stands of wild cereals in the Middle East a little over 10,000 years ago. This ecological change, which is demonstrable in pollen profiles from the region, may have considerable significance for the local sequence, but it cannot explain the origins of agriculture. Wheat, even if newly arrived in the Middle East, was certainly in existence elsewhere prior to this time. We are still lacking an explanation of why this resource was exploited by man (anywhere) only in the very recent past. In general, all climate-based explanations suffer from two problems. First, climate phenomena are reversible and repetitive; they cannot account for a single occurrence of an event or process that has demonstrated very little tendency for reversal. Second, climate changes are by their nature regional in scope and often of opposing direction in adjoining regions. Hence, they are inherently incapable of explaining parallel economic trends over a broad geographic region.

There are a number of other approaches to the problem of agricultural origins which have a potentially wider application because they are not so reliant on factors in the local environment. These are the hypotheses referred to above which explain agriculture primarily in terms of pre-existing cultural conditions. According to Braidwood (1960) the development of agriculture was based on increasingly intense use of local fauna and flora: the resulting familiarity with potential domesticates led ultimately to their manipulation. Braidwood would appear to be correct at least in a descriptive sense, not only for the Middle Eastern region in which he worked, but for other regions as well. One of the clearest trends in the archaeological record is the tendency of human populations to exploit intensively a variety of local fauna and, more particularly, flora just before they undertake domestication. Yet as an explanation of what occurred rather than as a description, Braidwood's account seems to fail. First, his explanation stresses emergent human knowledge or consciousness of the potential of domestication, when in fact there is good evidence that the principles of domestication are universally understood (see chapter 2). Man did not need education as much as he needed motivation. Moreover, Braidwood does not explain satisfactorily why this type of exploitation emerged when it did (or, as we now know to be true, why it emerged in so many places at the same time).

In a somewhat similar vein, Sauer (1952, 1958) and Watson and Watson (1969) have suggested that sedentism was necessary as a precondition for agriculture since it permitted not only familiarity and experimentation with local flora but also long-term observation and investment in vegetable resources. But this explanation, like Braidwood's, stresses knowledge rather than motivation. A further problem is that we now have evidence that sedentism is not in fact always a precondition of agricultural development, although it appears to be in many cases. More important, however, this model does not explain sedentism itself.

On a slightly different tack, one which focuses on the structure and habits of potential domesticates, Flannery (1965) has argued that the transport of favored plant species out of their customary environment to new regions would alter the selective pressures on the plant, resulting in the emergence of previously rare mutant types. This result could be accomplished either by the movement of human populations or simply by the transport of the favored species themselves through trade. Flannery was particularly interested in the latter alternative.

Lynch (1973) has expanded on Flannery's suggestion, noting that transhumant hunting and gathering might accomplish this result. He points out that transhumance involves tight schedules of seasonal movement which might force human groups to harvest certain resources either slightly before or, more important, slightly after their period of peak maturity. Lynch argues that the timing of the harvest might select for certain seed types. In the case of a late harvest, mutant forms which tended to retain their seeds instead of shattering would be selected, and this would be followed by immediate transport of the selected species to new regions along the route of migration.

Both suggestions again provide significant insight into the domestication *process*. Plant domestication is clearly linked archaeologically with movement of domestic forms out of their natural environment, and there is some evidence of both transhumance and early trade patterns associated with incipient domestication. However, we still lack an explanation of why these patterns themselves should emerge or why they should result in a general and synchronous worldwide adoption of an agricultural economy.

Another school of thought has postulated that the origins of agriculture might be found in mystical or ceremonial practices. E. Isaac (1970), for example, relates agriculture to the emergence of a new "world view" and new ceremonial practices which emerged at the end of the Upper Paleolithic. The latter hypothesis is, of course, largely untestable from archaeological remains; but, even if valid, it only describes the context in which agriculture developed rather than providing us with an explanation. In general, these models which explain agriculture in terms of other cultural variables add to our descriptive knowledge of the context in which agriculture appears, but they are not causal explanations. They do not account for the widespread acceptance of agriculture, since it is not yet clear why the cultural preconditions themselves would be achieved, much less why they would be achieved simultaneously on a worldwide basis; the postulation of previous cultural thresholds to explain the emergence of agriculture begs the question since we must simply seek further for the causes of these preceding cultural developments.

Perhaps the most powerful model yet put forward for explaining the emergence of agriculture is that offered by David Harris (in press). Harris's model deals effectively with some of the objections which have been raised concerning other theories although it too is subject to a number of theoretical and practical problems. He has expanded and generalized Binford's basic model, arguing that agriculture would emerge only under conditions of disequilibrium between population and resources. After eliminating or qualifying various other possible sources of disequilibrium he focuses on the growth of human population as a stress factor. He is restricted, however, by his assumption that human populations do not normally outgrow their resources. To resolve this dilemma, he argues that normal controls on population growth may become ineffective under conditions in which the mobility structure of a group is changing (when, in particular, group mobility is reduced and sedentism results). Changes in the environment are seen as inducing altered economic strategies which result in sedentism; sedentism in turn results in "abnormal" population growth or population pressure. Harris then considers a range of environmental and technological variables and develops a series of subsidiary models to demonstrate when population pressure would be likely to lead to domestication and when it would not. These models or "alternative pathways" prove to be an invaluable asset in understanding why agriculture developed in some regions slightly earlier or later than in others and why it never penetrated some areas at all (see chapter 7).

The basic weakness in Harris's model is that he has no satisfactory way of explaining why changes in group mobility patterns with such profound ecological effects occurred in so many parts of the world at roughly the same time. As noted above, Harris does call attention to the possible effects of environmental variables on group mobility and he notes the coincidence between post-Pleistocene climate change and the emergence of agriculture. But no convincing link is provided to connect post-Pleistocene climate change to sedentism or the emergence of agriculture on a broad geographic scale. Harris does, however, make one effective suggestion in this regard. He notes that the terminal Pleistocene extinction of migratory megafauna which is observable in many regions could have had a widespread tendency to alter economic strategies and promote sedentism and population growth. The suggestion is an attractive one, but it too is not without its difficulties. One problem, as Harris notes, is that the linkage in time (and space) between the extinctions and the emergence of sedentism is not as direct as one might wish. Also, this explanation begs the question of the causes of the terminal Pleistocene extinctions, a question which is very much subject to debate. Finally, like Binford's thesis, Harris's model focuses too heavily on specifically post-Pleistocene events and fails to recognize the relationship that exists between these events and patterns which are observable over a longer time. I would argue that it is more profitable to view the Pleistocene extinctions and the emergence of sedentism as symptoms of a common phenomenon (albeit interacting with one another) than to view one as the cause of the other.

A more satisfactory explanation of the widespread emergence of agriculture—and of the relationship between this process and the more general pattern of world prehistory—may be found if we put aside some of the traditional constraints of anthropological theory. We must reconsider the role of population growth and population pressure and view them as active forces that continually modify the ecosystem and human cultural response to it. Population growth is continuous and cannot be reduced to the series of occasional events envisioned by some of the authorities discussed. The growth of hunting and gathering populations up to some threshold or saturation level could have provided the stress which made it necessary for populations to begin artificial augmentation of their food supply. Moreover such a model could account for the fact that the stress was shared by a variety of human populations living in different environments and undergoing different kinds and rates of environmental change. It could even explain the fact that agriculture emerged in different parts of the world at roughly the same time, if it can be shown that mechanisms existed which were effective in distributing population pressure evenly from region to region (as I believe to have been the case). The hypothesis of continuous population growth has two further advantages. It accounts for the fact that archaeological evidence of disequilibrium and stress—population pressure—is not confined to the post-Pleistocene period (as various authorities have asserted) but is widespread in the prehistoric record. It also provides what I think is the most satisfactory explanation of the timing of the emergence of agriculture. Many authorities have called attention to the coincidence between the end of the Pleistocene and the emergence of farming. Though this is by no means without importance, there is a coincidence which I believe to be more significant—that between the end of the era of territorial expansion and the beginning of the period of rapid economic intensification as defined by the emergence of broad-spectrum economies and then of farming. It would appear that population growth, which first resulted in territorial expansion and the infiltration of unused ecological zones, gradually began to bring about an economic intensification when the potential for territorial expansion was exhausted.

There is, of course, a certain basic logic in considering population growth as a contributing factor in culture change. We know that human populations do have enormous growth potential (Birdsell 1958, 1968; Polgar 1972, 1975) and we know that population growth is a significant trend in human history. We know also that tendencies toward overpopulation are a key ingredient of evolutionary change in animal populations in which internal (intra-specific) competition based on excessive reproduction is the essence of natural selection. The main difference between biological evolution and human cultural evolution may in fact be that animals with relatively fixed ecological patterns respond to the stress of overpopulation primarily through high differential mortality and only occasionally through actual ecological change. Man, with his greater behavioral and ecological flexibility, may more often meet the same stress by adjusting his behavior and increasing his numbers (cf. Faris 1975).

The idea that population growth or pressure may contribute to technological change is of course not new. Demographic stress as a cause of domestication was suggested by Leslie White (1959:285). And a number of recent studies have attempted to show that population growth and pressure are contributors to the process of culture growth rather than simply results of technological change. (See Dumond 1965; Smith 1972a,b; Harner 1970; Spooner [ed.] 1972; Meyers 1971; Kunstadter 1972; Flannery 1969, 1973; Binford 1968; Harris in press; Cohen 1975a,b; Sanders and Price 1968; Carneiro 1967, 1970, 1972; and others.)

The basic discussion of the role of population growth as a determinant of technological change, and the study which prompted much of the later work, however, was the work of Ester Boserup (1965). It is Boserup who was primarily responsible for challenging the traditional Malthusian models of population growth and reversing our perspective on the cause and effect relationship between population growth and technological change. In her analysis of agricultural systems she argues that it is not so much technological progress as population density that determines what type of agriculture will be employed. She argues, in fact, that the relative efficiency of various agricultural technologies is largely a function of population density and that the various known technologies represent a continuous series of more or less elastic responses to growing population.

Despite these studies, and despite the cogency of Boserup's logic (if not always of her supporting data), the demographic stress model of culture change has met with little approval. It is implicitly or explicitly dismissed or criticized by a number of recent studies (Polgar 1972, 1975; Sheffer 1971; Bronson 1972, 1975; Hassan 1974, 1975a; Cowgill 1975a,b). More important, the approach is afforded only very limited applicability even by many of its supporters who explicitly or implicitly see demographic stress operating only under very particular conditions (Binford 1968; Harris in press; Flannery 1969, 1973).

I believe that this dismissal is unfortunate and premature and that this limitation is inappropriate. I intend to argue that population growth and pressure are contributing factors in the origins and growth of an agricultural economy worldwide. The prevalent models of human populations as systems that seek and maintain equilibrium should, I suggest, be replaced by models that stress the inherent growth of these systems. I do not intend to suggest that culture can be explained by any simplistic form of population determinism, although I am primarily interested in tracing the evidence for population pressure and its effects. Nor do I intend to suggest that population grows as an independent variable unaffected by other natural or cultural factors. There seems to be no reason to reject, for example, the assertion made by various authorities (see Sussman 1972; Hassan 1973; Dumond 1975) that the rate of population growth might have increased as a result of sedentism or other economic or ecological factors accompanying the onset of agriculture (although I think that this trend is probably overemphasized by the nature of archaeological preservation). My argument is simply that population growth (and population pressure) was a more ubiquitous and more significant trend among pre-agricultural peoples than is usually recognized.

Similarly, I do not rule out evidence that local cultural variables or climate changes may have been significant for determining the precise nature of local cultural sequences. I simply suggest that the widespread parallelism of prehistoric events indicates that *some common factor* is in operation in each of the widely divergent regions and that the patterns of events for the various regions are consistent with a picture of, continuous (although not necessarily steady or constant) population growth and population pressure.

In brief, I intend to argue that human population has been growing throughout its history, and that such growth is the cause, rather than simply the result, of much human "progress" or technological change, particularly in the subsistence sphere. While hunting and gathering is an extremely successful mode of adaptation for small human groups, it is not well adapted to the support of large or dense human populations. I suggest therefore that the development of agriculture was an adjustment which human populations were forced to make in response to their own increasing numbers. By approximately 11,000 or 12,000 years ago, hunters and gatherers, living on a limited range of preferred foods, had by natural population increase and concomitant territorial expansion fully occupied those portions of the globe which would support their life-style with reasonable ease. By that time, in fact, they had already found it necessary in many areas to broaden the range of wild resources used for food in order to feed growing populations. I suggest that after that time, with territorial expansion becoming increasingly difficult and unattractive as a means of adjusting to growing population, they were forced to become even more eclectic in their food gathering, to eat more and more unpalatable foods, and in particular to concentrate on foods of low trophic level and high density. In the period between about 9000 and 2000 B.P. populations throughout the world, already using very nearly the full range of available palatable foods, were forced to adjust to further increases in population by artificially increasing, not those resources which they preferred to eat, but those which responded well to human attention and could be made to produce the greatest number of edible calories per unit of land.

The argument is based on six central propositions. First, it will be argued that agriculture is not a single unified concept or behavior but an accumulation of techniques used to increase the range or density of growth of particular resources; that these techniques, most of which are used in various combinations by different "hunting and gathering" societies, represent no great conceptual break with traditional subsistence patterns; and that it is therefore not ignorance, but rather lack of need, that prevents some groups of people from becoming agriculturalists. Second, evidence will be presented to show that agriculture is not easier than hunting and gathering and does not provide a higher quality, more palatable, or more secure food base. Agriculture has in fact only one advantage over hunting and gathering: that of providing more calories per unit of land per unit of time and thus of supporting denser populations; it will thus be practiced only when necessitated by population pressure.

These first two propositions, which are not new, should be familiar to readers acquainted with the literature on agricultural origins, the recent literature on hunting and gathering economics, and particulary the literature dealing with labor costs, dietary quality, and population as they relate to agricultural origins (Lee 1968, 1969; Binford 1968; Flannery 1969, 1973; Spooner 1972; Harris in press; Sahlins 1968, 1972). The remaining propositions, however, mark a somewhat more abrupt departure from the existing literature.

The third proposition is that, contrary to contemporary equilibrium-system models of hunting and gathering populations (Birdsell 1958,

1968; Hayden 1972; Polgar 1972; Hassan 1975a; Binford 1968; Flannery 1969) which stress their demographic stability and their tendency to achieve equilibrium well below the carrying capacity of their resources, human societies have in fact grown throughout their history and have encroached progressively on their resources to the extent that the continuous development of new adaptive strategies and the continuous redefinition of ecological relationships were necessary.

The fourth proposition is that hunting and gathering populations enjoy widespread and highly effective mechanisms by which population pressure is balanced from region to region; that there is evidence of the operation of such mechanisms among prehistoric populations of hunters and gatherers; and that therefore it is not unreasonable to find a roughly synchronous building up of population pressure over very large portions of the globe, with the result that agriculture was "invented" or adopted by most of the world's population within the same fairly brief time span.

The fifth proposition is that the events leading to agriculture in the various parts of the world show a remarkable parallelism when they are viewed in a reasonably broad temporal and geographic perspective. Despite local variations, this parallelism seems to demand some common contributing factor acting in all these regions.

The sixth proposition is that the record of Paleolithic and Mesolithic man, as well as that of pre-agricultural man in the New World, can reasonably be read as indicating fairly continuous population growth and increasing population pressure in pre-agricultural contexts, and that in each case the adoption of agriculture appears to be only one in a long series of ecological adapations to increased population.

These propositions will be developed as follows: Chapter 2 will discuss a number of theoretical issues relating to the nature of hunting and gathering and agricultural systems. It will focus on an analysis of the conditions under which the "discovery" of agriculture is possible or probable and on those conditions under which an agricultural economy would be, and would be perceived to be, an improvement on a hunting and gathering way of life. Some of the theoretical issues concerned with population growth and population stabilizing mechanisms will be discussed, along with a theoretical model of a growing population and a consideration of the adaptive alternatives which such growth permits and necessitates. Finally, chapter 2 will consider the problem of population growth and cultural change on a broad geographic scale. Chapter 3 will discuss the nature of archaeological evidence bearing on the measurement of population growth and demographic stress, criticizing contemporary methodology and suggesting new lines of evidence which should be brought to bear. Chapters 4, 5, and 6 will review the evidence of demographic stress leading up to the origins of agriculture in the Old World and in North and South America. Chapter 7 will summarize the arguments and consider some of the problems and research priorities which the population pressure model generates.

2 THE THEORY OF POPULATION PRESSURE AND THE ORIGINS OF AGRICULTURE

In order to evaluate the role of population pressure in the origins of agriculture, it is necessary to deal with a number of theoretical problems concerning the nature of subsistence systems, the processes of economic change, and the nature of population growth. Some of these problems have received considerable attention in recent studies in anthropology. Although the arguments are not new, they are reviewed here because they are essential parts of the model which is being developed, and because they may be unfamiliar to those not conversant with the recent anthropological literature. Some of the problems which must be raised, however, deal with issues that are not yet satisfactorily resolved within anthropology. The latter will involve somewhat more controversy.

The Concept of Domestication

One of the key problems is to determine how the origins of agriculture should be conceptualized. As I suggested in the previous chapter, there have been two different approaches to this problem. One school has tended to perceive agriculture as a conceptual leap or an invention, and has emphasized research concerning the time and place and the circumstances under which this invention would have occurred. The other school has tended to assume that the knowledge of agriculture is universal and that what needs to be explained is not the availability of new knowledge but rather the process leading to the implementation of techniques which had presumably always been available. The distinction is a critical one for a population pressure model of agricultural origins (or for any model which assumes that agriculture was a response to stress). Such a model must assume that the concept of planting seeds was generally available, since it must be assumed that the implementation of agriculture depended *only* on need. But the distinction is also important for the model presented here in two other ways. First, if the concept of agriculture is simple and widely available to primitive man, then diffusion can largely be eliminated as a significant explanation of the rapid spread of agriculture. Twenty years ago diffusionists had no problem explaining the sudden, even explosive emergence of agriculture worldwide. Once invented, the new concept simply spread rapidly from group to group. But, if everyone knew about agriculture, and people in Europe and Africa, or in Peru, did not have to wait for messengers from the Middle East or Mexico bearing the news of the new discovery, then the problem of explaining the simultaneous adoption of agriculture is a significant one. In addition, the distinction between the two approaches has a significant bearing on the discussion below concerning the relative economic costs and benefits of hunting and gathering when compared to agriculture. As long as we assume that hunting and gathering groups are ignorant of agriculture then it is possible to assume that agriculture is the superior system and that those who avoid it do so out of ignorance. If it is conceded that most hunting and gathering populations know about agriculture, then it clearly must be admitted that there are at least some costs or disadvantages to domestication which lead people to avoid this system even when they are fully capable of implementing it.

There is a fairly widespread consensus now among anthropologists that the knowledge that plants grow from seeds is probably universal among hunters and gatherers and that this knowledge has probably been available to human groups since very early times, long predating its application in full fledged agricultural economies. For example, Flannery (1968:68) states:

We know of no human group on earth so primitive that they are ignorant of the connection between plants and the seeds from which they grow.

Similarly, according to Bronson (1975:58):

Deliberately growing useful plants was neither unique nor a revolutionary event. It probably happened in many places starting at an early date. This is not a complex idea or a difficult idea to develop. It is not beyond the inventive reach of any human being. We can be quite sure that activities resembling cultivation go far back into the Pleistocene.

There are numerous lines of archaeological and ethnographic evidence which support this position. First, analysis of ethnographic data on modern hunting and gathering groups not only suggests that they are aware of the concept of planting seeds; it also suggests that their lifestyle promotes and demands such awareness. The techniques of agriculture must be essentially self-evident to any hunting and gathering group. The independent discovery of the concept of domestication by several of these groups (or their prehistoric counterparts) would seem not only to be possible but to be inevitable.

In the first place, as has been pointed out by a variety of authors, survival for hunting and gathering groups is dependent on their intimate knowledge of the plant and animal communities they rely on. Such groups must know precisely the places where edible fruits, seeds, and roots are to be found and the conditions under which they best grow, as well as the feeding habits, movements, and ecological requirements of the game animals upon which they depend. William Allan (1965:4), for example, quotes D. F. Thomson of the University of Melbourne with regard to the Bindibu of Australia, who are, in his words:

> expert ecologists [displaying] a knowledge of the economic resources of their country far beyond that possessed by most white men. They have names for each type of country and botanical association and can name every tree and plant. They can also describe the food harvest or the fibre and resin that these will yield each season.

The Bushmen of the Kalahari display similar ecological expertise. According to Elizabeth Marshall Thomas (1959:10):

> Each group knows its own territory very well; although it may be several hundred square miles in area, the people who live there know every bush and stone, every convolution of the ground and have usually named every place in it where a certain kind of veld food may grow even if that place is only a few square yards in diameter.

Similar statements about the knowledge of hunting and gathering groups are echoed by other authors. (See, for example, Stewart 1956; Lee and DeVore 1968.)

What is particularly important from our point of view is that there is no significant difference between precise knowledge of the ecological conditions under which certain plants and animals thrive and the ability to help them thrive by modifying natural conditions to suit their needs. If one is aware that certain plants are typically to be found most abundantly under very specific conditions of light, moisture, or soil, it is not conceptually difficult either to plant or transplant them in that particular environment, or conversely to alter their existing habitat in the desired direction. There is in fact abundant ethnographic evidence demonstrating that such ecological knowledge does modify the behavior of hunting and gathering peoples vis-à-vis their food sources and that these modifications resemble incipient stages of cultivation. For example, the Andamanese Islanders, when digging wild tubers, are known to avoid harvesting during the season of new growth in order to assure the growth of the next year's crop. The same group is reported to protect the seed crop of other wild harvested plants as well (Heizer 1955:5). The Menomini Indians of Wisconsin are reported, when gathering wild rice, purposely to have allowed some of the rice to fall back into the water to insure a crop in the next year (Heizer 1955:5). Irrigation of wild food sources is reported among the Paiute (Steward 1929), and a number of authorities (Stewart 1956; Sauer 1952:11; Isaac 1970:18–19; Lewis 1972; Day 1953; Hough 1926; Gould 1971) have reported on the use of fire by nonfarming groups to promote the growth of desired plant species which are known to be good colonizers of recently burned lands, or to increase the biomass of game animals which can be supported by grazing on newly opened areas.

Similarly, Kroeber (1925:220) records the ability of the Pomo Indians to alter stream courses to create conditions favorable for the runs of certain fish, and in the same vein, Heizer (1955) reports conservation measures for wild game and fish species which are practiced by a number of Indian groups. He reports, for example, that the Klamath Indians of the Yukon, the Yukaghir of northeast Asia, and the Klallam Indians of Washington State during the historic period all employed care in the damming of salmon streams to avoid destroying their food resources by overexploitation. It was the practice of these groups either to avoid complete damming of the streams or else to restrict the time period during which the dams could be in place.

A study of the life-styles of hunting and gathering groups also suggests that, even if such ecological knowledge were not already part of the cultural repertoire of any group, intimacy with certain usable species and a knowledge of their habits would be thrust upon the people by the plants or animals themselves responding to the basic behavior patterns of the human populations. Moreover, the evidence suggests that this knowledge would obtrude itself upon such groups as early as they were mentally equipped to perceive and analyze the evidence of their surroundings in human fashion. We know that even nonhuman primates inadvertently accumulate around their home ranges gardens of their favorite plant species growing from seeds or vegetative parts dropped in the course of eating or in the animals' feces (Jolly 1972:59). Human hunting and gathering groups tend to gather gardens of their favorite foods near their houses in the same manner, quite independent of any attempts at purposeful cultivation. Such natural accumulations of preferred foods have been reported in the campsites of a number of hunting and gathering peoples (Schwanitz 1966:12; Heizer 1955).

We know furthermore that hunting and gathering groups often unintentionally propagate their food plants by accidental resowing of parts of their crop in the process of harvesting (Schwanitz 1966:12). We know that continued collection of certain edible species in the wild, quite apart from cultivation or purposeful selection, has selective effects paralleling many of those occurring under domestication (Heizer 1955:12). We know also that man, and even nonhuman primates, are often responsible for the inadvertent or purposeful transportation of species beyond their typical home ranges, which, as mentioned in the previous chapter, is one of the characteristic features of domestication (Heizer 1955:12).

Pre-agricultural man commonly created pathways, village clearings, dump heaps, and recently burned areas, which provided open habitats for colonization by new plants (Anderson 1952; Hawkes 1969; Heizer 1955:12), and many of man's cultivated plants originated as weedy volunteers, preferring these particular disturbed habitats although receiving no intentional help from man (Anderson 1952; Sauer 1952:71). We know even that desirable morphological changes such as the gigantism of edible parts and early maturity, usually considered signs of domestication, have occurred in plants in the vicinity of human settlement without any direct or intentional human participation (Isaac 1970:18).

Taken as a whole, this evidence indicates two very important things about the "concept" of domestication: first that any human group dependent in some degree on plant materials, possessing the rudiments of human intelligence, and having any sort of home-base camp structure (of the type associated with all men probably since *Homo erectus*) will be almost bound to observe the basic process by which a seed or shoot becomes a plant. Ignorance of this basic principle is almost inconceivable. In addition, however, this evidence demonstrates that agriculture is not a single unified concept or behavior but a combination of behaviors, any one of which may be either inadvertent or purposeful, including such things as the creation of clearings in which certain plants thrive; the enrichment of certain soils; the planting of seeds; the irrigation of plants; the removal of competing species; the practice of conservation measures; the transporting of species beyond their original ecological boundaries; or the selection of preferred types. None of these behaviors alone constitutes agriculture; taken together they *are* agriculture, yet all of them have been demonstrated to be practiced alone or in various combinations, inadvertently or purposefully, by "nonagricultural" groups.

This suggests that the conceptual break, or for that matter the operational distinction, between agricultural and nonagricultural practices is not very great. Populations cannot really be categorized as agricultural or nonagricultural (at least when seen in historical perspective). Rather there is a continuum in the degree of assistance different human populations offer to the species on which they depend (Higgs and Jarman 1969, 1972). The same population may very well offer differing degrees of assistance to various plants, depending on their importance, their scarcity, and their relative ability to survive unaided. This sort of continuum is reflected in Anderson's (1952) description of primitive gardens with their haphazard organization of crop plants and their graded distinction among the plants which are actively discouraged or removed, those which are tolerated, those which are utilized but not encouraged, those which are given varying degrees of assistance, those which are planted, and those which are carefully selected and stored for planting.

Such a continuum should also be reflected in the history of any group evolving toward dependence on agriculture. The group will presumably gradually bring to bear an increasing repertoire of techniques for assisting the growth of its selected plants and will employ these techniques with gradually increasing intensity, without ever self-consciously crossing any dividing line between gathering and agriculture. Thus, agriculture should be conceived more as a de facto accumulation of new habits than as a conceptual breakthrough.

The archaeological evidence, such as it is, also supports the contention that agriculture is not a difficult concept to develop, that it has in fact been developed many times, and that it was not primarily ignorance which prevented human populations from becoming agricultural sooner than they did. We now possess archaeological data which indicate fairly clearly that agriculture was developed independently in a fairly large number of places rather than diffusing from one or a few hearths of discovery. The archaeological record clearly shows that domestication occurred independently at least four times: Southwest Asia, Southeast Asia, Mexico, and the Peruvian Andes all display early evidence of domestication. The relative dates of these early centers, the distinctiveness of the first cultigens in each area, and the lack of convincing diffusion horizons connecting them, all indicate that these centers must represent separate developments. There are moreover a number of additional regions, including several in South America, Africa, Asia, and the western and eastern United States, where it is apparent from the botanical evidence that new species were brought under domestication. In fact, if we look at the botanical evidence concerning the precise point of origin of our various cultigens, the number of separate centers of domestication becomes almost infinite, even if we confine ourselves to the major recognized crop plants. Unfortunately, in most of these areas, the archaeological evidence is not yet adequate to establish whether cultivation was achieved independent of stimuli from other regions or was dependent on stimulus diffusion. I believe however that as our archaeological knowledge of poorly studied regions and obscure cultigens becomes more complete, the number of apparently independent hearths of domestication will grow rapidly. It is only very recently, for example, that work in Thailand (Gorman 1969, in press; and others) has established the relatively early date for agriculture in this region and at the same time shown on botanical grounds the independence of this hearth of domestication from that of the Middle East.

The real insight, however, comes from detailed analysis of the evidence for early domestication in the known primary centers themselves. In each of these centers, archaeological or botanical evidence or both suggest rather clearly the independent parallel domestication of a variety of wild races and species in a number of subregions. Thus, in the Middle East (chapter 4) domestication appears to develop over a broad area ranging from Iraqi Kurdistan to the Levant, and possibly extending to Anatolia and the Aegean region as well, within one general time span, rather than radiating out from any single center; moreover, the known distribution of the main wild cereals suggests that they were domesticated in different locations. Similarly, the archaeology of early domestication in Mexico (chapter 5) points to a number of regions evolving domestication at approximately the same time but utilizing different cultigens. Chang (1970) and Li (1966) have suggested that the East Asian cultigens (chapter 4) are clearly traceable to at least two centers, one in Southeast Asia and one in North China. In South America (chapter 6), although the archaeological record of early domestication is poor, the ecological range of the various cultivated species suggests enormously diverse origins. As Heiser (1965), Pickersgill (1972), and Harlan (1971) have all pointed out, one of the most startling facts about the cultivated plants of the nuclear area of the New World as a whole is the number of genera in which two or more geographically distinct species appear to have been domesticated separately, with different species appearing earliest in the archaeological records of different regions. There is even some suggestion that many of the New World domestic *species* may have been domesticated independently more than once from different wild races in different areas.

There is one further type of evidence in the archaeological record which fairly clearly supports the assumption of multiple independent developments of domestication technology. As will be discussed in considerable detail in subsequent chapters, the first appearance of agriculture in any region almost invariably occurs in an economic context of the type referred to as "Mesolithic" in the Old World and as "Archaic" in the New. This striking parallelism in the economic context in which early agriculture first occurs in different regions suggests that agriculture is in some sense a product of that context rather than simply a new technology grafted by diffusion on a haphazard series of recipient cultures. Moreover, the fact that the similarity of the underlying cultures is in the basic economic structure rather than in easily diffusible stylistic elements suggests that agriculture is a function of separate but parallel economic trends of long standing. That agriculture is demonstrably grafted on to Mesolithic economies by diffusion in *some* regions (notably Europe and coastal Peru) does not alter the fact that this type of economy is clearly conducive to agricultural development or the consequent high probability that such an economy fostered agricultural development independently a number of times.

The archaeological record also provides a series of other hints that the *concept* of domestication was not the key stumbling block in the development of agriculture. A number of authors (notably Sauer 1952) have suggested that cultivation techniques may have been applied first to plants whose value was magical, medicinal, or utilitarian rather than primarily dietary. If this is true, it would imply that agricultural techniques were known and used long before they were brought to bear on the food supply. Although the record of most such early nonfood cultigens has been lost (if, in fact, they ever existed) there is at least one plant, the bottle gourd (*Lagenaria siceraria*), that occurs both widespread and very early in the archaeological record of both hemispheres. In South America, at least, this species occurs in several regions well before any evidence is found of other cultigens. If these early remains were in fact domestic they would establish the existence and use of plant control techniques long before the emergence of economic agriculture. As Richardson (1972) and others have pointed out, however, there is no proof that these early forms were domestic.

Bronson (1975) has also speculated that cultivation, in the sense of the small-scale manipulation of a few edible plant species, may significantly have predated agriculture, in the sense of economic dependence on cultivated plants. Again, if true, this would imply that factors other than the difficulty of mastering the domestication concept were responsible for delaying the acceptance of agriculture as an economic strategy. In this case, the archaeological record provides strong supporting evidence. In many of the regions studied (chapters 4, 5, and 6) there is evidence of significant delay between the earliest evidence of cultivation (or even, in fact, full morphological domestication of certain plants, which is usually the earliest evidence of cultivation identifiable from the archaelogical data) and dependence on agriculture as a way of life. In addition, in Peru, in the eastern United States, and in much of Europe there is good evidence of a significant delay in the acceptance of agricultural technology, even when there is evidence that the techniques were available by diffusion from nearby regions.

Thus, both the archaeological and the ethnographic evidence appear to indicate that the knowledge of cultivation techniques is universal, or at least that such knowledge is significantly more widespread in time and space than the actual practice of agriculture as a subsistence strategy. If this is true, then we are forced to find some reason other than ignorance to explain why some people have remained hunters, and, conversely, some reason other than discovery or diffusion to explain why others became farmers.

The Advantages and Disadvantages of an Agricultural Economy

Prior to about 1960, hunting and gathering groups were commonly pictured as existing near starvation, struggling constantly to find adequate food resources. According to standard reconstructions of the period, the advantages of agriculture over hunting and gathering in improving the quality and reliability of the diet, reducing labor costs in the food quest, and providing leisure time were sufficient (and sufficiently obvious to recipient populations) to make acceptance of agricultural technology automatic once the concept was perceived. This model was an attractive one. Western science is imbued with a sense of its own progress; with a deeply rooted belief in the superiority of Western man; and with an abiding belief in the sanctity of hard work, particularly work with the soil. Descriptions of nonwestern (and largely nonwhite) populations tended strongly to emphasize the backwardness of their life-styles and the poverty of their existence.

An increasing number of studies of contemporary hunting and gathering populations, however, have tended to challenge these traditional assumptions. Although the sample of contemporary hunting groups is small, and although very few good quantitative scientific studies have been done, a good deal of evidence is accumulating which suggests rather uniformly that the diet of hunting and gathering populations (outside the Arctic) may be calorically quite adequate, and at the same time richer in food variety, vitamins, minerals, and above all protein, than that of agriculturalists. These recent studies suggest also that hunting and gathering involves activities widely preferred to those of agriculture and provides foods widely preferred for consumption to the main agricultural staples—grains and tubers; that the food supply of hunters and gatherers may be more reliable than that provided by agriculture; and that it may be obtained with as little, or even significantly less, labor than is necessary for agricultural production.

In fact we have gone through a rather abrupt reversal in the literature to a point where hunting and gathering is seen, somewhat uncritically perhaps, as a superior mode of life. We *may* be overemphasizing the quality of the hunting life now, just as previously we overemphasized its poverty. Cowgill (1975b) has warned that the hunting existence may not have been as stress-free as we now commonly picture it. Similarly Hayden (1975), citing Bartholomew and Birdsell (1953), warns that the economic well-being of a community may be more accurately measured in its reaction to periodic stress years than in its reaction to normal periods of plenty. Thus, the bulk of our observations, which presumably occur in "average" conditions, may be somewhat misleading. Moreover studies can be cited (Meggitt 1962; Steward 1938) reflecting at least occasional economic stress among hunters and gatherers.

On the other hand, there is no evidence to suggest that such stress periods are more frequent among hunter-gatherers than among farmers, and there is both logical and empirical argument to suggest, on the contrary, that periodic stress is more often a function of farming. Thus, despite warnings about the contemporary "bandwagon" emphasis on the quality of life among hunters, a certain optimism about their condition still seems to be warranted.

The contemporary literature contains a few excellent studies of hunting and gathering groups in which good quantitative data on work effort, calorie consumption, protein consumption, and other relevant variables are available. One such study is by Richard Lee (1968, 1969), who spent fifteen months between October of 1963 and January of 1965 living among the !Kung Bushmen of the Kalahari Desert in South Africa and made a detailed study of the work and dietary habits for one group over a four-week period. Lee found that during that time, which was neither the richest nor the poorest time of the year, caloric intake for the group averaged 2,140 calories per person per day, well in excess of the Recommended Daily Allowance (1,975 calories per person per day) for persons of comparable stature and activity. He further found that their daily average consumption of protein (93.1 grams per person per day) was nearly half again the RDA. Lee found in addition that, rather than being compelled to eat whatever was available, the Bushmen were in fact highly selective, drawing about 90 percent of their vegetable diet from only about 23 of 85 available edible plant species while consenting to eat only 54 of some 223 local species of animal known to them and getting the bulk of their animal food from only 17 of these. Lee also found that even among the species preferred for consumption the total available biomass was only partially consumed, the remainder being left to wither on the vine, or, more aptly, to rot on the ground. He found moreover that the human population was healthy, suffering less from kwashiorkor than neighboring agricultural peoples and showing no clinical-level symptoms of vitamin deficiency. He found that the Bushmen were relatively long-lived, having a proportion of adults over sixty years of age of nearly 10 percent (46 individuals in a population of 466), a percentage which compared favorably with primitive agricultural populations and in fact with most nonwestern populations. He found in addition that the Bushman food supply was surprisingly dependable. They did not suffer the annual hunger periods characteristic of many African farming groups, and even more surprising, during a particularly poor year when neighboring agricultural tribes were reduced to famine and dependence on aid from the United Nations, the Bushman economy was essentially unaffected.

With the wisdom of hindsight, none of these revelations should be as surprising as they were when first offered. A selective yet wide-ranging diet such as that typical of hunters and gatherers should be expected to provide a broad range of nutrients, a healthful diet, and a healthy population (Stini 1971). Conversely, Yudkin (1969), Barnicot (1969), Brothwell (1969), Stini (1971), and Hassan (1975b,c) have all offered arguments suggesting that agriculture and accompanying sedentism may bring with them a host of nutritional diseases including deficiencies in protein and in a number of vitamins and minerals. (For a contradictory view, see Sengal [1973] and Stephens [1975].) Both Barnicot (1969) and Dunn (1968) have also suggested that agriculture and concomitant sedentism will bring with them exposure to new parasitic diseases. Similar conclusions have been stated by a number of other scholars (Bates 1955; Haldane 1957; Hiernaux 1963; Polgar 1964; Angel 1972; Hassan 1975a,b,c).

Similarly, it is logical that hunters and gatherers would enjoy a degree of continuity and reliability in their food supply which is lost with the advent of agriculture. The Bushman diet protects them against seasonal shortages and periodic starvation. The natural biotic community from which the Bushmen obtain their food is buffered against "crop" failure in two senses. First, the very variety of the resources available guarantees that some foods will be available if others fail, or in seasons when others are not available; the complexity of the ecosystem buffers it against disaster. Second, the native wild plants and animals on which the Bushmen subsist are already preselected for their ability to survive whatever climatic extremes may occur in their particular environment. It should thus be clear that agriculture which both simplifies the ecological community and introduces foreign species will tend to undermine the stability of resources and introduce periodic ecological crises. This is particularly true, of course, if the natural vegetation is entirely removed and replaced by a single "clean" seed crop such as wheat. However, even tropical vegeculture systems, such as those described by Harris (1969, 1972), which mimic the variety of the natural environment, and which involve selective rather than total replacement of indigenous species, must surely lose some of the stability inherent in a natural ecological community of long standing.

What was particularly surprising in the Bushman example was the ease with which their food was obtained. Lee's records indicate that the working adults spent an average of about 2.5 days per week in subsistence activities, a working day being defined as about 6 hours. This suggests a total weekly work output of about 15 hours or an average of slightly better than 2 hours per day spent in the food quest. Lee further pointed out that such labor excluded adults over sixty (who, as he indicated, constituted fully 10 percent of the population) and, more surprising, excluded unmarried "children," which in this case meant all "girls" under the ages of fifteen to twenty years and "boys" of less than twenty to twenty-five. The total work force thus constituted only about 65 percent of the population. He further pointed out that it was only the female portion of the labor which was truly productive of calories. Men hunt, and on occasion actually return with meat from large animals; this is highly prized food, but in fact constitutes only a small portion (one-third or less) of the total caloric consumption. It should further be pointed out that Lee's observations were made on the Bushmen in what is considered a poor environment during what was considered to have been a particularly bad year. Lee's figures may be summarized for comparative purposes by noting that the productive adults work an average of about 780 hours per year in the total food quest. When comparing this labor effort to that of agricultural systems, however, it is important to note that only a little more than half of this labor (primarily that of the women) went into vegetable production. Thus the average adult participation in the gathering of vegetable calories (the closest available equivalent to agricultural labor) was only about 400 hours per person per year.

A study of Australian Aboriginals in Arnhem Land (McCarthy and McArthur 1960; McArthur 1960) also suggests that hunter-gatherers can obtain an adequate diet for a moderate labor investment. The study indicates that adults spent an average of 3.5 to 5 hours a day on food gathering and that the work was not particularly arduous. Two of the four groups studied had average daily caloric intakes in excess of 2,100 calories per person. Two groups had reported caloric intakes below recommended levels, but since all groups were reported to have regular access to more food than they actually consumed, it seems unlikely that this was a serious deficit. Protein consumption was high in all groups and the diets were otherwise considered well balanced, although, in the brief period of the study, intake of some nutrients was low in one group or another.

There are a number of other studies of hunting and gathering groups which confirm the image of good health, good diet, and relatively low labor costs among such populations. The Hadza of Tanzania are noted by Woodburn (1968a) to maintain high standards of health and nutrition with minimal effort, and, like the Bushmen, without either being forced to eat foods they dislike or threatening to exhaust their food supplies; they are reported to be as selective in their eating habits as are the Bushmen and to be as effectively buffered against "crop failure." Woodburn estimates that something less than 2 hours per day on the average is spent per person in the food quest, suggesting a total of about 750 hours per adult per year in the gathering of vegetable and animal foods. Again, presumably about half of this labor goes into vegetable food production. Further observations of a similar nature are available from a number of anthropologists dealing with different hunting and gathering groups: Bose (1964); Neel (1970); Turnbull (1965); Ackerknecht (1948); Maingard (1937); Clark (1951); Huntingford (1955); Allan (1965); and Marshall (1961). Moreover, these references are not limited to contemporary authors, who might be viewed as being in the throes of an anthropological fad or as studying people with particular twentieth-century well-being. Sahlins (1972), culling the older literature, has brought to light a number of older references, dating back in many cases to the mid-nineteenth century, which tell the same story: Grey (1841); Quimby (1962); Curr (1965); Hodgkinson (1845); Bonwick (1870); Spencer and Gillen (1899); Eyre (1845). Both Grey and Eyre are explicit (though not necessarily scientific) in documenting the number of hours spent in the food quest (presumably the quest for both animal and plant foods) by the groups studied. Eyre reports a figure of 3 to 4 hours per person per day (1,000 to 1,400 hours per year), and Grey suggests a figure of 2 to 3 hours per day (750 to 1,000 hours per year)—both for groups of Australian aborigines.

The labor costs for these various groups may be compared with figures, such as they are, available for labor inputs into primitive agriculture. Conklin (1957:151) indicates that the Hanunoo agriculturalists of the Philippines put about 1,200 hours per year per person into swidden cultivation, suggesting an average of something over 3 hours per person per day. But, as Sahlins points out (1972:34), this figure does not take into account labor devoted to gathering, raising of animals, or secondary activities related to tool maintenance and so forth. A. J. Richard's figures for the Bemba of Rhodesia as summarized by Boserup (1965:46) suggest that an average of 1 to 2 hours per day (365 to 720 hours per year per adult) is put into agricultural activity alone, again exclusive of hunting, collecting, or food preparation. Clark and Haswell (1966:116ff.) cite a number of studies on labor inputs in tropical Africa and elsewhere providing a wide range of figures for per capita annual labor inputs in subsistence agriculture, all of which suggest, however, that agricultural labor inputs are at least equal to and in most cases significantly higher than the labor of vegetable gathering attributed to the hunting and gathering groups discussed. In general they appear to suggest that anything less than 1,000 hours per year per male represents the lower limit of labor inputs in subsistence agriculture. They cite one study from Nigeria in which the average number of working hours in agriculture per adult male in two societies was 997 and 1,327 hours per year respectively. They cite a second study from the Calabar region of Nigeria where the men averaged 4 hours per day in agricultural work (or about 1,500 hours per year). Data from a case study in Nyasaland (Malawi) show that the number of hours worked per year in agriculture ranged from 400 to 900 for men and from 580 to 760 for women, while Haswell found in a study in Gambia that adult men and women averaged about 855 hours per year in agricultural work. In the village of Warwar in the Cameroons men averaged 4 and woman 5 hours per day in agriculture except for a three month period when both sexes averaged 10 hours per day, suggesting a total per adult expenditure of over 2,000 hours per year. A case study in Guatemala suggested that the average man worked about 750 hours per year in agriculture.

The smallest figures I have encountered for agricultural labor inputs are those quoted by Marvin Harris (1971) from Rappoport's study of slash and burn agriculture among the Tsembaga Maring of Australian New Guinea. Here the average investment per worker in agriculture is only 380 hours per year and the total average labor input in food production (including pig raising) only 780 hours per year. Among the Tsembaga, however, this average is calculated for everyone over ten years of age, totalling 146 of 204 persons, or about 72 percent of the population, a higher percentage than is gainfully employed among the Bushmen. (Harris, incidentally, used a calorie input/output ratio computation to demonstrate that swidden as practiced by the Tsembaga had nearly twice the labor efficiency of Bushman gathering. It is noteworthy, however, that if one combines Harris's figures for caloric inputs and outputs for swidden and for pig raising, the Maring's total efficiency figure is virtually identical to that which Harris computes for the Bushmen.)

In a study of a slightly different type, Carneiro (1968) compared the productivity of the hunting and agricultural portions of the economy of the Amahuaca Indians of the Peruvian montaña. He measured the number of man-hours required by each type of activity to produce a standard number of calories (one million). Hunting was found to be somewhat less productive than farming, requiring 795 man-hours to produce the same calories produced in 603 hours of farming; but, as Carneiro points out, the result is hardly compatible with the assumption that major labor saving is obtained by a switch to farming. He also observes that much of the agricultural labor is more intense than what is required in hunting, and he notes that the difference in productivity is probably due in part to the recent perfection of farming techniques and cultigens (implying that the difference between hunting and *incipient* cultivation would have been somewhat less). It should also be pointed out that the comparison was made in an environment (the tropical rain forest) which is one of the earth's poorest biomes with respect to the biomass of huntable animals (Bourlière 1963); and that the comparison is slightly misleading since it compares hunting (as opposed to gathering) with agriculture, whereas hunting is clearly calorically the least efficient portion of the hunting-gathering economy.

By comparing figures such as these, Sahlins (1968:86; 1972) concludes that our traditional formulae are inverted; that the amount of work per capita may actually increase with the evolution of our tech-

nology rather than the reverse. Bronson (1975), however, has challenged this conclusion, noting that the amount of good quantitative
data comparing labor costs among collectors and simple agriculturalists
is very small and hardly conclusive. Certainly, fair comparison is no
easy matter, as the data cited above should indicate. Quantitative data
are scarce and, even when available, are often not presented in a manner
which makes comparison possible. Different scientists measure different
things and report their findings in different forms. Some authors report
only the work done by men; others report for all adults but not for
children. Very few define the status of adulthood or note what proportion of the entire population consists of working adults or what proportion of the adult population is working on food production as opposed
to other craft specialties. Some authors count only the direct labor
costs of food production and harvest while others count time spent in
food preparation and tool maintenance as well. This is not simply the
problem of a poorly coordinated discipline. It is highly probable in fact
that no standardized methodology for measuring and presenting labor
statistics could be developed which would fairly compare hunting and
gathering with agricultural activities. The behavior categories are simply
not comparable; hence, no strict balancing of labor costs is possible. In
addition to food preparation and tool maintenance costs common to
both types of society, does one, for example, include the labor costs of
storage—a problem mainly confined to agricultural groups? Does one
take into account the fact that almost all known agricultural peoples
devote at least some fraction of their time to the production of surplus
food and that their labor costs therefore will appear disproportionately
high relative to actual food consumption? Or does one assume that at
least some part of this surplus production is necessary insurance against
crop failure and is thus a necessary labor cost even of the most basic
"subsistence" agriculture?

Given the range of these variables and the limits of the available data,
it is clear that little can be proved conclusively. We are left with impressionistic comparisons. Nonetheless, what data we have do tend to bear
out Sahlins's contention. Certainly there is nothing in the statistics presented to suggest that people would shift to agriculture to save labor or
to gain leisure time. The existing data are consistent with the assumption that labor costs increased rather than decreased in the transition to
agriculture. Again, with the wisdom of hindsight, this conclusion is

hardly surprising. Plant gathering at its simplest is nothing more than the harvesting phase of an agricultural economy, and as such, should represent only a fraction of the labor costs of a full agricultural system. It stands to reason that as a group takes on more and more activities to care for and promote the growth of certain plants, their labor costs will rise, even though this tendency should be offset in part by the reduction in the travel time associated with gathering.

Pure labor costs, however, as counted in man-hours, are not the only important variable. As Bronson (1975) has pointed out, there are a number of other somewhat more subjective factors involved in the decision to shift or not to shift to agriculture besides the simple counting of labor costs or caloric outputs. He suggests that such factors as *perception* of labor costs, the desire for security, and simple cultural preference in behavior and diet are factors to be considered. To these I might add simple conservatism, the desire of most societies not to change the status quo until forced or until the perceived advantages are so immense as to outweigh the basic cultural costs of change itself. Such subjective factors are, of course, impossible to predict for individual prehistoric societies, and no doubt these variables contribute some of the complexity that surrounds the archaeological interpretations of economic change. Nonetheless, certain objective realities and certain cross-cultural generalities enable us to predict in general how such factors will affect the choice. In the first place, many activities related to hunting and gathering are apparently commonly perceived as less arduous and more prestigious than is agricultural labor (Murdock 1968); hence the persistence of what Murdock refers to as "hunting mentality" among groups with mixed economies. Second, hunting and gathering labor is dispersed in time, but highly correlated with consumption rewards, whereas agricultural labor involves very high concentrations of labor in short spans of time long before any returns are achieved. It seems unlikely that groups will voluntarily put in such concentrated efforts so far in advance of their harvests on the abstract principle that in the long run they might put in fewer man-hours per year.

On the same general theme it is worthy of note that labor efficiency per se is apparently not particularly highly regarded among the Bushmen (and, by inference, other hunting and gathering groups). Their low labor costs are maintained despite the fact that labor minimization is by no means the foremost principle in their decision making. The effort involved in food gathering could easily be lessened by reducing the proportion of meat in the diet, or by the group being less selective and more eclectic in the foods they are willing to eat, or, conversely, less insistent on eating the full range of preferred foods. Apparently dietary variety, involving not only meat but also varied vegetable substances, weighs heavily against minimizing labor costs, and there is thus the strong suggestion that hunting and gathering populations may tend to avoid reducing the range of their diets with their spice and variety (as well as nutritional benefits) even in the face of possibly reducing overall labor costs. Since agriculture systems move in the direction of simplifying the ecosystem and reducing dietary variety, it is likely that this direction will be resisted even if some labor saving is found to be involved. If agriculture provided particularly relished foods, of course, this situation would be reversed. But, as has been observed by an increasing number of anthropologists and nutritionists, agriculture not only tends to cut down on the proportion of typically favored meat in the diet, but also involves concentration on foods, especially seeds and tubers, which are usually particularly *low* priority foods in terms of preference (Flannery 1973:307; Yudkin 1969:548).

Another point is worth noting. One of the advantages traditionally attributed to agriculture is the opportunity it supposedly affords for a population to settle down and become sedentary, thereby presumably reducing labor costs and reducing biological stresses on the human population. Certainly there are advantages to sedentism. Among the most important is probably reduction in the strains inherent in continuous mobility, particularly the burden on mothers carrying children in arms or *in utero*. As Sussman points out (1972; see below) the reduction in this strain may be a major factor in the increased rate of population growth accompanying sedentism. A reduction in mobility also presumably cuts labor costs associated with transport and minimizes risks of bodily injury associated with movement. Moreover, the reduction in mobility permits the accumulation of capital goods which may both enrich life and help reduce the labor costs associated with the food quest. However, as Hassan (1973) points out, hunting and gathering groups actually do not move all that much, and the strains, even those associated with bearing children, have probably been significantly overplayed. Moreover, much of the advantage that *we* perceive in sedentism may be a function of hindsight or, more important, may result from our own accustomed dependence on the capital goods that sedentism permits. There is considerable evidence that hunting and gathering groups did not perceive the same advantage in settling down. The abundance of foods available to these groups suggests that at the cost of slightly greater labor inputs in the food quest or somewhat reduced selectivity in their diet, they could, in many cases, remain sedentary or form larger aggregates than they actually do (Lee 1972b). The prehistoric record (see chapters 4, 5, and 6) shows that sedentism and large group aggregations were achieved in many parts of the world without agriculture, and often in fact without any evidence of significant new technology. Sedentism in many regions at first seems to have been based simply on increasingly intensive use of vegetable foods and coastal or riverine resources, which had previously been exploited only on a temporary or small-scale basis. The implication appears to be that, like agriculture, sedentism was within the technological capability of man long before he adopted it as a way of life.

Thus, sedentism does not appear to be inextricably linked to, or dependent upon, artificial control of the food supply. Nor, since it was apparently eschewed by both prehistoric and modern hunting and gathering populations who possessed the technological capability to settle down, should it be viewed as an essentially desirable end in itself. Since, as already noted, sedentism carried with it exposure to new parasitic diseases and restriction on the range of dietary sources, it may have been not so much a desirable end as a necessary evil. As will be discussed further below, I suggest that sedentism was itself a response to population pressure.

The only really important question remaining in comparing the relative affluence of hunter-gatherers and agriculturalists is to determine whether the relative affluence which appears so widespread among contemporary hunters and gatherers outside the Arctic Circle is truly indicative of the conditions under which such groups existed in prehistoric times. It is widely recognized that contemporary hunters and gatherers live in marginal environments (Lee and DeVore 1968:4-5), regions where they continue to exist by virtue of the lack of competitive interest on the part of their more powerful neighbors. From this it might be concluded that Pleistocene hunters with their choice of environments would have been even better off. It is not necessarily true, however, that areas marginal from the point of view of neighboring agriculturalists are equally poor in the eyes of the collecting populations that inhabit them. The telling argument, however, is that many of these "marginal" environments are marginal not only by the standards of contemporary farmers but also by the standards of prehistoric hunting and gathering groups as interpreted from the archaeological record. As is discussed below (chapters 4, 5, and 6), grasslands in both the Old and New Worlds, which supported the heaviest biomass of large game animals, tended to be the first portions of their respective continents intensively exploited by hunters, whereas desert regions and tropical forest areas, two types of environment characteristically inhabited by hunters and gatherers today, were heavily utilized only after the grasslands were fully occupied.

It is not clear to what extent the food quest has been eased for modern hunting and gathering groups by the availability of modern technology. None of the groups considered uses guns; most do possess iron, but there is little indication that the quality of tools employed contributes more significantly to their food quest than it does to that of the modern farmers, also possessed of iron, to whom they have been compared.

A more significant problem is the very real possibility that contact with modern populations may have made available techniques of food processing or recognition of foods which have widened the dietary scope of the collecting peoples in question. Knowledge of wild foods and their preparation is more likely to have diffused the other way, however. Also, outside contacts are hardly likely to account for the relative affluence of Australian aborigines lacking contact with more sophisticated economies when they were first described (as early as 1840).

It is also not entirely clear whether the mere proximity of other non-gathering groups of agriculturalists eases the burden of hunting and gathering groups or makes their lives more difficult than they would otherwise be. On the one hand, the proximity of agriculturalists may be preferable to head-to-head competition with other hunters, and neighboring farmers may help to provide conditions favorable for the propagation and survival of collected species. On the other hand, the proximity of farming groups may make life more difficult for hunters by restricting free wandering or by altering the natural environment in ways detrimental to the preferred wild species. The relatively dense agricultural populations may even provide stiffer hunting competition than would other hunting and gathering bands.

Many of these factors are worthy of further research. On balance, however, the widespread prosperity of contemporary hunting and gathering populations outside the extreme latitudes (but otherwise distributed over a very wide range of ecological and sociopolitical environments), suggests that Pleistocene hunters and gatherers enjoyed equal if not greater prosperity. I think that the inference can safely be made that prehistoric hunters, like their modern counterparts, enjoyed adequate, high quality diets, good nutritional standards, reliable, well-buffered food supplies, and abundant leisure. Since contemporary hunting and gathering groups appear to fare at least as well as, and in most cases far better than, their agricultural neighbors, who have the benefit of highly evolved crop plants and time-tested cultivating techniques, we may safely assume that Pleistocene hunters and gatherers would have fared at least as favorably when compared with early agricultural pioneers, who had not yet perfected either their cultivating techniques or their cultigens.

This in turn creates a new problem in the interpretation of agricultural origins. If agriculture provided neither better diet, nor greater dietary reliability, nor greater ease in the food quest; if it did not of itself confer the capability of sedentism, but conversely provided a poorer diet, less reliably, at equal or greater labor costs; why did anyone become a farmer? According to Lee (1968), Bushmen, who like other contemporary hunting and gathering groups know all about planting seeds, argue that this would be foolish since there is so much wild food available to harvest.

What, then, does agriculture actually accomplish? The various techniques that constitute agriculture have one property in common; they provide only one economic benefit: the ability to grow and harvest more food from a unit of space in a unit of time. As Flannery (1973: 307) has pointed out, the primary advantage possessed by the otherwise low-preference foods which constitute our major cultigens is their high density per acre or hectare and their positive response to human manipulation. In other words, agriculture permits denser food growth supporting denser populations and larger social units, but at the cost of reduced dietary quality, reduced reliability of harvest, and equal or probably greater labor per unit of food. If it is true then that agriculture is not a difficult concept but something readily available to hunting and gathering groups, and if it is true that its only advantage lies in the greater density of food produced, it follows that agriculture will occur only in situations where greater productivity per unit of space is required.

The Demand for Increased Productivity

Increased demand of the type which would promote agriculture is not necessarily synonymous with population growth. As Bronson (1975) has pointed out, the demand for greater productivity per unit of space can be stimulated in a number of ways, not all of which require population growth or even imply population pressure. Increased productivity might be stimulated simply by the desire of a group of people to obtain more of a particularly prized plant than was available naturally. However, while this explanation might well account for the initial cultivation of special medicinal or industrial plants or spices, it would hardly account for a general shift to the cultivation of plants whose utility was primarily caloric and which, as we have seen, tend to be low-preference foods. The desire of a population to consume at a higher rate or to accumulate surplus for purposes of taxation or exchange might also stimulate increased productivity.

Increases in population density requiring concomitant intensification of food resources might also result simply from cultural choices. A human population might concentrate for defense or for the purpose of expediting the exploitation of some desirable but spatially limited resource such as mineral deposits or water. Or it might concentrate as a result of centripetal social forces. The desire to create large aggregates of people for social, political, or even religious purposes could result in artificially high population densities in particular areas. Similarly, the evolution of new sociopolitical systems capable of maintaining order among larger aggregates might account for the emergence of high local population densities.

The demand for greater productivity might also be stimulated by changes in the environment which reduced either the natural density of particular wild resources or the total area available for exploitation. Thus, for example, Lanning (1967a,b) has linked the development of agriculture on the Peru coast to the decline of certain natural resources in the region which he considers to have been the result of climate change; and Gorman (in press) has noted that rising post-Pleistocene sea levels resulted in a significant reduction in the total Southeast Asian landmass, which may have caused an increase in population density and triggered attempts by local populations to increase the density of their food plants.

All of these factors are potential causes of stress on resources requiring increased productivity per unit of space, and all, alone or in combination, have undoubtedly contributed at various times and places to the intensification of resource utilization leading toward agriculture. But all are insufficient as explanations of a general worldwide shift toward agriculture. Many of these factors are reversible and in the long run might be expected alternately to increase and decrease density requirements. Others are occasional events which correlate only rarely with individual historical instances of agricultural origins. Others, such as concentration for defensive purposes or incentives to produce surplus, involve variables which, from what we know of the archaeological record, tend to occur only after dense agricultural populations have evolved. In other cases, where the archaeological data are not clear, logic would seem to dictate the priority of population growth. I suspect, for example, that cultural incentives to increase the local population aggregate and the development of political machinery to permit these larger numbers are more likely to result from, or interact with, growing population than to precede it. Improvements in political organization, it would seem, are more likely to grow out of trial and error attempts to cope with large group size than to be developed *in vacuo* and only later applied to larger aggregates.

Similarly, some of the environmental variables discussed above would function to increase stress only in the context of already high population pressure. Changes in climate or the reduction of inhabitable landmass such as occurred in many regions at the end of the Pleistocene would be significant only if population was already approaching critical densities. The combination of dense population and environmental changes might trigger the demand for agricultural production when these changes themselves were insufficient. The occurrence of this *combination* of events at the end of the Pleistocene might explain why the last glacial retreat had economic consequences which were not observed at the end of the previous glaciations. Even in combination with high population densities, however, climate changes at the end of the Pleistocene could only have been of regional significance, accelerating the development of population pressure in some regions and presumably decelerating it in others. Thus, allowing for local variations produced by natural or cultural variables, it would seem that there is only one factor which can account for the irreversible and nearly universal adoption of an agricultural economy: a general pattern of population growth to levels in excess of the densities which hunting and gathering economies would support.

Population Growth, Carrying Capacity, and Stabilizing Mechanisms among Hunters

It is at this point that the argument encounters serious opposition from a number of sources and begins to run counter to what is still commonly accepted in anthropology. I will argue that, historically, human populations have in fact tended to grow almost continuously, and that this growth has forced them more or less constantly to define new adaptive equilibria with their environments. This assumption is counter to a number of statements in the recent literature in both anthropology and biology, which suggest to the contrary that animal populations and human hunting and gathering populations tend to stabilize in numbers and to achieve stable (or at least roughly balanced, fluctuating) equilibria with their resources.

One of the questions which occurs immediately, of course (and one which has received a good deal of attention in the recent literature), is whether or not human populations *can* grow under the conditions imposed by the hunting and gathering life-style and whether or not this growth is limited totally or in part by biological or cultural constraints imposed by a nomadic existence. Polgar (1975) has noted that in the course of the evolution of *Homo sapiens* a number of changes occurred in female reproductive physiology and in child-care patterns which have significantly increased human potential for population growth in comparison to the growth potential of our primate ancestors. At the same time, however, various authors have noted that the exigencies of the hunting and gathering life-style may effectively limit population, or at least force human groups to adopt cultural practices whose cumulative effect is to eliminate or severely limit population growth. In the first place, infant and child mortality is known to be extremely high among hunting and gathering groups. Denham (1974), for example, estimates that average infant mortality among hunting and gathering populations may be in excess of 50 percent. In addition, various authors (see Birdsell 1968) have suggested that the rigors of carrying and nursing children may require the relatively wide spacing of births among hunter-gatherers. Nursing tends to suppress ovulation, and Lee (1972a) has suggested that the wide spacing of births among the Bushmen may be related to the suppressive effects of prolonged nursing. In addition, body weight is known to affect fertility, since body-fat composition affects both the age of menarche in girls and the regularity of ovulation in adult women (Frisch, cited in Coale 1974). Nancy Howell (cited in Coale 1974) notes that low levels of body fat among the Bushmen may result in irregular ovulation, particularly when nursing adds an additional drain on body resources, and that this, in turn, may help to account for the wide spacing of children. On a more mechanical level, it is possible that the rigors of bearing and carrying children under nomadic conditions may both increase the natural abortion rate and encourage potential parents to practice artificial population-limiting techniques.

Sussman (1972) has suggested in fact that the requirements of child spacing among hunting and gathering groups combined with high child mortality would effectively have prevented significant population growth among these populations. He calculates that women in hunting and gathering societies would have been forced to space out their children over four-year intervals to avoid excessive labor in child transport. The spacing would be accomplished by natural infertility resulting from prolonged lactation or from occasional infanticide or abortion. Sussman estimates that women in the Pleistocene would have been fertile for a period of only about sixteen years, from age sixteen until death, estimated (after Deevy 1960; Vallois 1961) to have occurred on the average not much after age thirty. He then calculates that such a woman would have an average of only slightly more than four live births in her lifetime, an estimate that corresponds well with observed birthrates (Carr-Saunders 1922; Krzywicki 1934; Neel et al. 1964). With natural infant and child mortality of approximately 50 percent, women would average only slightly better than two children who would themselves survive to adulthood. This reproduction rate, of course, would just allow replacement of the adult population or very slight growth.

According to this model, zero population growth might result as the accidental by-product of a child-spacing strategy necessary to insure the survival and well-being of the mother.

Denham (1974; see also Yesner 1975) has also argued that Pleistocene population was limited primarily by the exigencies of the hunting and gathering existence, although his interpretation of the mechanisms involved differs sharply from the model proposed by Sussman. Like Sussman, Denham notes the high rate of infant mortality common in hunting and gathering groups, but he takes issue with the assumption that the spacing of children was a major factor in population control. Denham argues that most hunting and gathering populations did in fact grow most of the time, but he argues that this growth was counteracted by periodic sharp reductions in local populations or even the extinction of the local group resulting from such factors as random variations in birth spacing and sex ratios, epidemics, natural disasters, or the failure of local resources. According to his reconstruction, the incidence of local population failure was sufficient in the Pleistocene to maintain the overall stability of the human population without recourse to cultural population controls such as infanticide.

Other scholars, however (Hassan 1973, 1975a; Hayden 1972; Polgar 1975; Birdsell 1968; Divale 1972), have tended to minimize the role of cyclical decline or natural disasters as population checks and, at the same time, to take a more optimistic view of potential population growth among hunter-gatherers. While not denying the existence of physiological spacing mechanisms, naturally high infant mortality, or the need for the spacing of births to protect the mother, this group has tended to assume that the growth limits imposed by these mechanisms are not so stringent as postulated by Sussman. These authors tend to assume that hunting and gathering populations are capable of more rapid growth than has been demonstrated for the Pleistocene, and they assume that during this period population growth was limited primarily by cultural means, including the institutionalized use of abortion, infanticide, and intra-specific aggression, whose function was not so much the spacing of children as the maintaining of populations at levels comfortably within the resource potential of their environments. Birdsell (1968) estimates, for example, that between 15 and 50 percent of all live births were eliminated by systematic infanticide during the Pleistocene as part of the cultural machinery for creating equilibrium between population and resources. Hassan (1975a) has suggested that an abortion or infanticide rate on the order of 25-35 percent would have been necessary to account for the difference between potential and real growth rates during the Pleistocene, and he suggests that implementation of these birth control techniques may have been determined primarily by the parents' perceptions of the relative economic costs and benefits of additional children rather than by the necessity of strict child spacing. Divale (1972) has suggested that a whole complex of population control mechanisms including systematic female infanticide and inter-group feuding may have been universal to human populations for at least the last 70,000 years.

My own inclination is to agree with those who stress the importance of cultural checks on population. I tend to discount the possibility that periodic disasters could have acted as a major factor in limiting population growth. I have no doubt that such disasters occurred and that local populations occasionally became extinct, but the relative health of hunting and gathering populations and the ecological stability of their food supplies seem to suggest that if anything such disasters would have been less frequent among hunting and gathering groups than among early farmers. Similarly, although random fluctuations in birth spacing or sex ratios may have occurred at the local level, it seems unlikely that these events could have had as significant a disruptive effect as Denham implies, since well-documented social mechanisms exist among hunting and gathering groups to equalize such irregularities among adjoining populations. I also tend to discount Sussman's pessimistic assessment of the growth potential of hunting and gathering groups. For one thing, it is not clear that any of the biological limiting factors which are mentioned are universal among hunter-gatherers or unique to that group of populations, nor is it clear that any of these factors regularly operate more effectively among hunting and gathering groups than among other populations. For example, as Hassan (1973) points out, it is not clear that the problems of child transport are really so rigorous that they would provide a major determinant of child spacing. Neither is it clear that the rate of infant mortality is particularly high among hunting and gathering groups in comparison to that of primitive farmers. In addition, the documented capacity of hunting and gathering groups to expand rapidly when faced with an open environment where space and resources are temporarily unlimited (Birdsell 1968) is inconsistent with the assumption that population growth is limited by any inherent biological incapacity to grow and is equally inconsistent with the assumption that the hunting and gathering life-style necessarily enforces low reproductive rates. Clearly hunting and gathering populations *can* grow; apparently under most conditions potential growth rates are dampened by cultural mechanisms.

The really important question with regard to a population pressure model, however, is not just whether human population can grow, but whether the postulated cultural control mechanisms will permit population growth in *bounded* environments, when such growth implies increased population density and consequently threatens the carrying capacity of local resources. Birdsell (1958, 1968) has suggested that hunting and gathering populations regularly stabilize at, or below, the carrying capacities of their environments and that except for minor fluctuations such populations show no tendency to exceed their carrying capacity. Following Birdsell's lead, a number of other scholars have recently affirmed that hunting and gathering populations actually stabilize at levels well *below* (equal to perhaps 20–30 percent of) the carrying capacity of their environments (see Lee and DeVore [eds.] 1968; Lee 1968, 1969; Flannery 1969; Woodburn 1968a; Balikci 1968; Wynne-Edwards 1962; Deevy 1960; Hainline 1965; Hayden 1972; Polgar 1975). The general theme of these studies is that human (and many animal) populations are regulated in their numbers not by Malthusian checks but by homeostatic social mechanisms which cause population growth to level off long before the food supply is threatened. This observation presents a serious obstacle to a population pressure model. If population in fact stabilizes below carrying capacity and for reasons other than the competition for scarce resources (cf. Benedict 1972), then population pressure on resources cannot build up and cannot cause economic change. Even Lewis Binford, who has himself proposed one of the pioneering formulations of a population pressure hypothesis for agricultural origins (1968, see chapter 1), assumed that population pressure could build up only under very special conditions because equilibrium mechanisms in human population systems were so widespread and so efficient. His statement (1968:326) can well stand as a model of this position:

> Most demographers agree that functional relationships between normal birth rate and other requirements. . . favor the cultural regulation

of fertility through such practices as infanticide, abortion, lactation taboos, etc. These practices have the effect of homeostatically keeping population size below the point at which diminishing returns from the local habitat would come into play. . . . These data [on contemporary hunting and gathering populations] suggest that . . . within any given habitat the population is homeostatically regulated below the level of depletion of the local food supply.

Binford also argues that human population cannot tend constantly to grow and put pressure on available resources, because if this were true then:

man would be continually seeking means for increasing his food supply. . . . There would be ubiquitous and constant selective pressure favoring the development of technological innovations such as agriculture which serve to make larger amounts of food available to the group. [But] *there is a large body of ethnographic data which suggests that this is not the case* (italics mine).

A statement from Flannery (1969:75) is also of relevance:

A growing body of data supports the conclusion stated with increasing frequency in recent years that starvation is not the principal factor regulating mammal populations. Instead, evidence suggests that other mechanisms including their own social behavior homeostatically maintain mammal populations at a level *below* the point at which they would begin to deplete their own food supply (italics his).

The widespread acceptance of this position has resulted in the frequent dismissal of population growth as a causative factor in cultural evolution, at least among hunting and gathering groups. Or, as in the case of Binford and Flannery's work, it has led scientists to what I believe are unnecessary extremes to explain "special instances" of population pressure. Such assumptions about the functioning of equilibrium systems in human populations, however, represent, I believe, both a theoretical error and a misreading of the empirical evidence.

In the first place, these conclusions (as regards our own species) are based on the analysis of certain modern human populations which I suggest represent a skewed sample of human groups. Quite clearly, contemporary hunting and gathering groups are anomalous simply by virtue of the fact that they are still hunters and gatherers. (They may even have genetic peculiarities which affect population growth. Stephens [1975] has noted, for example, that genetic factors as well as environmental conditions may affect age at menarche. Similarly, if genetic factors help to determine the storage of body fat, as seems evident, then the ovulation patterns of adult females may be in part genetically determined.) Moreover, traditional anthropology is largely defined as a social science by its emphasis on nonwestern "primitive" culture groups. Such groups quite obviously are preselected for their technological conservatism. If it is correct that technological change, at least in the food quest, is largely equatable with population growth, then such groups must also be preselected for the stability of their populations; they are by no means representative of the human condition. If one looks at the archaeological record (chapters 3-6) it would appear, despite Binford's statement, that there is in fact a large body of data which suggests that human populations are, and have been, continually seeking means for increasing the food supply.

A more serious problem, however, lies in the widespread misuse of the "carrying capacity" concept in its application to human populations. This concept implies the existence of fixed population ceilings related to the productive capacity of the environment. According to this model there is a specific fixed maximum level of consumption of any resource which the environment can tolerate. Consumption at or below this level is compensated for by the regenerative powers of the resource. Consumption above this level exceeds the regenerative power of the resource and results in the destruction of the system. For animal populations (such as horses, for example) with relatively immutable behavior patterns and comparatively well-defined, sharply bounded resources, the concept is of considerable importance. If the horse eats more grass than the grass population can regenerate, the system breaks down. Therefore it is incumbent on the horse population to maintain its numbers by some means other than outright competition and starvation below the level at which competition begins to destroy the grass supply.

Population pressure arguments for human beings are traditionally couched in similar terms. Population pressure is assumed to refer to the growth of the population to the point where it necessitates consumption levels beyond the carrying capacity of the resources, resulting in environmental degradation. The traditional counter-arguments (see Binford, cited above) are based on the observed failure of human populations to approach this critical level.

However, the carrying capacity concept is difficult to apply to human populations and, I believe, may have little relevance to human biology. At best, the concept can be used as a measure of the relationship between a population and its economic strategy at a particular point in time (cf. Zubrow 1971, 1975), but in no case should this measurement be construed as indicating a fixed ceiling on potential consumption or on potential population growth.

Hayden (1975) has documented the difficulty of dealing with carrying capacity in practical terms. He notes (following Birdsell 1968) that it is extremely difficult to measure the actual carrying capacity of any region at any point in time; and he also points out the practical difficulty of applying an essentially static concept to resources which vary in their availability through cycles of greater or lesser duration (see also Ammerman 1975; Yesner 1975). Finally, he notes that the carrying capacity of human populations is an elastic quantity geared, at least to some extent, to demand (see also Faris 1975; Street 1969).

It is the latter problem that I am primarily concerned with. In the first place, human populations have an extraordinarily broad range of potential foods because of their omnivorous capabilities and flexible (culturally controlled) patterns of food selection. Human groups rarely exploit all the resources available to them. Actual consumption is determined by cultural choices, which in turn are based on a variety of factors: food preferences, nutritional needs, prestige, labor costs, and activity preferences. Thus, population pressure need not take the form of exhausting resources or approaching "carrying capacity." As population grows, some preferred resources may in fact be exhausted. Others may be subjected to conservation measures of the type already documented among hunting and gathering populations. Still others may become increasingly hard to get, long before carrying capacity is reached. In any case, the result will be that people begin to transfer their attention to other foods. Thus, in human societies, population pressure may simply take the form of a progressive shift from preferred foods to less and less desirable resources.

Furthermore, we know, as documented above, that there are a large number of behaviors human populations can perform to *increase* the quantity of particular resources available in the environment—techniques such as burning off of ground cover, fertilization, creation of clearings, planting of seeds, irrigation, and so forth—which are widely practiced by nonagricultural peoples and presumably known to other populations who do not bother to use them. Thus, the quantity of any food available in nature is not fixed as far as man is concerned. It is a variable dependent in part on the amount of effort people put into promoting growth. (Some food resources of course are more variable and more responsive to human attention than others.) Population pressure, then, in addition to causing a population to shift to eating secondary foods, should also lead them to increase the amount of effort they put into promoting the growth of their food resources. A growing population could consume more and more food, but compensate, first by eating a wider range of foods and second by artificially increasing the quantities of wild food available. As population continued to grow, more and more techniques for increasing the food supply would be brought to bear until full-scale agriculture and, later, increasingly productive forms of agriculture were adopted.

In short, food ceilings on human populations are flexible; carrying capacity as a static limit inherent in the productive potential of the natural resources of a region has little meaning. The equilibrium in question is not that between man's population and his food supply, but that between man, his cultural preferences, his level of labor investment, and his resources as modified by his behavior. It is therefore quite natural that human societies rarely appear to approach "carrying capacity." Population pressure must be redefined. *It is here defined as nothing more than an imbalance between a population, its choice of foods, and its work standards, which forces the population either to change its eating habits or to work harder (or which, if no adjustment is made, can lead to the exhaustion of certain resources).* By this definition, population pressure can be seen to motivate technological change in the food quest without ever threatening carrying capacity in the absolute sense, without ever reducing the human population to starvation, and without threatening to break down the ecosystem.

This altered notion of population pressure, incidentally, answers one criticism which is often leveled at the concept of population pressure (or stress in general) as an explanation of the origins of agriculture. It has been argued, for example, by Carl Sauer (1952:20) that agriculture could not arise out of food shortage because people under stress do not innovate; they lack leisure time for experimentation with new techniques and cannot risk their scanty supplies on the uncertain promise of distant future rewards. It has already been pointed out, however, that experimentation probably has nothing to do with it. The techniques are already widely known; all that remains is to implement them. More important in the present context, however, is the realization that the pressure to change need not imply a crisis, but merely an awareness on the part of a people that they will have to change their eating habits or work harder at assisting the growth of their food plants in order to maintain their standard of living.

I do not intend to argue, incidentally, that *all* populations necessarily have grown through history. It is fairly well documented that effective population control mechanisms are known to essentially all contemporary cultures and that they are widely practiced (Birdsell 1968; Devereux 1955, 1967; Watanabe 1968; Balikci 1968; Laughlin 1968; Marshall 1962; Sussman 1972; Hassan 1973; Dumond 1975). If we include infanticide then it is certainly a reasonable inference that all prehistoric populations had the capability of restricting population growth as well. As already indicated, many scholars assume that these practices were extensively employed in the Pleistocene. Thus, some prehistoric populations may well have chosen consciously to limit population and maintain a high economic standard; others may have been limited by the application of unconscious spacing or fertility reducing mechanisms such as those outlined by Stott (1962) or Wynne-Edwards (1962); and some may even have been limited by inherent physiological factors such as a stronger than average tendency for lactation to suppress ovulation. Binford and others are accurate in their observation that many contemporary groups display successful population control mechanisms and achieve balanced equilibria. It is evident, however, that not all contemporary populations are so well adjusted. Many groups in the world today are growing and are tending to redefine the ecological problems of the human species as a whole. I believe that the archaeological record, discussed in subsequent chapters, indicates that whatever stable populations existed in the past similarly coexisted with growing populations, which required a pattern of more or less continuous ecological adjustment on the part of the species as a whole.

I also do not mean to deny the existence or the ecological impact of the kind of random, short-term fluctuations in local populations which have been described by Ammerman (1975; see also Denham 1974; Bronson 1975; Yesner 1975). Clearly, on a small scale, both demographic events and environmental variables can result in profound short-term changes in population density and structure. There are several reasons, however, for minimizing the importance of these fluctuations in the present context. First, as will be discussed below, I believe that mechanisms exist which rapidly and effectively level out such local fluctuations, minimizing their ecological impact. Their primary importance may in fact be the slight dampening effect they have on the overall rate of population growth. Second, possibly because of the efficacy of the leveling mechanisms and because of the gross nature of the archaeological sample, the impact of such fluctuations on the archaeological record is minimal. The overall pattern still appears to be one of slow population growth. Finally, while the occurrence of these fluctuations may be of significance for understanding the pattern of prehistoric events on a small scale, such random patterns of population change, which are highly localized, short-term, and reversible, can have no bearing on the general pattern of human economic evolution which is the focus of this work.

The Rate of Population Growth in the Pleistocene

It is still necessary to account for the apparently very slow rate of population growth and technological change during the Pleistocene and to relate this slow rate of growth to the argument being developed here. Several authorities have argued that the slow rate of population growth is incompatible with a population pressure model of technological change. Cowgill (1975b), for example, suggests that the very slow rate of population growth in the Pleistocene is inconsistent with the assumption that population tends constantly to increase. In a similar vein, Bronson (1975) has argued that when population does grow it is inherently too rapid a process to explain the very slow evolution of Pleistocene technology.

By all estimates, Pleistocene population growth *was* slow in comparison to rates observable in later periods. Cowgill's estimate of .0030 percent per year is roughly in keeping with other estimates that have been offered. Hassan (1975a) estimates Pleistocene population growth at an average annual rate of between .0010 and .0020 percent; Dumond (1975) has estimated a rate of between .0007 and .0015 percent; and Polgar (1972) has estimated Pleistocene population growth as less than .0030 percent. These estimates contrast with an annual rate of population growth during the Neolithic estimated at .1 percent (Hassan 1973; Hole et al. 1969; Carneiro and Hilse 1966) and with growth rates of 1 or 2 percent or more per year observed in contemporary history.

These estimates for Pleistocene population growth may be slightly low, but they must be of essentially the correct order of magnitude. The growth rate calculated depends in part on the estimate that is given for the earth's population just prior to the emergence of agriculture. A number of scholars estimate the terminal Paleolithic or Mesolithic population of the earth at between three and five million individuals, or provide density estimates of approximately .1 person per square mile, which implies a similar figure (Hassan 1973; Keyfitz 1966; Dumond 1975; Braidwood and Reed 1957; Deevy 1960). Such estimates are based either on the relative densities of archaeological sites for the period in question or on ecological assumptions about the size and density of the human population that could have been supported by hunting and gathering. I suspect that, whichever basis for calculation is employed, there may be a strong tendency to underestimate the terminal hunting and gathering population. I will argue in the next chapter that the archaeological sites of this period are systematically underrepresented; and I would also suggest that the prevailing assumptions about hunter-gatherer population dynamics may be skewing our estimates downward.

The earth's land surface is about 50 million square miles, not counting the Antarctic, the only major landmass which was uninhabited 10,000 years ago. The very territorial extension of man from Africa to the eastern United States and Tierra del Fuego implies that the land between these extremes was reasonably filled. Since hunting and gathering populations have been observed to expand territorially at population densities well below levels which their existing technology could support (Birdsell 1958), these intervening areas admittedly need not have been filled to capacity for this territorial dispersal to occur. On the other hand, if one accepts the evidence of high population pressure given in subsequent chapters and if one believes, as I do, that agriculture spread because the world was, in effect, saturated with hunters and gatherers, then a relatively large post-Pleistocene hunting and gathering population is implied. Lee and DeVore (1968) have estimated that the theoretical maximum population density for hunting and gathering groups in most circumstances is about one person per square mile.

Clearly not all of the inhabited globe could have supported population densities approaching this figure; but assuming that the saturation point for hunting groups had been reached, an early post-Pleistocene population for the world of approximately 15 million persons does not seem unreasonable. Even allowing for this significantly larger estimate of Mesolithic population, however, the rate of population growth over the preceding two million (or more) years could not have been significantly greater than the estimates which have been offered. Even with this more generous estimate, growth rates during the Pleistocene could not have begun to approach those observed in more recent phases of prehistory.

The slow rate of population growth in the Pleistocene is not incompatible with a population pressure model, however, unless one takes an all-or-none position with regard to population growth and assumes that population either achieves equilibrium or grows unfettered. On the contrary, a very slow but steady rate of population growth resulting in the gradual buildup of population pressure might occur under any of a variety of circumstances. It would occur, for example, if only a small number of growing populations were generating the population pressure for the species as a whole. It would also occur if there were only a slight preponderance of growth over decline in the fluctuation patterns of local populations. It might occur if populations were biologically constrained at levels near zero growth by any of the mechanisms discussed above (although, as discussed above, I consider this an unlikely possibility). Most important, I suspect, a pattern of slow but steady growth might occur under conditions where cultural population-control mechanisms are in operation but where such controls are less than a hundred percent effective, since de facto population control never quite lives up to the norm. Various scholars (Woodburn 1968b; Hassan 1973; Dumond 1975; Cowgill 1975a,b) have pointed out that, despite certain pressures the community can bring to bear, birth control and infanticide decisions are often a private matter for parents alone to decide, rather than a matter of strict group policy. Parents make their choices based on their perceptions of a whole range of relevant variables. Moreover, as Hayden (1975) argues, strict population control for the group as a whole requires significant effort and expenditure of energy, and groups will tend to spend only as much energy in this regard as is required for their survival. He notes also that there will be a natural tendency for structural degeneration to occur in the population control system due to accidents in cultural transmission or other sources of deviance, and that this tendency will be counteracted only to the extent that deviance is selected out. He suggests that the strictness with which such systems are enforced will be a balance between the risk of starvation (and, I might add, hard work or lowered economic standards) on the one hand and the excessive energy costs of overly strict enforcement on the other. Under such conditions, slight population growth might easily occur despite a cultural norm favoring population stability. In fact, natural selection among human cultural systems might favor the evolution and persistence of slightly imperfect population equilibrium mechanisms. The total absence of a population limiting system might be disastrous to a population since it might lead to the destruction of the environment. Or too rapid population growth, even if it did not threaten the environment, might require rapid behavioral adjustments which would threaten the integrity of the culture system. Any such system presumably has a limit to the type and rate of behavioral change it can absorb without losing its identity. Population growth forcing cultural change in excess of these limits would be destructive. On the other hand, a perfect equilibrium system might be equally disastrous in the long run. Such a system, if it resulted in total demographic stability, might result in cultural stagnation in the midst of a highly competitive and changing world. Populations whose equilibrium systems were too perfect and lacked flexibility may have tended to be eliminated historically or pushed into marginal environments by those populations that allowed continued slight population growth and as a consequence continued to grow technologically.

The ideal population control device may be one whose tolerance of deviation from the norm of stability is geared to environmental circumstances and to the capacity of the system to absorb cultural change. The same population could then allow slow or rapid growth or even maintain total stability of numbers, depending on its immediate problems, without needing to overhaul its value system. Under normal circumstances, growing population among hunter-gatherers in a bounded environment will result in either reduced quality of diet, harder work, or both, since at this level of technology the problem of increasing demand is not offset by any concomitant increase in efficiency. Unlike modern productive strategies, hunting and gathering strategies usually have little to gain by increasing the scale of their operations (Zubrow 1975). Under such circumstances population growth should be extremely slow. All people suffer if population density increases; therefore they will bring strong social sanctions to bear on those who reproduce in excess. Tolerance of growing population would then be geared to the rate of change that people will tolerate in their life-style. If they will tolerate no change, population growth may be stopped entirely; more likely, it will be limited to the level at which change is not perceptible over the short run or perhaps even within the life span of any one individual. In this manner, a population in a bounded environment may well tolerate very slight changes in subsistence strategy based on occasional or very slow population growth, yet invoke strict application of cultural norms for population control when the rate of change seriously threatens contemporary standards.

On the other hand, if an expanding population can move into open territory of an ecological type similar to or superior to its present habitat and requiring little or no change in exploitative strategy, quality of diet, or labor costs, little heed may be paid to family planning; population growth may well approach its maximum rate, doubling or tripling in each generation. If, however, the colonized territory is of a new ecological type, with an unpleasant climate requiring harder work and providing poorer food, or simply necessitating a new life-style, toleration of population growth will again be geared to the rate of culture change the group will accept.

A Model of Population Growth on a Small Scale

Let us consider for a moment the model of a hunting and gathering population and look at its alternatives. If the Kalahari Bushmen are in any way typical, we can assume that such a group can obtain an adequate and nutritious diet with a minimum of work by exploiting the wild resources within a loosely defined radius of its camp. Usually such exploitative systems seem to support less than one person per square mile (Lee and DeVore 1968b:11), so that the size of the group is limited by the distance people are willing to walk to obtain food. Among modern hunting and gathering groups an exploitation radius of a few miles seems typical. A radius of 6 miles, such as that estimated by Lee (1968) for the Bushmen, will bring approximately 100 square miles within the area of exploitation. It is thus characteristic of such groups that the population of a single camp is rarely more than 100 people and is typically closer to one-half or one-quarter of that figure (Lee and DeVore 1968b:11). Depending on local conditions, the group may move once or more during the year to exploit new areas as preferred resources become scarce around their old campsites, or they may move from place to place for any of a host of other reasons.

The high quality of the diet and the low labor costs involved can be maintained as long as the population is constant. But this nice balance may be threatened if population tends to grow beyond these limited figures. It is my contention of course that for many, if not most, human populations, this tendency has been a constant problem to be dealt with. Increased population threatens the group either with a decline in the quality and quantity of food available or with an increased work load (per capita) or both. I think we must assume that the people involved would be capable of realizing, in practical if not in theoretical terms, that the more mouths they had to feed the harder the gathering process would be. The group then has several alternative solutions to its dilemma. First, it can choose to limit its population. This is the strategy most in keeping with retaining both the quality of the diet and the low labor costs. It is not, however, the only solution; nor, I suggest, is it the one that has prevailed historically. There are several alternatives. A group can increase the radius it exploits (which involves increased labor costs in travel), or it can send some of its members off to form new daughter camps elsewhere. These two alternatives in combination are of course the manner in which hunting and gathering groups populated the world. The group may also send some of its people to neighboring, less populous camps. It can search harder for increasingly less readily available food sources within the exploited area (again implying greater labor costs); or it can turn to resources that have hitherto been avoided: less preferred or less nutritious foods; foods which, while equally nutritious and palatable, are harder to find or to prepare; food resources whose gathering involves demeaning (low prestige) activities; or foods occurring in microenvironments within its area that have previously been considered unproductive, dangerous, unhealthy, taboo, or simply unpleasant. The group might also be forced to respond to growing population by beginning to store wild produce from seasons when food is plentiful for use in seasons when it is scarce. If, in the course of further population growth, they gradually became heavily dependent on stored foods harvested in one or a few seasons of the year, they might actually be forced to settle down in the vicinity of their storage bins. Alternately, if they were accustomed to moving seasonally between desirable but relatively scarce resources and less preferred but readily available resources, they might eventually be forced by growing population to settle permanently in the vicinity of the more plentiful resource, which hitherto they had relied on only in emergencies or during the lean season.

We can in fact assume that some combination of these alternatives would be tried and that, as population continued to grow and pressure mounted, more and more of these strategies would come into play. All such responses, however, are inherently limited in their adjustive capacity. An increase in the area exploited from a single village or the budding-off of daughter colonies would eventually abut against territorial constraints. Such constraints might be obvious ecological boundaries such as shorelines or mountain ranges; they might be formal political boundaries; or, if no such strict political boundaries existed, territorial constraint might take the form of competition with other groups which made the new territories as uninviting as the old. Similarly, working harder to find wild foods within the same area, settling for less desirable resources, or storing foods which are plentiful in particular seasons are only temporary solutions to growth, since population will soon tend to outstrip the new standards just as it did the old. At this new level, the decision might again be made to limit population and stabilize the economy. For those who do not take this option, however, and continue to allow population to grow, the only other alternative is eventually to begin artificially to increase the density of desirable crops within their gathering radius by the application of one or more of the techniques which together constitute agriculture. As population grows, more and more of these will be brought to bear; but any new technique will occur only when the people decide that the labor involved in it is the least of evils. Each technique represents added labor costs, and will only be undertaken when these costs are outweighed by the disadvantages of retaining the old methods.

All of this of course allows for a wide range of latitude in group decision. Cultural differences in the *perception* of labor costs, shortage, or the desirability of certain foods mediate between population growth in the absolute sense and population pressure as a force for culture change, but these factors would appear to be nothing more than minor variations on the general theme. More important, perhaps, are the cultural factors weighed in the choice of strategies. Are there good rational reasons for wanting or needing larger population despite the costs? Are there cultural norms lending prestige to large families or requiring that each family have, for example, at least one son? Are there attitudes preventing the use of certain effective population control techniques? Do the joys of parenthood outweigh the perceived additional labor costs of population increase? Is it considered harder to walk farther to collect food, to move the village more often, to store food, or to weed the wild "garden"? How important is it to have the meat of large mammals in the diet?

It is clear given these alternatives that populations have an enormous choice of possible strategies. Some in fact have chosen (or been forced by biological limitations) to limit population at densities appropriate to hunting and gathering. Some populations just as clearly have not. All these variables, however, are regulated ultimately by the one overwhelming factor of competition. If we assume that various groups of hunting and gathering people made different decisions in these situations, that some limited population while others did not, how do we explain the scarcity of hunting and gathering groups? The answer is that at each stage those who for one reason or another chose to allow population to grow and responded to need by devoting extra labor to the food-gathering process were able to compete successfully for space with those groups that chose to remain small or for genetic reasons could not grow. Thus, even if only relatively few populations allowed population growth and compensated with intensification of their resources, they would over a period of time replace the more conservative groups in all but the most marginal environments. Moreover, awareness of such competition may very well have been one factor involved in the original choice of many populations not to limit population. Thus the population has a choice of several adaptive strategies at any point in time, but only one strategy—the growth of population and the consequent intensification of resources by application of agriculture techniques—is viable for most groups in the long run. The other choices either provide only temporary solutions which delay but do not eliminate the beginnings of agriculture, or they lead to evolutionary dead ends, since such populations cannot ultimately compete with those utilizing agriculture.

Incidentally, this model of expanding population may help us to understand the relationship between animal husbandry and plant domestication. It is my intention as much as possible to avoid complicating the discussion with a consideration of animal domestication, but some comment on the place of this activity in the evolution of human economic systems seems warranted. The problem is that, unlike plant domestication, which has clear benefits for a human population in terms of increased production of calories, there is no evidence that animal domestication at a primitive level increases the overall production of animal biomass. Bronson (1975) argues in fact that the replacement of wild game by domestic animals would probably have the reverse effect. How then would animal domestication relate to a model of increasing population pressure? The answer, I think, lies in the effects of population pressure on human geographical distribution. Wild animal populatons may be more efficient than domestic ones in their overall use of space, but their habits are not amenable to generalized human exploitation once severe territorial competition among human groups restricts the area available to each group or once individual groups become committed to sedentary behavior involving the storage of seasonal resources. Since vegetable foods provide the main sources of calories under conditions of intense population pressure it is these resources which will increasingly dictate the habits of most human groups. The domestication of animals, I suggest, is primarily a means of subjecting them to the requirements of an exploitative system geared more and more to intense exploitation of spatially limited vegetable resources (whether or not these resources themselves are yet to be considered "domesticated"). The only other alternative for human population, as demand for calories and protein continued to grow, would be to eschew the vegetable resources and follow the herds in the fashion of nomadic pastoralists; but this alternative would be available only to a very few.

Population Growth and Pressure over a Large Geographical Area

Before we proceed to apply this population growth model to world prehistory, one other major issue must be dealt with. What is the relationship between population growth on a local scale and technological growth patterns which demonstrably cover a very large portion of the globe? Bronson (1975) has objected to a demographic stress model, arguing that independent local growth of population does not proceed evenly but fluctuates wildly, and therefore cannot be related to technological progress, which does not show the many reversals noticeable in population patterns on a local scale. He has also pointed out that if population growth in individual groups were responsible for changes in food technology, we should expect to see enormously wide chronological variation in the origins of agriculture in various parts of the globe.

The answer to these problems may be found in large part in the competitive success of denser populations and the tendency of competition itself to promote broad parallelism in the development of population density and, therefore, in the evolution of agricultural technology. Since there is a tendency for the denser populations to expand at the expense of the less dense, population density relative to environment will always tend to even out. This will be particularly true once agriculture has developed in any region. Relatively rapid expansion of agricultural populations may help to build up population pressure in adjoining areas, and this fact may in part account for the tendency of agriculture to be adopted over a wide region. However, the competitive success of an agricultural economy cannot account entirely for the chronological parallelism observed in the origins of agriculture, and it certainly cannot account for the widespread parallelism in the world-wide buildup in population pressure prior to the development of agriculture (which will be discussed in subsequent chapters).

The answer to both of the dilemmas presented by Bronson may lie at least in part in the widespread occurrence and apparently great effectiveness of population equilibrium mechanisms, not within each local population, but *between* adjoining populations of hunters and gatherers. The existence of these mechanisms suggests that population growth during the Pleistocene may have been geographically a very diffuse process; that local fluctuations would have been balanced out; and that widespread regional parallelism in the buildup of population pressure during the Pleistocene might be expected.

Turnbull (1968:132) has called attention to what he refers to as population "flux," or changeover of personnel, between local groups of both Ik and Mbuti pygmy hunting and gathering groups. Similarly, Woodburn (1968b) and Lee (1968) document a very loose type of local group structure among the Hadza and the !Kung Bushmen respectively whereby local groups as sharply defined human aggregates do not exist.

Rather, individuals are free to move from group to group, partake of local resources, and participate in whatever cooperative social efforts occur wherever they are. Lee and Woodburn both indicate that the turnover is in fact considerable. Moreover, it has been widely noted that in such simple hunting and gathering bands, conflict is widely resolved by group fission (Lee and DeVore 1968). One result of this fluidity is that local endogamy is minimized and local groups show little tendency to diverge culturally or genetically from their neighbors (Yellen and Harpending 1972). A second result, however, is that population density and pressure tend to be equalized from region to region (Ammerman 1975; Williams 1968; Lee and DeVore 1968). Given the widespread occurrence of such flexible group structures among hunting and gathering groups, this mechanism may in fact be a powerful force for equalizing population pressure over broad areas.

Even in the presence of more formalized group arrangements, for example, among the Australian aborigines, similar mechanisms may exist. Service's (1962) model of the exogamous patrilocal band, which he claims to be the basic group structure of hunting and gathering populations both now and in the Paleolithic, places a high premium on the movement of women between local bands. At a higher level of organization, Birdsell (1958, 1968) has called attention to the very high frequency (14 percent) of "intertribal" marriages among the Australians, referring not to band exogamy but to marriage outside the larger cultural and linguistic unit. Similarly, Owen (1965) has noted the importance of band exogamy and points out that rules governing exogamy combined with low population densities would necessarily result in a very high incidence of extratribal marriage. The fact that such marriages constitute the movement of women (the productive portion of the population) from one area to another suggests that one function of such movement may well be to equalize demographic stress (cf. Kunstadter 1972).

Strong evidence demonstrating the effectiveness of these density equalizing mechanisms comes from Australia in the work of Birdsell (1953). Birdsell studied rainfall figures and tribal areas for a series of 123 "dialectical tribes" of Australian aborigines living in ecologically uniform territories, but otherwise distributed over a large portion of the continent. He found that a negative logarithmic relationship occurred between mean annual rainfall and tribal area, and that there was a very high coefficient of curvilinear correlation (r=.8). Since the tribal groups studied were all of approximately the same size (numbering about 500 individuals), Birdsell in effect established a correlation between rainfall (the main population limiting factor in the desert environments studied) and population density. From this he drew two conclusions: first, that there was some mechanism by which human populations achieved balanced equilibrium with their environments; and, second, that the populations were stabilized approximately at the carrying capacity of their environments.

It is possible, however, to place a somewhat different, and, I believe, much more direct interpretation on the same data. Birdsell may not have demonstrated the existence of a mechanism by which a local population achieves balance with its own environment, but rather the existence of a mechanism by which a series of widely distributed populations equalize population pressure among themselves. Another way of stating his results is to notice that, correcting for rainfall, all 123 groups have approximately the same population density in similar environments. It seems unlikely that such even densities would be achieved by a series of populations adapting independently to local conditions even if the conditions were similar, since cultural differences in the interpretation of environment would lead to the establishment of different density equilibria. What is indicated instead is that these groups have to some degree common perceptions of environment and share mechanisms to move people around to equalize density and population pressure from region to region. Nothing in Birdsell's data suggests that any one group is in equilibrium with the carrying capacity of its environment, since the "carrying capacity" is in fact not measured (and is, I believe, unmeasurable). All the data show is that the densities of the various groups relative to their resources, as measured by rainfall—or, if you prefer, the population pressure levels of the various groups—are balanced *with one another*.

What is of primary significance is that Birdsell's data suggest that equilibrium systems governing the geographical dispersal of human populations can be, and are, effective on a huge scale. In his example, the better part of an entire continent is embraced in one vast population flux network. Since similar density relationships have been demonstrated in Micronesia (Hainline cited in Birdsell 1968) and among the Shoshone of North America (Vorapich cited in Birdsell 1968), it is evident that Birdsell's data are not unique. It would appear that among hunting and gathering populations, widespread and effective mechanisms operate to keep population pressure balanced over very broad areas.

This does not mean, of course, that population pressure is always in perfect balance from region to region, since these mechanisms would hardly be perfect. More important, however, it must be established that the existence of such mechanisms does not at all imply that population density per se will even out. What tends to balance out from region to region is population pressure as measured by a people's perception of the relative difficulty of obtaining resources and the relative quality of the resources obtained. Just as Birdsell's populations are adjusted to relative rainfall, we might expect populations in a variety of environments to show markedly different densities while experiencing approximately equal degrees of stress. Since, given flux, people will tend to gravitate to areas of high rewards and low labor costs, and since by moving they will tend to decrease population stress in the old area and increase it in the new, there will always be a tendency for cost-reward ratios to equalize from area to area.

One of the most important consequences of this system of population flux, incidentally, is that events or conditions on one part of a continent have a rather direct effect on cultural choices made in other locations. There is, for example, a feedback system between territorial expansion in one location and intensification of resource use or the exploitation of secondary environments in another. An open frontier will not only draw population from immediately adjacent areas; it will also tend to reduce population pressure throughout the culturally homogeneous territory which shares in the flux system. Conversely, barriers to migration will have the effect of concentrating population in secondary environments or enforcing efforts to increase food supplies not only at the edge of the expanding population but throughout the system. Thus there is an inverse relationship between the possibility of expanding into new territory in one region and the necessity of utilizing secondary resources or even taking up farming in another. Moreover, since expansion and intensification of resource use are to some extent alternative strategies, we should expect to see each taking place primarily when the other becomes difficult; the rate of intensification should be inversely correlated with the capacity for territorial expansion and should accelerate rapidly as expansion becomes more difficult.

Such a system of population flux might be theoretically capable of distributing population pressure evenly enough so that groups throughout the world would be forced to adopt agriculture within a few thousand years of one another. And the flux system described might also account for the fact that agriculture developed at the end of the period of territorial expansion not only along the frontiers of expansion but in other regions as well. There is a serious problem, however, in attempting to establish the actual existence of such a system during the Pleistocene. The best evidence is probably ethnographic analogy: the widespread occurrence of various flux mechanisms in modern groups suggests that they probably operated in the past as well. However, there are more direct hints that such systems may have been operating in the Pleistocene and that they may have broken down only very slightly before the advent of agriculture.

One type of evidence concerns human biology. The fact that most "racial" variation in human beings is in the form of clines rather than in the form of sharply bounded trait clusters would seem to indicate that relatively open mating systems have operated throughout most of human history, supporting the assumption that people have moved freely from group to group.

The other type of evidence is the presence of widespread artifact style-horizons and long-lasting stylistic groupings which may define the geographic extent of population flux networks. The problem here is to decide how to interpret such widespread tool complexes or horizons as the "Acheulian," the "Mousterian," the "Stillbay," or the "Clovis," and how to interpret the widespread, synchronous or roughly synchronous development of certain specialized flint-knapping techniques, such as the use of prepared cores or the manufacture of blades or microliths. Such groups may represent true cultural groupings implying the actual movement of people and the diffusion of ideas between local groups. In this case, I think, the tool complexes might plausibly be interpreted as representing population flux networks.

A number of archaeologists have in fact suggested that the stone tool assemblages defined by archaeologists do have sociocultural reality and that their extent defines the boundaries of prehistoric social interaction spheres within which people and information moved freely. Having noted the fluid structure of Bushman bands (see above), Yellen and Harpending (1972) suggest that the homogeneity of Late Stone Age

Wilton assemblages in South Africa is a product of this fluid structure which created open networks of exchange and prevented the development of sociocultural isolates, thus preventing the emergence of separate stylistic traditions. As they note, however, Wilton industries are only homogeneous over a fairly small region of southern Africa and they interpret the stylistic boundaries of the Wilton assemblages as reflecting the limits of this particular prehistoric social system.

Both Collins (1969) and Isaac (1968, 1972) have suggested that, at an earlier time level and over a much broader geographical range, the Acheulian tradition also represented a real sociocultural unit. Isaac in particular suggests that the distribution of stylistic variation within the Acheulian tradition reflects the operation of an open system of social interaction and informational exchange and he suggests in fact that the Acheulian represents the residue of a pattern of social movements of the type described by Owen (1965, see above). In studying the range of variation in Acheulian assemblages from different sites, Isaac makes several key observations. He notes that the Acheulian tradition as a whole is remarkably homogeneous and conservative, but that there is a good deal of variation in the specific permutations and combinations of tools which occur at specific sites. Most of the variation, he concludes, is random, displaying neither patterns of geographical clustering nor patterns of functional specialization. He suggests that the variation primarily represents local idiosyncratic interpretation of the main cultural tradition, a kind of cultural "drift." He notes, however, that the cultural drift never seems to lead to the emergence of distinct traditions. Apparently the tendency of local groups to develop idiosyncratic variants is constantly counteracted by some homogenizing influence. Isaac concludes that the local group apparently did not persist long as a cultural isolate but was in fact part of an open social network which lacked mechanisms to prevent the flow of information from group to group. He even suggests that it may have been this tendency for local variation to be cancelled out which helped to promote the overall conservatism of the Acheulian tradition. Isaac then goes on to contrast the homogeneity of the Lower Paleolithic with the regionally more restricted distribution of Late Pleistocene and Holocene tool traditions. The contrast results, he argues, from changes in the nature of the communications networks which tended to produce closed systems. Isaac even argues that population density may have been one factor governing this transition.

This interpretation has gained support from a number of studies published recently by Martin Wobst (1974, 1975, 1976). Arguing that Paleolithic archaeology has lacked a firm grounding in meaningful social units, Wobst has attempted mathematically to simulate the size and scope of social interaction spheres and networks among hunting and gathering groups. In his models he has attempted to determine the size and distribution of mating networks which would have been capable of providing mates and insuring reproduction among Paleolithic populations. Wobst's most interesting conclusion, for present purposes, is that, at low population densities, closed mating networks (such as Birdsell's "dialectical tribes" or Steward's "maximum bands" which tend to be endogamous cannot function. At low population densities a local group could not afford to restrict itself to potential mates available within the maximum band or tribe since, except for those located near the center of the tribal territory, travel distance to potential mates would be prohibitive. Instead, each local group would be forced to seek its mates more or less symmetrically from adjoining populations. The result would be a series of overlapping mating networks rather than a closed mating system. A closed system such as the dialectical tribe would emerge only after a certain population density threshold was reached, such that each local band could afford the selectivity inherent in tribal endogamy. Only at this density, Wobst postulates, could signs and symbols of tribal membership emerge to reinforce endogamy and only at this stage would barriers to gene flow, information flow (and population redistribution) emerge. Wobst, an archaeologist working in Europe, contrasts the homogeneity of mid-Pleistocene assemblages with the heterogeneity of Late Pleistocene assemblages, and he suggests that indications of the development of closed mating systems emerge only relatively late in the European Upper Paleolithic, when ritual symbols reinforcing band membership proliferate in the archaeological record and become increasingly localized. Starting with theoretical propositions very different from those of Isaac, Wobst thus arrives at a reconstruction of prehistoric processes which is essentially identical to that which Isaac has offered.

Not all scholars agree, however, that the tool complexes defined by archaeologists imply sociocultural unity. Coles and Higgs (1969) explicitly warn against the assumption that stylistically similar archaeological assemblages necessarily represent culturally linked communities; and David Clarke (1968) has suggested an explanation for these similarities which is at variance with the cultural unity model. Clarke defines a "technocomplex" as a series of parallel adaptations to similar ecological problems by groups having a common trajectory or common evolutionary background. Such groups are not visualized as culturally homogeneous, and clearly no movement of people or flux is required to sustain this model. It should be pointed out, however, that the pattern of intertribal marriage described by Birdsell and Owen might create a situation similar to that envisioned by Clarke where some parallelism between adjoining, but culturally and linguistically diverse groups would be maintained.

A third possibility is that these archaeological "cultures" represent patterns of stylistic evolution that converged either by chance or by the process of adaptation to similar ecological situations. Whether or not chance can be invoked depends primarily on the degree of similarity involved and the extent to which stylistic similarities are dictated by raw material or by certain cross-cultural regularities in human conceptualization and manipulation. It might be possible to argue, for example, that the handaxe was an "obvious" development of simple flaking technique, being nothing more than the extension of bifacial flaking around the entire perimeter of the stone. The Acheulian style as a whole, however, seems to be too complex to have been derived in various locations strictly by chance. It can also be argued that chance, independent invention—even if guided by common adaptive or perceptive principles —would be unlikely to create an industry of Acheulian type only within one continuous region of the Old World. Such independent invention would make more sense if tools of the same type could be shown to be widely scattered in space or time.

Likewise, whether or not the similarities can be explained by convergent evolution through adaptation to similar activities or life-styles depends on the degree to which certain life-styles can be shown to require and to inspire specific tool types. Attempts at functional analysis of specific tool assemblages (Binford 1972; Binford and Binford 1966) assume that there is a fairly strict correlation between form and function even among the most primitive of stone tools. Implicit in such analysis, therefore, is the assumption that similar stone tool assemblages could reflect common activities rather than social ties. Isaac (1972) has pointed out, however, that there is very little evidence to support most of the functional interpretations which have been put forward. He notes, for example, that at least among Acheulian industries there is little evidence of correlation of tool types with particular types of non-artifactual remains or with particular environmental niches suggestive of specific activities. On a slightly different tack, Collins (1969) has pointed out that there is very little ethnographic evidence to support the assumption that people undertaking the same activities will independently develop stylistically similar tools to perform them. On balance it thus seems to me that a reasonably strong (although admittedly not conclusive) case can be made in support of the assumption that widespread tool traditions do in fact represent the operation of large, open cultural networks maintained by the free movement of people between local groups.

One other complicating factor must still be considered, however. Some of these so-called tool traditions may exist only in the minds of archaeologists. The "Stillbay" complex in Africa has been challenged as a meaningful descriptive category by any number of recent scholars. Similarly, Fitting (1970), while not denying the existence of a Clovis horizon, questions whether it really embraces all of the artifacts which are commonly attributed to it. Clearly any estimate of the extent of prehistoric population flux networks must depend in part on a critical screening of the various presumed tool horizons. Although I have admittedly not done such a screening, but am relying largely on tool horizons as described and criticized by other synthesizers of archaeological data, I will argue that there are sufficient well-established, wide-ranging tool horizons in the Pleistocene and post-Pleistocene history of the various world regions to suggest that population flux networks were widely distributed both in space and in time. One of the striking patterns encountered in the prehistory of each world region is that, regardless of when a particular continent was first occupied, the earliest inhabitants, spread throughout the continent, seem to have produced artifacts very similar in style. In each continent, it is only shortly before the emergence of agriculture that strong regional differentiation in artifact styles emerges. I would argue that in each case we are witnessing the existence of large, culturally homogeneous units (presumably involving networks of population flux) which break down into isolated regional cultural groups just prior to the beginning of agriculture.

Whether or not these flux mechanisms could have operated across the bottlenecks separating the continents, however, is more problematic.

There are stylistic horizons linking Africa and Eurasia in many different periods. Similarly America and Asia appear to be stylistically linked at more than one time (Willey 1966), and North and South America (Willey 1971, Lynch 1974, 1976) are similarly connected. But it is difficult to imagine this process operating on a sufficient scale between the New and Old Worlds to account for the synchronous buildup of population in the two hemispheres. As I shall point out in subsequent chapters discussing the various regional histories, several environmental factors appear also to have affected rates of population growth bringing the New World into synchronization with the Old.

3 THE ARCHAEOLOGICAL MEASUREMENT OF POPULATION GROWTH AND POPULATION PRESSURE

Before proceeding to analyze the archaeological evidence from various regions of the world, it is necessary to spend some time discussing the methods archaeologists employ to measure prehistoric populations. I have already suggested that some of our traditional reconstructions of prehistoric events should be reconsidered. It is my contention also that standard archaeological techniques for reconstructing prehistoric populations may contain sources of error and significant biases. I think that these limitations in our existing methodology may account in large part for the tendency to underestimate the extent of population growth in the Pleistocene and the importance of population pressure as an incentive for economic change during that period.

The first, and probably the most serious, problem is the fact that at present there is essentially no methodology in archaeology designed to identify, far less to measure, the elusive concept that we refer to as "population pressure." Almost all of our established procedures deal with the measurement of relative or absolute population size or population density (size relative to geographic area); none deals directly with the relationship between population size and resource utilization patterns. The recent literature (see, for example, P. E. L. Smith 1972a; Harner 1970; Hayden 1975; Cohen 1975b) offers a number of hints as to types of evidence which might be used to measure prehistoric population pressure, but to my knowledge no systematic methodology for dealing with this concept has ever been devised.

The problem is further complicated by the fact that our standard methods for reconstructing prehistoric population size and density are themselves fairly crude. They provide only a very rough, and often

A slightly different version of this chapter has been published in *American Antiquity* (Cohen 1975a).

unreliable, indication of the direction and magnitude of demographic change. Moreover, for reasons which will be discussed, these methods are particularly weak in measuring the size and density of nonsedentary populations, and thus are especially insensitive as indicators of demographic trends during the period prior to the development of agriculture. In order to defend the population pressure model, therefore, it will be necessary first to discuss some of the problems inherent in the standard archaeological methods of population measurement, and, second, to propose a number of additional types of evidence which can serve as indicators of population growth and, more particularly, of population pressure among nonagricultural groups.

Consider for a moment the standard techniques used by archaeologists to estimate the size and density of prehistoric populations. Our techniques are all built around the same basic model (see Cook 1972). First, a count is made of some parameter which is assumed to be representative of population at a particular point in time. Any number of different parameters can be used. Counts typically are made of the total number of houses or rooms in a village; the total number (or total acreage) of sites from a particular horizon or time period; the total number of shellfish or animal bones contained in single layers of a certain site; the total number of grindstones found in a village; the total number of acres under irrigation at a certain time; and so forth. At this point the method has variants; the counts may be used in one of two ways. First, they may be converted directly to absolute population estimates for the period in question by establishing the ratio of actual persons to the units counted. If five people are known or estimated to have occupied a particular type of dwelling, then population is estimated as the number of houses multiplied by five; if a cultivated acre of land supports an average of two people, then population is estimated as the number of cultivated acres multiplied by two; conversely, if one person consumes 50 shellfish in a day, then the number of shellfish divided by 50 determines the number of man-days of shellfish consumption.

The alternative method is to use the parameters to establish the relative size of the total population in two different periods. Thus, if there are twice as many houses occupied in one period as in another, twice as much acreage cultivated, twice as many grindstones used, or twice as many shellfish consumed, population is assumed to have doubled between the two periods. This latter method permits us to establish the relative size of population even if no absolute estimate of population is available for any period. Often, however, at least one of the periods in question provides evidence that permits us to estimate absolute population size. The absolute population figures for the remaining periods can then be calculated proportionately.

This method of population reconstruction (in either of its variant forms) makes three central assumptions. First, it assumes that the relevant parameter is fully preserved and recovered (if absolute estimates of population are to be made), or that preservation and recovery of the parameter, if not total, is at least constant from period to period (if only relative population size is to be estimated). Second, it assumes that the ratio of units of the parameter to actual persons is known or at least that these ratios are constant. Third, it assumes that the various units of the parameter counted are strictly contemporaneous, or at least that they all fall within a narrowly defined time period of known length. Ideally we wish to know exactly how many rooms were occupied or acres cultivated at a particular point in time or exactly how many shellfish were consumed in a single day. Usually we settle for knowing the relative number of rooms occupied, acres cultivated, or shellfish eaten within two broadly defined periods of roughly comparable length.

This basic methodology for estimating prehistoric population, with its two variants, enjoys an extremely wide vogue. It is, essentially, the only method we have for dealing with populations for which no census data are available. The method nonetheless has a series of significant flaws. The first important problem is that each of the three basic assumptions involves some inaccuracy, and as a result the method always includes some minor degree of error. The archaeological record is only a fractional sample of past human behavior and its residue. The archaeologist certainly cannot count on total recovery, nor can he presume with any assurance that the degree of sampling error is the same for all the periods studied. There will always be some differential preservation or recovery from various periods. In addition, any assumption that there is a fixed ratio between a unit of the parameter and a unit of total population will obviously involve some error. Families vary in size, for example, so that there will never be a perfect correlation between the number of houses occupied and the number of persons in the population. Moreover, average family size may well change from period to period without any obvious change in house size, so that the populations of two different periods need not be strictly proportional to the house or room count for the two periods. Finally, absolute contemporaneity between units counted can almost never be reliably established, nor is absolute dating sufficiently accurate to permit us to define with any precision the interval of time the units span. We can never determine with any assurance exactly how many houses in a village were occupied at any one point in time. And, if we are dealing with a more general measure such as the relative number of rooms or houses occupied during two prehistoric periods, we cannot control precisely for such variables as the length of the two periods being compared, the length of time each room or house was occupied during each period, and so forth.

These sources of error are widely recognized and appreciated. Despite the popularity of this general method, most archaeologists do not attempt to define prehistoric population with great precision, nor would they argue that their reconstructed curves of population trends represent anything more than a very approximate indication of the shape of population growth and decline.

The second major problem with the methodology is far more significant. Aside from these constant minor sources of error, there are conditions, not always readily perceived, which can render results so completely misleading that even the general shape of the reconstructed population curve is wrong. If, for example, there is a major change in the conditions of preservation in an area, comparisons of the numbers of units of any parameter for two different periods may be meaningless. For example, as will be discussed in more detail in subsequent chapters, Butzer (1971:536) argues that the apparent decline in population in post-Pleistocene Europe, based on the observed decline in the number of preserved archaeological sites, is probably nothing more than a change in the conditions of site preservation and has no demographic significance at all. Similarly, if a major shift occurred in the cultural patterns governing the utilization of any of the items counted—a shift that affected either their preservation and eventual recovery or their numerical relationship to people—the reconstructed population curve could be very misleading. For example, the movement of a population from an area of good preservation and easy recovery to one of poor preservation and difficult recovery might very well give the appearance of a major population decline when in fact none had occurred. Likewise, if a culture changed from providing one grindstone for each nuclear family to providing one for each female over the age of twelve, the number of grindstones might increase considerably, giving a false impression of significant population growth.

The third major flaw in this methodology, and the one most relevant from my point of view, is that it is essentially incapable of dealing with nonsedentary populations. All the potential sources of error which are relatively minor when one is dealing with permanent villages become magnified enormously when one deals with temporary campsites. In the first place, temporary campsites tend to contain relatively little imperishable material. They are frequently not preserved, and they are easily overlooked by an archaeological survey. As a result, the archaeological sample of temporary campsites is likely to be much smaller and subject to far greater sampling error than is a sample of sedentary villages. In addition, with temporary campsites it is much more difficult to establish contemporaneity or measure the duration of occupation than it is with villages. Whereas two permanent villages of the same general time-span can often be determined to be either contemporary occupations of separate populations or sequential occupations by a single population, temporary campsites utilized during one or more seasons of one or more years may represent almost any number of different populations. Similarly, establishing a ratio between persons and artifacts is more difficult in temporary campsites than in permanent habitations. Do the outlines of three houses in a temporary campsite represent three families living together or the reuse of the site by a single family on three occasions? Unless the outlines actually overlap one another, there may be no way to tell. Do six grindstones in two campsites represent six families, or three families who left their grindstones behind when they changed their location?

The result of this confusion is that this methodology, which is essentially designed for sedentary sites, can be applied to nomadic groups only with great care—if, in fact, it can be applied at all.

The most serious problem occurs, however, when this same basic methodology is used to compare pre-agricultural and post-agricultural populations in any region. In this case, the results are very badly biased. Agriculture correlates roughly with sedentism in most regions of the world, and it is obvious that sedentism, whether or not it results in actual population growth, will result in marked increases in the number of sites recorded by archaeologists. Permanent sites are simply much more likely to be preserved and to be found by archaeologists than are temporary campsites. We are left with the impression of a population explosion accompanying the development of agriculture, but this may result from nothing more than the preferential preservation and discovery of agricultural villages. Even if population growth does accelerate when populations become sedentary and begin to farm (and I am willing to concede that such is probably the case), it seems probable that this acceleration will be overemphasized in the archaeological record by the disparity in site preservation. The problem is thus a compound one: on the one hand, a methodology poorly adapted to the study of mobile groups tends to minimize our appreciation of group dynamics in preagricultural societies; on the other hand, we witness an abrupt (but possibly artificial) increase in the number of sites correlated roughly with the beginning of agriculture. The combination clearly leads to a strong bias toward the Malthusian assumption that population tends to remain stable until liberated by agriculture, after which revolutionary growth results.

Some of this theoretical bias may be eliminated if we consider a number of additional types of indicators of population growth. Although these are not quantitative indicators in the sense discussed above, they are, I believe, equally valid as measures of population change in a region (and particularly of population pressure on that region) as is site or artifact density. In fact, since these new methods do not rely primarily on *quantitative* data except in the broadest sense, it is probable that they are less subject to problems of sampling error than are the strictly quantitative types of estimates already discussed. These methods do have their problems, however, and must be used with care. Like all archaeological data, the indicators used by these methods are subject to some sampling error and can be considered reliable only if results can be replicated in many separate instances. In addition, many of the indicators which will be disucssed are subject to systematic biases in preservation. In some cases differential preservation can be expected to mimic the appearance of valid historical trends. Considerable care will have to be applied to distinguish patterns resulting from preservation from those which are of real significance.

The additional indicators to which I refer are changes in settlement patterns, food refuse, and food-related artifacts. If the archaeological record demonstrates certain changes in subsistence strategy, and if the changes demonstrated can reasonably and convincingly be assumed to result from population pressure (or imbalance between a population and its resource gathering strategy), then the evidence of economic change itself may be taken as an index of population pressure.

Changes in the subsistence economy can, I believe, be taken as evidence of population pressure as long as certain conditions are met. First, of course, the change in the subsistence economy must be in the direction of increasing the total caloric productivity of the region under consideration. Second, if the changes are to indicate economic stress they should occur in the absence of the diffusion of new or complex technologies or the appearance of new resources in the environment; it should be clear that the altered economy is not simply a response to new opportunities. Third, it should be established that the economic shift occurs in the direction of utilization of resources which are in some manner less desirable than those they replace. For example, the new resources may involve increasing rather than decreasing per capita labor costs. Or the new resources may be calorically productive but otherwise less nutritious than the resources they replace. The new foods, though highly productive, may be foods that most people (in cross-cultural survey) consider unpalatable; or they may require extraction and preparation techniques that people, cross-culturally, consider to be particularly odious, distasteful, or demeaning forms of labor. (The latter condition might be indicated in the archaeological record simply by the fact that the resources were previously unused, although both the foods themselves and the technology for exploiting them were clearly available.)

In short, when a change in subsistence patterns cannot be shown to depend on newly available opportunities, and when the shift appears to be in the direction of calorically dense but otherwise less desirable resources, I believe that we are justified in assuming that such shifts represent compensation for demographic pressure. The following is a list of types of archaeological evidence which may be interpreted as evidence of such situations. The list is derived from the theoretical model of expanding population presented in chapter 2. Since many of these occurrences are capable of alternate explanations in particular cases, we should hope to find several such indicators occurring together to be sure that they represent population pressure. Proving that population pressure results from actual population growth poses another problem, since, as discussed in the previous chapter, disequilibrium between population and resources may well be caused by a variety of factors other than population growth per se. In order to prove actual population growth, we should hope to find these stress indicators occurring in situations where alternative explanations can be ruled out (where, for example, pollen profiles show no climate change), or occurring in such a general manner, widespread both in time and in space, that only a general explanation such as population growth can account for the observed pattern. The list is as follows:

1. When it is possible to isolate the exploitative cycle of a single group making its annual round, evidence that the range covered is increasing (i.e., that people are traveling increasing distances for food) should indicate population pressure. For example, if it can be shown that the travel distance from the base camp to outlying camps (other than special purpose camps unrelated to the food quest) is increasing or that food resources are being transported to the base camp from greater distances, it is a reasonable assumption that the population is encountering increasing difficulty in supporting itself on preferred foods available near its home base. Care must presumably be taken to eliminate the role of nonfood resources or other variables such as political factors affecting the pattern of movement.

2. When a group expands into new ecological zones and territories, population pressure may be assumed, especially if expansion takes place into areas or latitudes which demonstrably present new adaptive difficulties such as extreme heat, cold, high altitude, disease, or danger of predators. The emphasis here is on the word *expansion;* mere migration is insufficient. It should be established that the group both occupies its original zone and has moved into additional regions. One special example of this trend would be the adoption of scheduled transhumance by populations formerly inhabiting only a single zone. Such an economic change might be indicative of the increasing difficulty in obtaining preferred resources during certain seasons. Lynch's (1973) model of agricultural origins (see chapter 1) would fit this concept nicely. If the movement of transhumant groups has become so tightly scheduled that certain resources must be harvested out of season, then the group presumably faces significant difficulty in feeding itself in one or even a few selected locations. Here again, however, care must be taken to evaluate the role of nonfood resources or political factors as motivations for movement.

3. When the inhabitants of a region become more eclectic in their exploitation of microniches, utilizing portions of the environment—such as deserts, coastal areas, or forests—which have previously been ignored, while continuing to exploit the old niches, demographic pressure may again be assumed. It is not necessary, incidentally, that any one group span all of these niches, so long as the aggregate effect among groups of neighboring populations is toward more complete use of available space. Such a pattern might of course emerge by chance in any one region, but a wide-ranging and consistent movement into previously ignored microenvironments would seem to be significant.

4. Similarly, when human populations show a shift toward more and more eclectic food gathering patterns, shown by reduced selectivity in the foods eaten, it may be argued that they are demonstrating the need to obtain more calories from the same territory in order to feed denser populations. As in the index above, this eclectic pattern need not appear in the economy of any one culture group so long as it characterizes the inhabitants of a region as a whole. Individual human groups might even conceivably increase their concentration on particular resources, but as long as the overall tendency among the groups in a region is toward fuller utilization of all resources, population pressure can be assumed.

5. When a group increases its concentration on water-based resources relative to its use of those that are land-based, especially when the resources are shellfish whose exploitation is independent of the invention of any new technology, this shift may be viewed as resulting from demographic necessity rather than choice. (See Harner 1970; Evans 1969; Cornwall 1964; Clark 1952.) This need not imply that fish or shellfish become the dominant food. Parmalee and Klippel (1974) have pointed out that shellfish are probably not as significant as a source of food for human populations as they often appear to be in the archaeological record, nor are they as good a potential resource base as is often assumed. The point here is that shellfish are evidently low-prestige resources of last resort for a great many human populations, so that *any* increase in their utilization is probably significant. In individual cases, of course, the exploitation of shellfish may result from cultural preferences or from environmental changes such as modifications of local coastlines; but a widespread increase in their importance presumably indicates a significant stress on other resources.

6. When a group shifts from eating large huntable land mammals to eating smaller mammals, birds, reptiles, and land molluscs, demographic stress may again be assumed. Large mammals make up a relatively small portion of the local biomass in any region (Deevy 1968), but they are apparently a highly favored food in most cultures (Murdock 1968; Clark 1970). Conversely, the smaller fauna are less desirable, low-prestige items (Binford 1968), but they make up a relatively large portion of the animal biomass. A shift in favor of the consumption of the smaller fauna clearly represents the sacrificing of quality for quantity. Such a change might occasionally occur by cultural choice, but a widespread trend in this direction would appear to be a significant indicator of population pressure.

7. When a group shifts from the consumption of organisms at high trophic levels to the eating of organisms at lower trophic levels (in particular, when it shifts from meat to plant foods), population pressure may again be assumed. This change in diet will increase the consumable biomass, but it runs counter to both widespread prestige values and widespread food preferences (Clark 1970; Yudkin 1969; Murdock 1968). As with other indicators discussed above, this change may occasionally happen by chance, and its reliability as a measure of population pressure depends in considerable degree on the extent of the geographic area over which the trend can be observed. This indicator, too, is particularly sensitive to interference from differential preservation since the relative proportions of meat and vegetable foods in the diet as reconstructed archaeologically depends almost totally on the quality of preservation.

8. When a shift occurs from the utilization of foods requiring little or no preparation to foods requiring increased amounts of preparation in the form of cooking, grinding, pounding, leaching of poisons, etc., population pressure is again indicated. Such practices clearly expand the range of edible foods, but at high labor cost. This index is particularly amenable to archaeological analysis since many of the tools for food preparation are highly distinctive and imperishable. Erroneous identification of food processing tools may, however, be a serious source of error.

9. When there is evidence of environmental degradation suggesting human efforts, particularly through the use of fire and land clearance resulting in the maintenance of subclimax vegetation, it may be argued

that larger human populations are increasing their interference in natural ecosystems to augment the productivity of their preferred foods. One of the primary problems here, of course, is to distinguish between environmental changes caused by man and those occurring naturally. As will be discussed in subsequent chapters, there is in fact considerable controversy about the relative importance of man as a modifier of the environment, and in particular about the role of man (if any) in the creation of open environments.

10. When skeletal evidence of malnutrition increases through time, it may be argued that demographic stresses are resulting in reduced quantity or quality in the diet available to each individual. Hayden (1975), although he is not concerned with population pressure in quite the sense that the concept is used here, does suggest that the relationship between a prehistoric population and its food supply can be gauged through the study of skeletons by measuring the actual incidence of morbidity or mortality resulting from malnutrition. The techniques for identifying malnutrition from skeletal evidence have been provided by a number of other studies (see Garn et al. 1969; Acheson 1959; Garn et al. 1968; Jones and Dean 1956). For an actual application of this measurement of stress, see Cook (1975).

11. When the size or quality of individuals exploited from a particular species shows a steady decline through time (when, for example, the size of molluscs in one or several shell middens decreases), it may be argued that human populations are consuming resources beyond their carrying capacity, resulting in the degradation of the exploited population. This index is particularly vulnerable to misuse since a variety of environmental factors might affect the size and health of exploited species. The key to its successful use would be the elimination of such alternate explanations by careful control of other related variables.

12. When an exploited species disappears from the archaeological and fossil record, it may be argued that the species was exploited beyond its carrying capacity. Here again, however, there is an enormous problem in distinguishing between extinctions caused by human agency and those resulting from other causes. This is another source of major controversy, and will be discussed in detail in subsequent chapters. One point to note, however, is that the disappearance of an exploited species from whatever cause will presumably result in disequilibrium and increased population pressure.

The preceding types of evidence are relatively straightforward and self-explanatory. In addition, I would like to suggest two other kinds of evidence which may also indicate population pressure, although in both cases the argument is somewhat more tenuous and controversial than those discussed above.

13. It has already been pointed out that hunting and gathering cultures are characterized by great fluidity in the structure of their local groups. People move back and forth easily from camp to camp. I have argued that one result of this fluidity will be widespread homogeneity of artifact styles and have called attention to a number of far-ranging artifact style-horizons in the Pleistocene which I believe are indicative of this kind of movement. Conversely, local specialization in artifact styles should be indicative of relative isolation among populations. Local isolation, I suggest, will result from a combination of two forces. First, if desired resources are scarce, the group will be increasingly jealous of its local resources and tend to become closed to outsiders. Second, scarcity, as discussed in chapter 2, will require a group to invest more and more labor in the future productivity of its resources. Since available food thus becomes more and more a function of the group's previous labor investments, the group will again become increasingly jealous of outsiders. Thus, increasing scarcity of resources should result in the gradual breakdown of the system of open population-flux. If this is true, then regional specialization of artifact styles may itself be an indicator of population pressure. (For an alternate explanation of this phenomenon, however, see the work of Wobst [1974, 1975, 1976] discussed in chapter 2.)

14. I would argue also that sedentism and the practice of artificial food storage may indicate population pressure, particularly if they are not linked to the availability of new resources or technologies, but rather are combined with intensive exploitation of old resources by traditional means. As has already been discussed (chapter 2), sedentism, despite certain advantages, implies high labor costs in the collecting of many food items and reduced dietary variety. The incidence of disease rises, and labor costs in the preparation of food for storage and in storage itself grow. Sedentism increases the threat of the loss of stored foods by rotting or rodent action; intensifies the danger of expropriation by other human groups; and, by tying it to a particular location, greatly increases the vulnerability of the population to exploitation and

enslavement by other groups. I suggest that sedentism *in most cases* occurs, not because of newly discovered resources which *permit* year-round residence in a single location, but rather because of the decline of resources associated with other parts of the traditional annual cycle, or because of territorial impingement by other groups. Either of these events would make it necessary for a group to stay in one place and stretch the resources of a particularly productive season to cover those other periods of the year when seasonal foodstuffs are no longer sufficient. In defense of this argument I would point out that the early history of sedentism (see chapters 4, 5, 6) is closely linked with the exploitation of resources (water-based resources and vegetable foods) which appear to have been largely ignored by earlier human populations in the same environments where sedentary communities later emerged. Moreover, these foods are today widely regarded as low-prestige "necessary" foods rather than as desirable ones. This picture is much more consistent with the assumption that human populations settled down out of necessity than with the assumption that sedentism was the strategy of choice.

It may seem paradoxical that population pressure can be cited as the basis both for sedentism and for seasonal transhumance, but I suspect that such is in fact the case. Under ideal conditions of low population pressure, human groups presumably adopt a fairly informal pattern of small-scale movements, which maximizes their harvest of the most favored resources without committing them either to the problems of sedentism or to the rigors of tightly scheduled seasonal movements. Increasing population pressure might well force human populations either to begin relatively rapid movement or to assume a sedentary posture, depending on the distribution of resources and other human groups on the landscape. In some environments, particularly those characterized by migrating game populations or rapid seasonal progression of resources (as in mountainous terrain), pressure would eventually force human groups to move seasonally. In regions where abundant but low-quality foods are available seasonally, population pressure might force a group to rely more and more heavily on these secondary foods and ultimately to settle down and develop storage facilities for their maintenance.

As I have tried to indicate above, none of these lines of evidence is alone sufficient to demonstrate population pressure, since all are susceptible to alternate explanations. But several occurring together, all indicating behavior that runs counter to the manner in which we would expect human populations to act by choice, surely must represent attempts to adjust to the need for more calories or to the loss of traditional food sources. Moreover, if these behaviors occur widely enough in time and space, and thus separated from particular events of climate change or other localized variables, population growth would seem to be the only plausible explanation.

If we now retrace the Pleistocene prehistory of the various regions of the world, watching not only for changes in site density, but also for changes in settlement patterns and economic strategies of the type outlined above, a picture of prehistoric culture change emerges which differs significantly from our traditional models. In particular, population growth appears to be much more continuous, and population pressure much more significant as a motivation for culture change, than is usually assumed.

4 THE CASE FOR THE OLD WORLD

If we review prehistoric events in the Old World in a broad enough perspective, noting changes in subsistence and settlement strategy of the type outlined in the previous chapter, a reasonable case can be made that agriculture is the culmination of a buildup in population pressure which embraces the whole hemisphere. Although there are significant local variations resulting from local climate phenomena or random cultural trends, the buildup of population pressure is observable in almost all the major regions of the Old World, and the observed patterns are surprisingly alike. This growth in population pressure appears to be synchronous throughout the Old World, suggesting that population flux mechanisms were acting with considerable effectiveness. Most important, perhaps, the buildup seems to be a process of considerable duration. Indicators of population pressure of the type described begin to accumulate well back in the Pleistocene, suggesting that the process of economic change that culminated in agriculture was not simply a post-Pleistocene event.

Very briefly, the archaeological record suggests that agriculture occurred throughout the Old World (1) when human population had completed an expansion from tropical to temperate and finally arctic latitudes; (2) when, starting with a high degree of selectivity in his choice of niches in each latitude, man had begun to exploit an increasing number of different habitats; (3) when the density of occupation sites had increased in most regions; (4) when the selectivity of man's diet had been greatly reduced and broad-spectrum consumption had become the rule; (5) when aquatic resources were being increasingly exploited almost everywhere; (6) when there was a growth in the consumption of small animals rather than big game animals; and (7) when increasing quantities of vegetable foods, including those involving complex preparation, were being eaten. In addition, although possibly less significant, agriculture appears to have occurred in a context in which many people had already become sedentary and in which artifact styles had already demonstrated a marked trend toward greater regional differentiation.

The Geographical Expansion of Old World Populations

The geographical expansion of early human populations is one of the easiest of these trends to document since in this case the evidence is somewhat less subject to poor preservation and to sampling error than is the evidence of the other economic patterns which must be considered. *Australopithecine* (or very early *Homo*) populations of the Lower Pleistocene, judged by the distribution of known fossils, were confined to limited portions of Africa or (depending on the dating and interpretation of certain fossils) to Africa and tropical southern Asia. The distribution of Olduwan tools presumed to have been made by the men of this period is very slightly broader, suggesting that, by the end of the Lower Pleistocene, these populations may have occupied all latitudes in Africa and extended as far north as the Mediterranean coast of Europe and the Middle East. This evidence provides a fairly clear idea of the general area of the globe in which early human evolution took place. It also indicates the type of ecosystem in which early man participated, and, by inference, the type of climate to which man is ideally suited and in which he would have preferred to live.

By the end of the Middle Pleistocene, something more than 100,000 years ago, man had extended his range of occupation significantly by moving into higher latitudes. This can be determined from the distribution of Acheulian and contemporary tools and from the distribution of fossils of the *Homo erectus* type. He now inhabited much of temperate Europe, including southern England, parts of France, and central Europe; the southern portions of the Caspian Sea region; and eastern Asia approximately as far north as Choukoutien, near Peking, while continuing to occupy Africa and the tropics. In the period between 100,000 and about 40,000 years ago, man further expanded the northern boundary of human settlement, entering for the first time such regions as central Germany, southern Poland, the southern Russian plain, the Iranian plateau, Turkmenia, and Uzbekistan. Between about 40,000 B.P. and the end of the Pleistocene, modern man further extended the range to include northern Europe as far as southern Scandinavia, a good deal of Russia, and Siberia at least to 61 degrees north latitude. At about the same time populations began to colonize the New World, as well as Japan and Australia. This expansion continued even into the beginning of Neothermal conditions. The northern expansion of man at this time was permitted in part by the exposure of land surfaces previously covered with ice, but there is also evidence of the settling of northern environments which had previously been used only sparsely or not at all, despite the fact that they had long been open and had even provided fairly rich resources for human exploitation. Portions of northern Russia, for example, which previously had largely been ignored or exploited only very lightly were now more densely settled. In western Europe, much of Scotland and Ireland (portions of which had been free of glaciation and rich in deer and reindeer in the late Pleistocene) were occupied for the first time, along with most of the Scandinavian peninsula (J. G. D. Clark 1962, 1967, 1969; Clark and Piggott 1965; Butzer 1971; Butzer and Isaac 1975; J. D. Clark 1970).

For some reason, this northward expansion of population in the Pleistocene is rarely considered when the role of population pressure is evaluated; it is rarely seen as part of man's cultural response to his own population growth. Yet there would seem to be little question that the movement represents an increase in overall population; and it would seem probable, too, that the motivation for expansion is in some way related to an imbalance between human populations and their selected resource base within their traditional territories. Human populations may migrate for any of a number of reasons, and it is probable that the actual pattern of migration by various human groups was one of random movements in response to a variety of cultural and biological imperatives. Yet one consistent vector in this random movement seems to have been toward the colonization of previously unoccupied territory. Moreover, this expansion cannot simply be equated with the unhindered colonization of "open" environments of the type described by Birdsell (1958, 1968) since in this case territorial expansion would have involved significant cultural costs. Along the northern frontier, this expansion involved crossing a great number of ecological barriers and the encountering of new ecological problems, the most important of which, of course, was the problem of coping with increasing cold. Our experience with both animal and human populations suggests that such expansion into new niches on a broad front is most likely to be explained by competitive displacement from old niches or by other stresses which render the older niches intolerable. In essence, such expansion should take place only when expansion and the problems it entails represent the *easiest* way of coping with new problems or the best way of approximating the old status quo (Romer 1933). In this sense, such northward expansion is very much an indicator of population pressure or imbalance between populations and their existing exploitative strategies, which required the utilization of latitudes previously defined as undesirable.

In this context, it should also be noted that the use of fire (and later of clothing) is probably explainable as part of a population pressure model. One of the primary functions of fire, of course, is to generate artificial heat. It thus permits expansion into new, colder regions that previously could not have been occupied. It would seem inappropriate, however, to consider fire as an "invention" that fortuitously "permitted" man to move farther north. One would hardly expect people to migrate into new, colder regions just because they had invented fire and wanted to try it out (or because they wanted to wear their new winter clothes). It would seem more plausible to assume that the use of fire and clothing were adaptations necessitated by enforced migrations into colder climates.

The known distribution of archaeological evidence for the early use of fire appears to support this contention. If the mastery of fire were a fortuitous discovery of mid-Pleistocene man, we might expect it to be shared throughout one of the culturally homogeneous regions which characterized Africa and Eurasia during this period; or, if no mechanisms for widespread cultural sharing existed, we might expect the use of fire to be randomly and sporadically distributed among the Acheulian and non-Acheulian populations of the time. In fact, however, early evidence of the use of fire occurs with at least two different stone tool traditions (as defined below), but it is found only in the northern, temperate portions of the occupied world. During the Middle Pleistocene there is evidence of fire in Europe and temperate Asia at sites such as Terra Amata in France (de Lumley 1975), Torralba and Ambrona in Spain (Howell 1966; Freeman 1975), Hoxne in England (West and McBurney 1954), Choukoutien in China (Oakley 1961a), and Vertesszöllös in Hungary (Kretzoi and Vértes 1965), some of which, at least, are estimated to be 400,000 years old or more. In Africa, surprisingly, fire is not known from most mid-Pleistocene sites. Here it is not in evidence until very

late or terminal Acheulian sites, such as Kalambo Falls, which has provided dates ranging from 60,000 to 200,000 B.P. (Clark 1970, 1975;
Isaac 1975b; Fagan 1974; but cf. Sampson 1974). In fact, most of the
early evidence for fire in Africa seems to be related to the post-Acheulian
industries of the early Upper Pleistocene within the last 100,000 years
(Oakley 1961a; Clark 1970). The fact that the earliest evidence of fire
clusters so strikingly along the northern margins of human expansion
suggests that the mastery of fire was not a recent "discovery" but
rather a technique necessary to the populations that penetrated these
new geographical zones.

It should be noted, incidentally, that this northward and eastward expansion of the human population cannot be viewed simply as a function
of the withdrawal of the glaciers. In part, of course, northward expansion was related to the retreat of the glaciers, which permitted expansion
and even drew populations northward as the climatic zones to which
they were adjusted and the animal and plant populations they exploited
all migrated north. On the other hand, much of this movement, most
notably the expansion of Mousterian and Upper Paleolithic populations
into northern Eurasia and Sibera, took place during a period of glacial
advance and must have involved migration in direct opposition to the
displacement of ecological zones caused by the expanding glaciers.
Moreover, in some cases, at least those involving Neothermal expansion
in northern Europe and Russia, as indicated above, expansion occurred
into unoccupied regions which had previously been free of glaciation
and supported edible game resources.

Finally, it should be noted that the expansion of population was not
simply a wave of colonization of open environments by groups on the
northern (and eastern) fringes of an otherwise relatively stable human
population. In all of the regions which have been reasonably well
studied, there is evidence that this expansion was accompanied by secondary expansions into previously little-used areas, filling in the inter-
stices between former centers of population. Simultaneously overall
increases in population density and modifications of subsistence strategy
were occurring. These trends can best be described if we start with a
brief summary of the nature and distribution of Acheulian and contemporary assemblages and then trace the subsequent history of human
occupation in various regions.

The Ecology of Mid-Pleistocene Populations

The Middle Pleistocene is a period of marked cultural homogeneity and extremely slow cultural evolution. However, the precise definition of cultural traditions within this period (from approximately 1.5 million years B.P. to approximately 100,000 B.P.) is a matter of some dispute (cf. Butzer and Isaac, eds., 1975; Isaac 1975b). All authorities appear to agree on the existence of a widespread and homogeneous Acheulian technocomplex extending throughout Africa and southwestern Eurasia, although not all would agree that this technocomplex implies cultural unity. In addition, however, there are a number of other tool assemblages (such as the "Tayacian," the "Clactonian," and the "Developed Olduwan") which have been defined by various authorities, based largely on the absence or scarcity of the typical Acheulian handaxes and other bifacial tools. These assemblages appear to be contemporary with the Acheulian tradition and largely overlap the Acheulian in its geographic distribution; in some cases they even alternate with Acheulian assemblages in separate layers of the same site. These industries are interpreted by some archaeologists as representing separate cultural traditions or even biologically distinct populations; they are viewed by others as representing simply functional or even random variations on the basic Acheulian tool kit. The debate concerning their significance is still unresolved. The only clear pattern that emerges is the almost total predominance of assemblages without handaxes on the northern fringes of human occupation in Europe and central Asia and the exclusive occurrence of such assemblages in eastern Asia. For simplicity, therefore, I will follow Collins's (1969) twofold classification of Middle Pleistocene industries, recognizing only an Acheulian complex in Africa and southwestern Eurasia and a Clactonian-Charentian complex distributed north and east of the area of Acheulian occupation.

One of the key problems with which we must deal is to evaluate the economic strategies of the people of this period. Much of the interpretation of subsequent economic history depends on how the adaptations of these early groups are reconstructed. The sites of the mid-Pleistocene have been considered to represent the first evidence of truly evolved hunting behavior, and, in much of the pre-1970 literature, Acheulian industries were characterized as representing an economy fairly specialized in the hunting of big game. A number of archaeologists (J. D. Clark

1970, 1975; Kraybill in press; Deacon 1975; Isaac 1969, 1971, 1975a,b),
however, have recently begun to question this stress on hunting. They
cite recent work on hunting and gathering groups (Lee and DeVore,
eds., 1968) which shows that contemporary "hunters" typically eat
much more vegetable matter than was previously thought, and in fact
derive most of their subsistence from such foods. Clark (1970:76,100)
has even suggested that Acheulian hunters, like their modern counterparts, probably derived between 60 and 80 percent of their subsistence
from vegetable foods.

However, it may be no safer and no more accurate to attribute a modern hunting and gathering economy to mid-Pleistocene populations
than it is to assume that these populations were strictly big game hunters.
In fact, although there are serious problems of interpretation involved,
there is considerable evidence to suggest that the economic strategies of
modern hunting and gathering groups are of quite recent origin (cf.
Washburn and Lancaster 1968). Moreover, given the observed preference
of human groups for meat, one might anticipate that at least by the
mid-Pleistocene, when the taste for meat and the ability to hunt had
both apparently evolved, but when the human population was not yet
large in proportion to its prey, meat would have provided a relatively
large part of the diet, since the desire for meat would not yet have been
offset by increasing labor costs resulting from scarcity. Even Clark
(1975), who argues that vegetable foods were more important than the
archaeological record indicates, suggests that meat was the preferred
food in the mid-Pleistocene.

One of the problems involved in interpreting the economies of these
traditions is that the functions of the stone tools recovered are only
poorly understood. Like other tools of questionable purpose, they have
typically been assigned to formal categories for classificatory purposes.
Too often in the earlier literature these categories were assigned functional definitions rather indiscriminately, the functions typically derived
from activities related to hunting in correspondence with the prevailing
stereotypes. The assumed predominance of hunting as an activity is in
part a result of nothing more than the persistence of these "functional"
classifications. More recent work in the classification of the tools has
tended to minimize functional assumptions and to concentrate on neutral, descriptive terminology. More important, the presumed hunting
"function" of many of these tool types has been questioned. Kraybill

(in press) has suggested that many of the tools usually presumed to be associated with hunting are more profitably to be analyzed as tools for vegetable processing. For example, she argues that the spheroids, sub-spheroids, cuboids, and polyhedrons reported as early as Bed 1 at Olduvai (where of course they predate the Acheulian) and reported in varying percentages from a wide variety of Acheulian sites are in fact tools for vegetable processing. At the Kalambo Falls site in southern Africa, which is one Acheulian site with good organic preservation, wooden shafts interpreted as digging sticks have been recovered (J. D. Clark 1969). The latter, if correctly identified, would also indicate the gathering of vegetable resources.

Whether either of these identifications of vegetable processing tools is correct, however, is an open question. In fact we simply do not know how most of these tools were used. Because of the significant time gap between mid-Pleistocene populations and modern hunting and gathering groups, ethnographic parallels for the use of such tools are limited and unreliable. In addition, use-pattern analyses of the type popularized by Semenov (1964) have not yet provided sufficient information to permit reliable indentification of tool use patterns. But one clear pattern does emerge from the analysis of the tool assemblages. Tools which can more reliably be associated with intensive processing of vegetable foods and which are common (although admittedly not universal) among modern hunting and gathering groups—grindstones, manos, nutting stones, mortars and pestles, for example—are all conspicuously absent. Since some modern hunting and gathering groups make significant use of vegetable foods without using such tools, or at least without making them out of stone, their absence at a particular location does not prove that vegetable foods were not eaten; on the other hand, their general absence in the Middle Pleistocene in comparison to their widespread proliferation in later prehistory certainly suggests that at least some classes of vegetable foods (and perhaps by extension all vegetable foods) were not as significant in the diet as they were later to become. In particular it seems likely that small grains requiring bulk processing were not widely exploited.

The interpretation of mid-Pleistocene economies is further complicated by the poor preservation of organic remains (Isaac 1968, 1971). As Isaac points out, organic remains of any sort are not as common at sites of the period as is usually believed; and where preserved at all, the remains may be significantly skewed. The faunal portion of the economy will clearly be better preserved than the floral; and large animals will presumably be better preserved than small. To a great extent, therefore, the apparent hunting orientation of these populations could be a function of differential preservation.

To be sure, at some sites there is ample evidence both of the success of mid-Pleistocene man as a hunter and of his fairly specialized concentration on certain species of big game. At Olduvai (Bed II) the articulated remains of a number of ungulates, especially the bovid *Pelorvis*, suggest hunting employing some kind of driving technique (L. S. B. Leakey 1951, 1957). Similarly, a concentration of a large number of giant baboons associated with Acheulian artifacts at Olorgesailie in East Africa (Isaac 1971) appears to represent at least an episode of fairly specialized hunting. And, at Torralba and Ambrona in Spain, elephants, rhino, and a variety of other large ungulates were apparently driven (with the aid of fire?) into swampy ground where they could be dispatched. At these sites the fauna recovered indicate a fairly clear selection pattern favoring the larger animals (Howell 1966; Freeman 1975; Freeman and Butzer 1966). Also in Europe Acheulian layers are reported with heavy concentrations of reindeer bones rivaling sites of later periods (Bordes and Prat 1965).

In South Africa, Sampson (1974; see also, R. Mason 1962) reports the occurrence of Acheulian sites near *neks* or gaps within long ranges of hills, locations which he suggests were chosen as natural funnels for migrating game. Sampson (1974) in fact is one scholar who argues that, at least in South Africa, the Acheulian economy as a whole was geared to a limited range of large game species. He has collected the faunal lists available from mid-Pleistocene sites in South Africa, and he notes that, although the lists are long and include a fairly broad range of different sized animals, there is a fairly pronounced numerical concentration on the larger game species, a concentration not found in sites of subsequent periods. As will be noted again in a later context, Sampson even suggests, following Martin (1966, 1967a), that this specialization resulted in the extinction of some of the large African mammalian forms.

Such heavy concentration on large mammals, however, is by no means typical of all sites of the period. Many sites contain no preserved bone and some scholars have argued from this negative evidence and from the related distribution of tool types that such sites represent a different portion of the economic cycle, perhaps one geared to plant

processing (Isaac 1975a; Binford 1972). Moreover, where bone is preserved it does not always come from large animals. Small mammals are
reported from many of the South African sites (Sampson 1974), although in many cases the associations, particularly those involving
rodents, are not firmly established. The fauna of the Olduvai living
floors spanning the Olduwan and Acheulian traditions include a very
broad range of vertebrate forms: amphibia, fish, reptiles, and even
some birds, as well as large and small mammals. In addition some land
snails have been found (L. S. B. Leakey 1951, 1965; M. D. Leakey
1967; Isaac 1975b). In France, Terra Amata shows a similar range of
vertebrate fauna extending from deer, elephants, boar, ibex, and aurochs
to rabbits, rodents, and fish; and there are even some marine shellfish
(de Lumley 1969). At Vertesszöllös in Hungary, rodents are found in
considerable quantity (Vértes 1966).

   On the other hand, there are no sites which document *extensive* exploitation of microfauna in the Acheulian tradition (Isaac 1971. Note
that Vertesszöllös is a non-Acheulian, or Charentian-Clactonian, site.)
Moreover, aquatic resources, whether fish or shellfish, are extremely
rare at sites of the period whichever tradition they belong to. These
items are occasionally reported in small quantities, as at Terra Amata or
Olduvai, but shell middens or evidence of extensive dependence on
aquatic foods are unknown from the Early and Middle Pleistocene
(Isaac 1971, 1975b; Klein 1974). The lack of such evidence is unlikely
to be explained entirely by poor preservation since shell is one of the
more durable organic substances and is quite commonly preserved in
geological contexts of the same time period. It is possible to attribute
the lack of aquatic resources, as Isaac (1971) suggests, to the fact that
these were gathered away from the camp and the refuse scattered rather
than concentrated at the site. Yet this explanation seems implausible as
a method of accounting for such a widespread and consistent lack of
shell refuse. Moreover, as Isaac himself points out, there is evidence that
Acheulian camps were moved to their resources rather than the food
transported to the camp. A more serious possibility is that evidence for
shellfish consumption has been destroyed by changes in sea levels which
have buried coastal exploitation stations from these early periods.
However, here two counter arguments may be offered. First, the scarcity of evidence for the use of aquatic resources does not refer to the
coasts alone but also to the apparent failure of mid-Pleistocene popula-

tions to make significant use of riverine and lacustrine resources. Certainly river profiles and lake levels underwent considerable alterations in the Pleistocene, which might have resulted in the destruction of some evidence; but both are subject to unsystematic local variables and, unlike changes in sea level, it is highly unlikely that they resulted in the *systematic* elimination of the evidence. Moreover, there are Acheulian sites related to coastlines where tectonic alterations of the land surface have more than kept pace with eustatic rises in sea level, so that the old coastlines are preserved. Such relationships are preserved, for example, near Casablanca (Biberson 1961) and in Angola (J. D. Clark 1966), but there is no evidence of the significant use of coastal resources even in these locations. In addition, in many regions of tectonic rebound where coastlines are well preserved, Acheulian assemblages are absent, suggesting that in general Acheulian populations were not attracted to coastal environments (Coles and Higgs 1969).

The relative absence of the remains of vegetable foods in the archaeological data may be somewhat more a function of preservation. At Kalambo Falls, where waterlogged deposits have preserved organic remains (but where we are dealing with a terminal, and possibly atypical, Acheulian population), vegetable remains include local fruit and palm nuts in small quantities (J. D. Clark 1969). Similarly at Choukoutien (Black 1933) hackberry seeds have been recovered. Moreover, although human populations can subsist entirely on game (if the stomach contents of the game is included), there is no reason to assume that any population given its choice would do so. Fruits and nuts in particular are reasonable variety-foods which we might expect to find even among peoples otherwise largely dependent on game for their livelihood, and it seems probable that some variety of vegetable foods was a regular part of the Acheulian diet. On the other hand, neither the kinds of vegetable foods encountered nor the quantities recovered seem to imply anything like staple dependence on these foods. When we combine this with the observed absence of the tools used for the processing of typical modern vegetable staples, it seems probable that vegetable food *was* in fact a significantly smaller portion of the diet for mid-Pleistocene populations than for modern hunters.

Thus, two of the main ingredients of the modern eclectic hunting and gathering diet appear to have been scarce or absent from the diet of Acheulian peoples. The scarcity of aquatic resources seems to be an unquestionable historical fact. The scarcity of vegetable staples, although clouded by the problem of poor preservation, also appears to have real significance.

In the case of the Acheulian assemblages, there is at least one other line of evidence which supports the assumption that relatively heavy emphasis was placed on the hunting of big game. This evidence comes from the distribution of the sites themselves. Over the entire Afro-Eurasian landmass, Acheulian sites are selectively linked to biomes noted for their relatively great biomass of large grazing animals. In Africa, for example, Acheulian sites are not found in the forested regions of the continent or in many of its deserts. They occur only in areas of grassland, wooded savanna, or semi-desert (or occasionally in what are now true deserts, during periods when conditions appear to have been moister or in places where localized sources of permanent water such as wadis, springs, lakes or marshes were available (J. D. Clark 1967, 1970, 1975; Howell and Clark 1963; Deacon 1975; Isaac 1971). These are precisely the areas where the largest concentrations of large game animals, the gregarious herbivores, were to be found (Bourlière 1963). Clark (1975) infers from this distribution that the availability of game was in fact the primary consideration in the placement of Acheulian camps.

In Asia, in addition to their avoidance of cold latitudes, Acheulian populations appear to have confined themselves to biomes roughly analogous to those they exploited in Africa, using river valleys and open hill and plateau country in what are now semi-arid or subhumid environments (Collins 1969; Clark 1975; Butzer 1971; Bar-Yosef 1975). In Europe, Acheulian populations also chose open environments involving alpine grassland and loess steppe. The European sites are rarely related to woodland, and in the few cases where such a linkage is known (for example, in northern France and southern England), there is a clear preference for open woodland areas (Butzer 1971:454; Collins 1969).

The case is not the same for sites of the Clactonian-Charentian tradition in temperate Eurasia. The sites of this group are typically more forest oriented than their Acheulian contemporaries (Collins 1969). Collins has suggested that this orientation toward biomes poorer in big game may suggest that these populations were somewhat less directed toward big game hunting than were the Acheulian groups. On the other hand, except for occasional concentrations of small mammals such as the rodents at Vertesszöllös, there is little direct evidence of an economy markedly different from that of the Acheulian tradition, and ethnographic analogies suggest that hunter-gatherers in temperate latitudes would, if anything, have been more dependent on game than their tropical counterparts (Isaac 1975b). It may be important that forests seem to have been utilized primarily in temperate latitudes and not in the tropics. Temperate forests are not markedly deficient in large game whereas tropical forests are among the world's poorest biomes in this respect (Bourlière 1963).

In sum, mid-Pleistocene populations appear to have been markedly careful in their choice of biomes to exploit. And although the traditional picture of these populations as big game hunters is probably somewhat oversimplified by poor preservation, there is good reason to believe that they concentrated significantly more on game and made significantly less use of aquatic and vegetable resources than did their descendants.

Subsequent history in Africa, Europe, and most of Asia, however, demonstrates a gradual movement out of these selected environments, accompanied by shifts in consumption away from big game toward a variety of other plant and animal resources.

The Ecology of African Populations in the Later Pleistocene and Early Holocene Periods

The precise order and nature of the changes in economy and settlement patterns that occurred in Africa in the Late Pleistocene are difficult to describe. One problem is that the transition from the Acheulian to subsequent industries appears to have occurred over a span of time just prior to, or coeval with, the outer limits of carbon-14 dating, so that the absolute chronology of the transition—and even the relative dating of some of the industries described—is difficult to ascertain. The picture is further complicated by the persistence of remnants of a somewhat simplified chronological and climatological model which originated prior to the first radiocarbon revolution. Lacking both carbon-14 dates and the detailed glacial chronology which enabled European prehistorians to erect a rough chronological framework before carbon dating was available, African scholars relied heavily on stylistic horizons (many now discredited) and on simplistic notions of climate change to date archaeological assemblages. In the African literature prior to about

1960, relative dating depended in large part on the assumption that industries could be correlated with continent-wide occurrences of increased moisture or "pluvial" periods, usually assumed to correspond
with glacial advances in Europe. Although climate change in Africa
during the Pleistocene is of course well established, the old model of
pluvial periods is now essentially discredited (Flint 1957, 1959; Cooke
1958). Nonetheless, many of the assumed archaeological correlations
which were based on these pluvials still persist. It is only recently (see,
for example, Sampson 1974) that a number of the old style-horizon
names and cultural correlations have begun to give way as a result of
the reexamination of the artifacts and the availability of large numbers
of radiocarbon determinations. It is very difficult, therefore, to apply
any sort of shorthand nomenclature to the Upper Pleistocene industries of Africa. What follows is necessarily a fairly loose and generalized
description of the observable trends. The precise relative chronological
positions of the various industries described and the terminological
contradictions among different scholars are left for the synthesizers of
African prehistory to unravel.

In Africa, during the entire time span of the Acheulian industries,
despite an increase in the density of sites (and therefore presumably of
population), the distribution of human groups appears to have remained
essentially as it was described above (Howell and Clark 1963; J. D.
Clark 1967, 1970; Isaac 1969). Then, beginning sometime between
100,000 and 60,000 years ago, a somewhat more complex picture begins
to emerge. One of the clearest trends is the expansion of human populations into new environmental zones. Although the grassland areas
remain occupied, there is now evidence for the first time of the expansion of human populations into the equatorial forest fringes and of the
occupation of desert or semi-arid regions of the East African Horn and
northeast Africa. This period yields the earliest evidence for the continuous occupation of regions such as the Congo forests, Somalia, and
the Nubian Nile. Subsequently, during the Upper Pleistocene, essentially all of the African continent was occupied (J. D. Clark 1967, 1970).
At the same time, though there is considerable controversy about how
many cultural traditions should be recognized and by what names,
there appears to be general agreement among African prehistorians that
the cultural uniformity of the Acheulian gave way to a variety of regional industries. At first the continent seems to have been divided into

a few broad cultural zones, but, in the course of the Upper Pleistocene and Holocene, increasingly fine regional variation emerges.

Some of the Upper Pleistocene variations in tool manufacture appear to reflect new economic activities geared to new environments although in many cases such functional analyses must be considered tentative. The "Sangoan" industries and their cultural descendants in central and southern Africa, for example, have been described as tool assemblages designed for woodworking, and thus seen as adaptations to the newly penetrated forest environment (Lee 1963; J. D. Clark 1965). They are also seen as implying the increased use of vegetable foods (Clark 1960). Davies (1968) suggests that the picks that characterize the Sangoan industries should be interpreted as tools for digging roots and tubers, but there is no direct evidence of such use and no reason to assume that they were any more specifically adapted to this purpose than were the handaxes of the Acheulian.

A potentially more significant trend in the stone tool industries of the Upper Pleistocene is the emergence of grindstones. In southern Africa grindstones are recorded in a number of different archaeological contexts beginning as early as about 49,000 B.P. These stones become both more numerous and more widespread on the African continent during the course of the Upper Pleistocene (Kraybill in press; Sampson 1974; J. D. Clark 1960, 1969, 1970; Louw 1969; Mason 1962). One of the problems in interpreting the economic significance of grindstones, however, is that until the stones occur in great quantity or assume modern shapes, or until they are found in direct association with the material being processed, it may be difficult to distinguish stones used for processing vegetable foods from those used for industrial purposes. Some, but not all, of the earlier specimens are covered with ocher, which would imply that they were not used to process foods. It is only toward the end of the Upper Pleistocene, as discussed below, that the numbers of grindstones encountered, their shape, their degree of use, and their association with other tools or organic remains give clear testimony to the importance of vegetable processing.

During the Upper Pleistocene, perforated stone rings interpreted as digging stone weights also appear. These, like grindstones, assume somewhat greater importance in the course of the Upper Pleistocene and become truly significant only in Late Pleistocene and early post-Pleistocene assemblages (Sampson 1974; Seddon 1968; Clark 1950, 1969;

Cooke 1963; R. Mason 1962; Gabel 1963; Posnansky and Cole 1963). If the interpretation of the function of these artifacts is correct, they, like the grindstones, should indicate the increasing importance of vegetable foods in the diet.

It is also early in the Upper Pleistocene that the first evidence for economically significant use of aquatic resources occurs in Africa. At caves at the mouth of the Klasies River in South Africa (dated apparently just at the outer limits of carbon-14 dating and thus reported as anywhere between 50,000 and 80,000 years old), remains of seals, marine birds, fish, and shellfish occur in an economy still centered largely on land-based hunting. Seals, shellfish, and penguins are reported in significant quantities, although fish and flying birds are still rare (Klein 1974). At the opposite end of the continent, at Haua Fteah in Cyrenaica (McBurney 1967), exploitation of coastal resources is reported at about the same time.

Corresponding roughly with the end of the Acheulian industries in Africa there is also a significant change in the composition of the fauna hunted by man and of the fauna recorded in paleontological contexts. In South Africa, for example, many of the larger game species which predominated in Acheulian sites disappear. Martin (1966, 1967a) records the extinction of some 50 genera of large African fauna corresponding roughly with the end of the Acheulian complex. At some Upper Pleistocene sites a tendency toward the specialized hunting of modern migratory herd animals occurs, analogous on a small scale to the pattern observable in European sites of the same period (Cooke 1963). The overall tendency seems, however, to have been in the direction of more eclectic hunting patterns. Although the samples are small, Late Pleistocene sites in Africa on the average have a wider range of associated fauna than their Acheulian predecessors (Merrick and Pastron 1969) and a more intensive unselective use of varied microenvironments is to be inferred (Sampson 1974:213).

There is one further hint that the scope of the food quest may have expanded significantly at the beginning of the Upper Pleistocene. As I have indicated, the earliest use of fire in Europe and Asia, along the northern fringes of mid-Pleistocene human expansion, seems to have been related to the need for warmth that came with the penetration of colder latitudes. Fire is first recorded in Africa, however, only in late Acheulian assemblages, and it is not widespread until the emergence of early post-Acheulian industries. Its appearance at this late date might be best explained if it is assumed that in Africa fire was first used primarily to expand the range of edible foods. Fire increases man's access to food by rendering edible portions of the (primarily vegetable) biomass which are indigestible or poisonous in the raw state (Barnicot 1969:527; Davies 1968). Fire and the use of grindstones may both have been developed to permit processing of otherwise inedible vegetable foods. Alternately, as indicated in chapter 2, fire may have been used to alter the environment in ways conducive to greater productivity or availability of wild foods.

The African Upper Pleistocene is thus characterized by a number of indications of economic expansion. The reasons for these economic changes, however, are not entirely clear. One possibility is that the changes reflect altered climate conditions in Africa at a time corresponding roughly with the onset of the Wurm glaciation in Europe (J. D. Clark 1960, 1970). For example, the Sangoan assemblages of the early Upper Pleistocene have been seen as relating to a phase of increased moisture which resulted in the expansion of the tropical forest environment (Lee 1963). There is evidence of climatic change at this time (Bakker and Clark 1962). Moreover, since parts of the Sahara which had previously been occupied by Acheulian populations were temporarily abandoned at the beginning of the Upper Pleistocene (J. D. Clark 1967), some redistribution of population in response to changing climate seems to have occurred. On the other hand, except for this temporary movement out of the Sahara, the new distribution seems to reflect less a population shift than an expansion of the inhabited territory. Moreover, interpretations about the direction, scope, and economic importance of the climate changes recorded have been questioned by a number of scholars (cf. Deacon 1975).

The invention and spread of new tools and new technologies may also account in part for the expanding range of environments occupied. The invention and diffusion of grinding equipment in southern Africa, for example, may have played a part in this economic transition. Similarly, the use of fire may have been instrumental in permitting expansion of human populations into new habitats. On the other hand, the very uneven distribution of fire in the Pleistocene suggests, as indicated above, that the use of fire did not diffuse but rather emerged as a specific response to local need. In a similar manner, the rise of marine oriented economies at opposite ends of the African continent in different cultural contexts would appear to involve no technological diffusion but rather parallel and independent responses to adaptive problems.

The extinction of many of the large mammals hunted by Acheulian people may also have contributed to this economic transition. This extinction could be wholly or in part a function of altered climate, but both Sampson (1974) and Martin (1966, 1967a,b) argue that it resulted largely from the efficiency of Acheulian man as a hunter of big game. Deacon (1975) takes issue with the "overkill" model, questioning the extent and the efficiency of Acheulian hunting, but as Isaac (1969) points out the archaeological samples are in fact too small to permit the question of Acheulian overkill to be resolved.

Whatever else is involved, one clear feature of this transition period is the overall growth of population. In addition to noting that the density of Acheulian sites increases through the Middle Pleistocene, J. D. Clark (1967, 1970:81, 108) argues that there is a definite increase in the number of sites on the African continent as a whole at about 60,000 B.P. or shortly thereafter. It seems to me that the most plausible explanation of the changes that occurred is that the growth of Acheulian populations eventually forced them both to expand into new habitats and to begin to broaden the spectrum of their dietary habits (Coles and Higgs 1969:109). Since the increase in site densities continues throughout the Upper Pleistocene (J. D. Clark 1967; 1970:137), accompanied by intensification of many of the trends described, a more or less continuous growth of population appears to be in evidence.

In the terminal Pleistocene and early Holocene in Africa, there is again an increase in the density of sites over the continent (J. D. Clark 1967; 1970:148). It is also during this general time span, when geographical expansion into new regions was no longer available as a means of adjusting to increased population, that intensified gathering of small animals, aquatic resources, and plant material becomes most evident. These trends are particularly noticeable in the lower Nile Valley, where after about 17,000 B.P. a striking number of indices of high population density and significant population pressure occur. As described by various sources (Wendorf 1968; Wendorf and Said 1967; Wendorf and Schild in press; Wendorf et al. 1970; Reed 1960; Irwin et al. 1968; P. E. L. Smith 1966) a number of groups appear in the Nile Valley between about 17,000 and 14,000 B.P. which combine the old economic patterns centered on the hunting of large mammals with an increasing dependence on the hunting of aquatic animals, including hippopotamus, crocodiles, and turtles, in addition to fishing and the gathering of freshwater molluscs. Beginning sometime before 14,000 B.P. a new element appears in the context of these riverine occupations. Grindstones in large numbers, showing signs of heavy use, occur at approximately this date in three regions of the lower Nile, at Tushkla near Abu Simbel, at Kom Ombo near Aswan, and at Isna near Luxor. The number of these stones and the degree of wear that they exhibit implies significant economic dependence on the harvesting and processing of grasses, as these authors point out; moreover, many sites also contain backed bladelets with "sickle gloss," implying their use in the harvest. There is also extensive evidence of burning, suggesting human interference in the natural ecosystem.

These sites display a number of other indicators of high population densities and high population pressure. First, the sites on the Nile appear to form a very tightly knit mosaic of cultures which maintained their individuality in spite of close proximity and the opportunity for easy communication. It has been suggested that their distribution is indicative of carefully maintained territorial divisions suggesting fierce competition for resources. It has also been noted that burials which have been recovered from some of the sites of the region display a remarkably high frequency of skeletal traumata, which may be wounds incurred in the processes of competition and territorial defense (Wendorf 1968; Clark 1971). The clustering of burials in fairly large cemeteries also indicates relatively large and stable settlements.

There is some question as to precisely what grasses were being harvested and processed by these groups and which, if any, of their cereals had been domesticated. J. D. Clark (1971) suggests that grasses such as the genera *Echinochloa* and *Eragrostis* may have formed part of the economy. Wendorf and Schild (in press) have suggested the possibility that wheat and barley were present in the wild state in the lower Nile during the Late Pleistocene, even though their early Holocene wild distribution as reconstructed by Harlan and Zohary (1966) does not include this region. In support of their contention they note that pollen from Isna dated at about 12,500 B.P. has been identified as that of barley, although this identification is considered inconclusive. The same

pollen profile provides the only evidence yet uncovered to suggest that
any cereal was domesticated at this early date in Africa, and it is (as the
authors themselves indicate) poor evidence at best. The pollen identified as barley undergoes a very rapid and significant increase in frequency at about 12,500 B.P., which *may* be indicative of some sort of
human aid in the dispersal of the plant; but there is no accompanying
evidence of morphological domestication. In fact, there is no clear evidence of domestication until a much later date. This lack is somewhat
surprising in view of contemporary developments in the Middle East
(discussed below), and it may be that the absence of domesticates is a
function of preservation and exploration geared toward the smaller
special-purpose camps of the area rather than the main (agricultural?)
settlements (Vermeersch 1970; Wendorf et al. 1970).

It is very important to note, however, that even if upper Egypt provides the most striking evidence of population pressure and the intensification of resource use in Africa at the end of the Pleistocene, it is by
no means the only region of the continent to show such trends. Both
the use of aquatic resources and other small animals and the increased
use of vegetable foods requiring grinding are widely attested in Africa
at dates beginning between 10,000 and 12,000 B.P. In the Late Pleistocene and early post-Pleistocene (although generally at somewhat later
dates than in Egypt) there is, for example, widespread evidence of the
use of aquatic and land molluscs as sources of food. The association of
the Capsian epipaleolithic industry with huge shellmounds of terrestrial
snails, along with the bones of lizards, tortoises, rodents, and occasional
large mammals, is eloquent testimony of an economic shift in northwest Africa (Vaufrey 1955). Along the northwest coast of the continent
among peoples of the Oranian tradition there is a parallel shift to the increased use of marine fish and shellfish, and this shift is also found
among coastal populations of southern Africa (Louw 1960; Goodwin
1938; Fagan 1960; Schrire 1962). Sampson (1974) argues that the intensive exploitation of coastal resources in South Africa begins with the
Oakhurst complex between about 12,000 and 10,000 B.P. This trend
continues and intensifies through the more recent Wilton and Smith-
field assemblages, in conjunction with the decline of large game as a
resource at some sites and the concomitant increase in the importance
of small land fauna (Klein 1974). Coastal fishtraps associated with the
Wilton culture are widespread on the South African coast (Sampson

1974:311). And J. D. Clark (1970:172) has related the accounts of early European explorers describing the exploitation of stranded whales and seals by historic Late Stone Age populations, although to judge from the faunal lists in South Africa such exploitation was not widely practiced before the end of the Pleistocene.

There is evidence that fish and other aquatic resources assumed increased importance in the diets of populations in the vicinity of freshwater sources on the continent. The sudden predilection for freshwater shellfish and fish observed in the Nile Valley is also documented in the lakes in the southern and central Sahara, where there is evidence of communities of fishermen using fishhooks of bone and hunting aquatic animals after about 8000 B.P. (Arkell 1949, 1953; Wendt 1966; Monod 1963). Similar communities are known also on the shores of Lake Rudolph in East Africa by about 7000 B.P.; at Gambles Cave on Lake Nakuru; and at Ishango on Lake Edward in the western Rift Valley of East Africa during the ninth millennium B.P. (Robbins 1967, 1974; Oakley 1961b; deHeinzelin 1957).

Similarly, increases in the use of grinding equipment and greater attention to vegetable foods near the end of the Pleistocene are fairly widespread. J. D. Clark (1950) reports, for example, that the Nachikufan industry from southern Africa, dated beginning about 10,000 B.P., shows an emphasis on plant processing, as evidenced by the high frequency of upper and lower grindstones as well as by the use of digging stick weights. Similar artifacts are reported from Oakhurst and Wilton industries in South Africa (Sampson 1974). In addition, Wilton sites such as Melkhoutboom Cave and the Gwisho site give evidence of significant exploitation of wild plant species through the preservation of organic remains (Deacon 1969; Gabel 1965).

In the Maghreb, the epipaleolithic sites dated to about 12,000 B.P. and after, already noted for their emphasis on land and marine molluscs, also contain grinding equipment and blades showing silica sheen which are assumed to be sickles (Camps-Fabrer 1966). And these industries, like those of the Nile Valley, appear to have supported fairly large and sedentary populations to judge by the size of accompanying cemeteries. At the Dungal Oasis in southwestern Egypt, sites with large numbers of grindstones and with sickle blades are reported at about 8000 B.P. (Hobler and Hester 1969), and at the same time similar grass harvesting and processing assemblages are widespread in the Sahara, which was apparently somewhat wetter than at present (Clark in press; Hagedorn and Jakel 1969; Fauré et al. 1963).

Although the buildup of population and of population pressure can be reasonably well documented on the African continent, the history of plant domestication and its precise relationship to this pressure is not entirely clear. A strong case can now be made for the indigenous domestication of a number of food plants in sub-Saharan Africa, including the following: sorghum (*Sorghum bicolor*); African rice (*Oryza*); various millets and other grasses (*Pennisetum, Eleusine, Brachiaria, Eragrostis, Digitaria*); a variety of minor pulses; tuberous crops including yams (*Dioscorea* spp.); a number of oil-seed crops, including the oil palm (*Elaeis*) and sesame (*Sesamum*); and a range of other crops including okra (*Hibiscus esculentus*) and watermelon (*Citrullus lanatus*). (For a complete listing see Harlan 1971; Harlan et al. in press; Purseglove in press.)

The distribution of these crops is of interest since they appear to have been domesticated in a number of distinct regions rather than in any single center. Individual species or genera often show signs of having been domesticated independently more than once. In addition, many of these crops are highly localized in their distribution even today (Harlan 1971). The pattern indicates that the act of domestication occurred independently a number of times rather than resulting from diffusion or from a single act of discovery. It is striking, however, that virtually all of these indigenous African food crops appear to have originated in a single broad geographical zone extending across the continent south of the Sahara and north of the equator (Harlan 1971; Harlan et al. in press). This pattern would seem to indicate that some common geographical factor was involved in the domestication process in the various regions. It has been suggested by various authorities (Clark in press; Ellis in press; Munson in press; Harlan et al. in press) that, in a context of high population densities, fluctuations in the southern margin of the Sahara Desert, which would upset existing ecological equilibria and displace human populations, might have been a major factor in generating population pressure on the savanna zone south of the desert, triggering domestication in this region.

It is not clear, however, when domestication occurred in sub-Saharan Africa, nor is it clear whether domestication occurred independently or only after the diffusion of domestication technology from the Middle

East. Traditionally, agriculture has been viewed as arriving from the Middle East, spreading first to Egypt and thence traveling by any of various routes westward and southward across or around the Sahara. (For a review of possible routes of diffusion, see Smith [in press], Ellis [in press], Munson [in press].) For a number of years, however, some scholars have argued, using linguistic data, tool distributions, and botanical data, that agriculture may have considerable antiquity in particular portions of sub-Saharan Africa and may have arisen independent of outside stimuli (cf. Murdock 1959; Portères 1950, 1951). The most famous of these assertions has been that by Murdock that an independent center of domestication involving approximately 25 food and fiber plants would be found along the upper Niger in west Africa dated to about 7000 B.P. More recently, the suggestion has been put forward based on the interpretation of certain tool forms that some type of vegeculture involving yams may be of considerable antiquity in west and central Africa, dating at least to the end of the Pleistocene and possibly extending back as far as the Sangoan industries (Davies 1968; Harris 1972; Coursey in press). Some of these early centers of domestication, such as that postulated by Murdock, are now at least partly discredited; and others, such as the one suggested by Davies and Coursey, are moot for lack of firm evidence. Yet there is in Africa, as on other continents, a growing tendency to move away from strict diffusionist models and to seek, at least in part, indigenous processes and local explanations for the origins of domestication (Clark in press; Smith in press; Harlan et al. in press). At the same time, there seems to be a continuing body of opinion based on indirect evidence that sub-Saharan agriculture may yet prove to have a history of five millennia or more, even if it cannot be pinned to any particular center (cf. Clark in press; Shaw in press; Ellis in press; Munson in press; Purseglove in press). One piece of evidence supporting the assumption of relatively great antiquity for sub-Saharan agriculture is the fact that the Sahara Desert is known to have undergone a period of increasing dessication and expansion beginning about 5000 B.P. (Bakker in press). If in fact the expansion of the desert acted as a trigger for domestication, we might expect agriculture south of the Sahara to have begun to emerge at least by this period.

The archaeological evidence unfortunately is still very spotty owing in part to poor preservation on much of the continent and in part to the underdeveloped state of prehistoric research. At best, the evidence provides a set of minimum dates by which agriculture can be shown to have been established in any particular region. But the direct archaeological evidence does not firmly establish an antiquity for African agriculture sufficient to prove its indigenous origins, nor does it provide a convincing picture of cultural diffusion linking the Middle East and the southern boundaries of the Sahara. In general, the earliest firm dates for agriculture in Africa tend to cluster toward the northern and eastern portions of the continent, suggesting the ultimate derivation of agriculture from the Middle East. But preservation is sufficiently spotty to render this pattern less than convincing and there are some indications that agriculture, although beginning early in northeast Africa, may have spread from the upper Nile or the Sahara to Egypt rather than moving in the opposite direction.

Reference has already been made to the possibility that very early domestication may have occurred along the Nile in Nubia by 12,000 B.P., much earlier than agriculture is known to have begun in Egypt. As indicated above, however, the evidence for domestication at this time is weak. The possibility has also been suggested (Hobler and Hester 1969) that early agriculture was associated with the grindstone and sickle complex at the Dungal Oasis in southwest Egypt, discussed above; but again the case is weak for lack of direct evidence. Slightly better evidence is available at Amekni in the Hoggar Massif in the Sahara, where two pollen grains of cultivated cereals have been dated between 8100 and 6500 B.P. and identified as *Pennisetum*. These occurred along with other pollen which may be that of wheat (*Triticum*) (Camps 1969, 1971). The pollen identifications are considered questionable however (Shaw in press).

Despite the flurry of recent activity in North Africa, in fact the earliest certain evidence of domestication in Africa still comes from the Fayum Oasis in Egypt, where actual grains of emmer wheat, barley, and flax have been recovered and dated to the middle of the seventh millennium B.P. (Caton-Thompson and Gardner 1934). At this date and in this location, the cultigens are generally presumed to have been derived from the Middle East. Similar finds from Badarian deposits in Egypt in the sixth millennium B.P. (Arkell and Ucko 1965) and from Merimde at a slightly later date seem to demonstrate that wheat and barley agriculture was well established in Egypt from this period on.

Outside Egypt, the record is even spottier. Fragments of oil palm (*Elaeis guineensis*) at Shaheinab in the Sudan, dated to 5300 B.P., may indicate the early domestication and transport of this sub-Saharan species (Arkell and Ucko 1965). Pollen from Meniet in the Hoggar dated to the mid-sixth millennium B.P. has been interpreted as representing domestic cereals because of its size (Hugot 1968), but the identification is uncertain as a result of the discovery of wild grasses with equally large pollen in the Sahara (Shaw in press). At Adrar Bous, in the Aïr region of the Sahara, pottery impressions have produced a single grain of *Brachiaria* dated to about 6000 B.P. and one of cultivated sorghum dated to about 4000 B.P. (Clark 1971). This scattered evidence suggests that agriculture involving a number of indigenous grasses may have been fairly widespread in the Sahara at least by 5000 B.P. (Munson in press), but most of the evidence must be considered questionable.

In Dhar Tichitt in Mauretania, where there is a better archeological record, it has been possible to trace the evolution of a farming economy from a hunting and fishing economy over a period of a few hundred years just prior to 3000 B.P. (Munson 1968, 1970, in press). In early phases of the sequence, hunting and fishing appear to predominate in the economy, while seed processing as evidenced by grindstones and seed impressions in pottery is only a minor activity. Subsequently both grindstones and seed impressions increase in frequency, suggesting the growing economic importance of a range of harvested wild grasses. At about 3000 B.P. a further shift is noted, this time toward the intensive use of a single genus, *Pennisetum,* specimens of which now show morphological signs of domestication. This economic transformation occurs in conjunction with a decline in the fishing potential of the region, and, as the author concludes, the increased emphasis on grass harvesting and the emergence of domestication seem to have resulted from attempts to find a substitute for the declining aquatic resources.

The Dhar Tichitt sequence appears to provide the earliest clear-cut evidence of African agriculture outside Egypt, but as such it is surprisingly late. As Munson (in press) points out, the sequence of gradual economic transition from generalized harvesting to specialized harvesting of a grain displaying morphological change is strongly suggestive of the evolution *in situ* of domestication—except that the transition is very rapid and very late in comparison with such sequences in other parts of the world. The latter considerations lead Munson to conclude that the

Dhar Tichitt sequence probably represents the intrusion of already domestic *Pennisetum* into an economy already geared to more generalized grass harvesting.

South of the Sahara the known archaeological evidence is all comparatively late, and claims for early agriculture—whether based on tools, linguistics, or botany—remain largely unsubstantiated by firm archaeological data (cf. Shaw in press). In West Africa, the earliest presumptive evidence of cereal cultivation comes from Iwo Eleru Rockshelter in southwestern Nigeria, where microliths with silica gloss have been dated to about 5000 B.P. (Daniels 1969). Unfortunately, while the gloss probably indicates the cutting of grasses, it need not indicate the harvesting of domestic cereals. Similarly, the pick and hoe-like forms cited by Davies as evidence of early vegeculture are not conclusive. At Ntereso in northern Ghana, impressions of *Pennisetum* have been found on potsherds dated to about 3500 B.P. (Davies 1968), but both the identifications and the stratigraphy of the site have been questioned by various authorities (Munson in press; Shaw in press). Similarly, the sequence from the K6 Rockshelter at Kintampo, Ghana, displays the seeds of hackberry (*Celtis*), oil palm (*Elaeis*), and cowpea (*Vigna unguiculata*), an indigenous pulse, in layers beginning at about 3400 B.P. (Flight 1970). However, the material is not considered proof of domestication (Shaw in press; Munson in press).

As summarized by Shaw (in press), Munson (in press), and Ellis (in press), firm dates for cultivation in West Africa south of the Sahara are generally on the order of 1500 B.P. or less, and the evidence suggests only that agriculture involving sorghum and probably *Pennisetum* and *Eleusine* was well established by this date. Sorghum is known, for example, from levels dated to about 1200 B.P. at Niani in Guinea and in levels dated to about 1000 B.P. at Daima in northeastern Nigeria.

In eastern and southeastern Africa, despite indirect arguments suggesting greater antiquity for agriculture, particularly in Ethiopia (Munson in press; Simoons 1965; Doggett 1970), the archaeological record as summarized by Shaw (in press) again displays no firm evidence of domestication before 1500 B.P. (partly perhaps because the archaeological record for Ethiopia, as for much of East Africa, is very poor). At Engaruka in Tanzania, sorghum has been recovered from all levels of the site, beginning at about 1500 B.P. At Inyanga in Rhodesia, dated to approximately 1200 B.P. sorghum, *Pennisetum*, *Eleusine*, and a number of melons and pulses have all been recovered. Sorghum and possibly *Pennisetum* have been recovered from Mwamasapwa in Malawi in levels dated to about 1000 B.P., and cowpeas have been reported from Nkope, also in Malawi, dated to about 1200 B.P. Sorghum has also been found at Isamu Pati and Kalundu in Zambia at about 1000 B.P., as well as at the Ingombe Ilede site in Zambia dating back about 1,300 years. In general, in Africa south of the equator, the spread of agriculture seems to have correlated (though not so perfectly as was once thought) with the spread of iron-using Bantu-speaking populations within the last 2,000 years (Shaw in press).

The European Sequence

In Europe the sequence of human occupations during the Upper Pleistocene displays a number of trends very similar to those already encountered in Africa. Overall increases in the density of sites are accompanied by geographical expansion of the territory occupied by human populations and by increasing regional specialization of human cultures. In the economic sphere, a system basically oriented toward hunting is supplemented increasingly by the use of coastal and riverine resources and, to a lesser extent, by the processing of vegetable foods. There appear to be two primary differences between the European and African sequences: first, in Europe, to a far greater extent than in Africa, the focus of hunting became increasingly concentrated on certain big game species; second, vegetable foods appear to be somewhat less significant in Europe than in Africa. These differences appear to be related to the particular environmental conditions of the two regions. Upper Pleistocene man in Europe survived in an environment modified by the expansion of the Wurm glaciation, which created a local environment similar to that found today only in more northern latitudes. As a result, the inhabitants of Europe adopted economies which in many ways parallel those of the most northern peoples today. The latter live in environments poor in edible vegetation and as a result are strongly dependent on animal and fish resources. Their habits are often strongly geared to the habits of particular game species on which they depend. At the same time their dependence on vegetable foods, by necessity, is among the lowest of known human groups. Signs of an adaptation of this type are evident in Middle and Upper Paleolithic Europe.

The Mousterian assemblages (associated with Neanderthaloid skeletons) which emerged in Europe shortly after 100,000 B.P. occur in somewhat greater numbers than those of the preceding industries, and at the same time they reflect a far greater degree of local stylistic or functional variation than did the industries of the Middle Pleistocene (Klein 1969a,b; Valoch 1968; Bordes 1953, 1961a,b; Sulimirski 1970; Freeman 1966; Binford and Binford 1966). At the same time Mousterian populations extended the known range of human occupation to the north, penetrating the European tundras in the earliest recorded human use of subarctic and arctic environments (Butzer 1971:463). Aside from this territorial expansion and some minor changes in economic activity, however, the Mousterian populations seem largely to have continued the ecological patterns focused on big game hunting which characterized earlier groups. There are sporadic reports of grinding equipment from Mousterian assemblages, such as from Molodova in the Ukraine and from Cueva del Castillo in Spain, which would appear to indicate some increase in the processing of small vegetable foods; but such reports are rare and there is little indication that these tools represent a major modification of the overall economy (Chernysh 1961; Klein 1969c; Freeman 1964, 1966; Sulimirski 1970). Similarly there is evidence that shellfish and other aquatic resources were exploited at a number of coastal Mousterian stations such as the Monte Circeo Caves in Italy and the Devil's Tower in Gibraltar (Clark 1948, 1952; Barker 1975; Evans 1969); and there is evidence for the small-scale exploitation of birds, small mammalian fauna, fish, and shellfish at inland Mousterian stations such as the Salzgitter-Lebenstedt site in Germany (Tode et al. 1953). The use of these resources may indicate some expansion in the scope of the food quest, and the exploitation of some, notably the birds, also may point to increased hunting proficiency (Butzer 1971: 468, 471). But these resources seem to have served as minor or occasional foods in a predominantly hunting economy, just as they did in contemporary early Upper Pleistocene sites in Africa.

Aside from the expansion in occupied territory, the most important trend in the Middle Paleolithic in Europe may be the increased attention paid by human populations to individual herds of migratory animals, foreshadowing the very specialized hunting patterns of the Upper Paleolithic. Concentration on particular gregarious species such as reindeer, mammoth, bison, or red deer is reported from a number of Mousterian locations (Movius 1953; Tode et al. 1953; Klein 1969a,b; Barker 1975; van Campo and Bouchud 1962; Bordes and Prat 1962). But while such specializations often approach 80 percent of the total fauna taken by Mousterian hunters, they rarely come close to the concentrations of 90 percent or more on individual species that are often observed in Upper Paleolithic sites (Mellars 1973). Moreover, the total kills are rarely as large as those of the Upper Paleolithic, and such concentration does not appear to be as common as it is in later sites.

The specialized hunting patterns appear to have involved some degree of adjustment to the habits of the prey. At the very least, the specialized exploitation of migratory herds would seem to require a pattern of seasonal migrations geared to the movement of the animals. As Butzer (1971:467) has pointed out, the number of temporary campsites with Mousterian tools exhibiting specialized exploitation of migratory herds seems to support this assumption of seasonal movement. The development of artificial shelters such as that tentatively identified at Molodova on the Dniester River (Klein 1969a,b) may be an adaptation to this newly required mobility. In a similar vein, Vallois (1961) notes the marked homogeneity of physical type among various groups of Neanderthal men in western Europe and concludes that such resemblances reflect widespread movement and contact between groups, presumably related to annual movements of human hunters in pursuit of game.

On the other hand, not all of the Mousterian sites appear to represent temporary or even seasonal camps. Mousterian cave sites in southwest France, for example, seem to have been occupied on a year-round basis, as judged by the antlers and dentition of reindeer taken (Bouchud 1954). The regional complexity of Mousterian industries in France (Bordes 1953, 1961a,b) would also appear to reflect stable, territorially defined, semipermanent populations. But the conclusion is only valid if we assume with Bordes that these variations reflect different culture groups; if they are simply functional (Freeman 1966; Binford and Binford 1966) no such mosaic is implied.

It should also be pointed out that many, if not most, Mousterian sites display a more generalized hunting economy than that pictured above. Klein (1969a,b) notes, for example, that Mousterian sites in some portions of European Russia, notably in the Crimea, display a fairly generalized hunting economy. Similarly, in much of Europe Mousterian hunters seem to have retained a generalized hunting style in environments which later supported specialized hunting by Upper Paleolithic groups (Barker 1973, 1975; Higgs et al. 1967; Mellars 1973).

After about 35,000 B.P. the Neanderthaloid Mousterian populations of Europe were replaced by biologically modern populations with a new tool technology based on the extensive utilization of blades. The new groups, however, seem largely to have continued the demographic and economic patterns and trends that had already been established. As indicated above, Upper Paleolithic populations further extended the northern boundary of settlement in Europe, and there is considerable evidence that the overall size and density of the European population increased. There is some debate (see, for example, Mellars 1973; Narr 1963; Vallois 1961) as to whether, or to what extent, individual population aggregates had increased in size over those of the Mousterian, but there seems to be very widespread agreement that the number and density of sites increased significantly in comparison with the previous period and that the increase in site density continued throughout the Upper Paleolithic itself (Nougier 1954; Bordes 1968; Sonneville-Bordes 1960, 1963, 1965; Klein 1969a,b; Butzer 1971; Mellars 1973; Sulimirski 1970). Mellars (1973) has noted, for example, that in the Périgord region of France there are more than five times as many cave and rockshelter sites dating from the Upper Paleolithic as are recorded from a Middle Paleolithic period of comparable length. Butzer (1971:481) has estimated that there may have been as much as a tenfold overall increase in population between the two periods. The trend toward increasing regional variation in culture, begun by the Mousterian peoples, seems also to have continued (Sonneville-Bordes 1963).

As among the Mousterian cultures, there is little evidence of the use of plant foods in the Upper Paleolithic. There is occasional reference to such foods or to the grinding tools possibly used to process them in studies of Upper Paleolithic sites (Chernysh 1961; Klima 1962; Sulimirski 1970; Freeman 1964; Clark and Piggott 1965), but it is not clear to what extent the grinding tools were used for food processing. There is thus little evidence that plant foods were a significant resource.

The evidence for the use of coastal and riverine resources, however, is quite substantial, indicating a significant increase in the importance of these foods in comparison to their minimal use in the Middle Paleolithic. Moreover, it is fairly clear that the importance of aquatic resources increased throughout the last 35,000 years of the Pleistocene, foreshadowing the highly developed use of these resources in the early post-Pleistocene Mesolithic cultures. Fishhooks, spears, and harpoons all appear for the first time among cultures of this period; and fishbone, both of freshwater and saltwater species, is fairly common in the refuse. Mellars (1973), for example, notes the occurrence of salmon, trout, carp, pike, bream, dace, and chub from Upper Perigordian, Solutrean, and Magdalenian sites in southwest France, and suggests that the riverine orientation of sites, particularly in later phases of the Magdalenian occupation of the region, may indicate increasing economic reliance on rivers as sources of food. J. G. D. Clark (1948) has called attention to the extent of fishing activities in Hamburgian and Ahrensburgian sites. A number of other authors (for example, Sonneville-Bordes 1963; Smith 1964) have also commented on the quantity of fish remains in Upper Paleolithic sites, particularly in the later cultures. It is noteworthy, however, that no mention occurs of deep-sea fish, suggesting that boats were not yet in use, at least not for economically significant fishing.

The Upper Paleolithic interest in coastal resources also includes the exploitation of marine mammals, which occurs for the first time in this context. The hunting of seals and walrus is reported and the distribution of tools such as harpoons may indicate that marine hunting had become an important activity (Sturdy 1975; Møhl 1970; J. G. D. Clark 1952). In addition shellfish are encountered in the archaeological refuse of the period. In most cases shells occur only in small quantities and a noneconomic motive is suspected (Mellars 1973). In other cases shellfish seem to have been of some economic significance (J. G. D. Clark 1948, 1952), and in a few locations, such as in the Grotto del Cavallo of the Uluzzian Upper Paleolithic in Italy, it is even possible to identify something approaching an economic specialization in their use (Whitehouse 1968).

Some authorities suspect that coastal exploitation may have been somewhat more extensive in the Late Pleistocene than the archaeological record reveals. For example, Newell (1973) notes that along the coast of Europe, wherever conditions are favorable for the preservation of Late Pleistocene coastal sites, there appears to be a fairly continuous record of coastal exploitation spanning the Late Pleistocene and early Holocene periods. In contrast, where such Late Pleistocene evidence is lacking, its absence is often traceable to specific conditions of poor preservation such as the submergence of the coast or the subsequent deposition of thick marine-layers.

This expansion in the scope of the resources used is also observable in the attention paid to birds and small mammals. Ptarmigan, grouse, and hare, with a variety of other small game, occur in some Upper Paleolithic sites in quantities that indicate more than casual gathering (P. E. L. Smith 1964; Mellars 1973); and it may be that in addition to the focus on hunting and fishing activities considerable attention was being paid to the trapping of small game and to fowling.

As in the Mousterian sites, however, the bulk of the diet seems to have come from large, gregarious animals. The Upper Paleolithic is characterized above all by the massive size of many of the kills and by the extraordinary concentration of the hunters on particular species. Concentrations of reindeer or mammoth bones constituting better than 90 percent of the fauna taken are not unusual. At one famous site, an Upper Perigordian occupation at Solutré in the Rhone Valley, the remains of approximately 100,000 horses have been excavated. In general, however, reindeer appear to have been the preferred quarry in much of western Europe; mammoth predominate in central Europe; reindeer and horse in central Russia; and bison in southern Russia (Butzer 1971: 477; Klein 1969a,b).

The significance of this pattern of increasingly specialized hunting, which begins in Mousterian sites and extends through the Upper Paleolithic, is not easy to assess. The Upper Paleolithic, with its relatively dense population and its flowering of cave art, is often pictured as a period of abundant resources providing a very high standard of living (see, for example, Butzer 1971:481). There is no question that the Upper Paleolithic economy was geared to an environment that was at least periodically rich in big game. Neither the extent of the kills recorded nor the degree of selectivity would have been possible in any environment other than open grassland or tundra supporting massive herds of gregarious animals. (It is noteworthy, for example, as Klein [1969a,b] points out, that specialization by Mousterian hunters typically occurred in open environments while more generalized hunters occupied forested regions.)

The herds may, in some locations, have provided a rich enough game base to permit hunters to be highly selective among an abundance of resources. David (1973) has suggested that the Noaillian population of southwest France may have lived in just this manner. Moreover, in regions where the herds were available throughout the year, these resources may have provided a rich and easily captured food supply permitting human hunters to eschew the use of other resources. There are in fact regions and sites in Europe where specialized hunting occurs in conjunction with year round or semipermanent occupation (Bouchud 1954; Klima 1954, 1962).

On the other hand, there are a number of reasons to assume that in most cases specialized hunting patterns represented not so much an enjoyment of plenty as an adjustment to the increasing difficulty of feeding dense populations by any other means. One important point is simply that today hunting focused on individual species does not occur in game-rich environments but rather appears to be confined to regions where the particular species hunted is the only available source of food, or at least the only source of meat in significant quantity. Such specialized hunting does not often occur today in Africa, for example; rather it is well documented on the American Great Plains and in arctic regions, where herds of buffalo and reindeer respectively provide the only reasonably large aggregates of meat. There are some problems in applying this model to the late glacial hunters of Europe, however. One is that an analogous, if less marked, tendency to specialization is noticeable in Africa at the same time although similar ecological conditions cannot be inferred. Another problem is that there is evidence in Europe itself that other animals were present and were taken. Most of the sites of the period, in addition to their specialized prey, seem to include a wide variety of other mammals, large and small, in very small numbers; cattle, horses, red deer, Roe deer, chamois, ibex, and others (David 1973). Nonetheless, the implication of such heavily specialized hunting would seem to be that, even if present, these other resources could not be relied upon to feed the population without massive supplementation from the slaughter of the herd animals.

The most telling argument is simply that in most cases the hunting of the herd animals would have involved considerable difficulty which would only have been undertaken for lack of other alternatives. Observations on modern reindeer herds, for example, suggest that although they are very easily taken at particular times, they are a very difficult source to rely on as a staple. Reindeer are easily killed, both because of their gregarious habit and because the animals are relatively unwary and are easily approached. Moreover, they are safe to hunt because they do not stampede or react aggressively to hunters. The major problem is that large herds form irregularly and only for short periods and that the reindeer migrate at a fairly rapid pace. Thus, although occasional massive kills are possible, reindeer can be killed in most locations only for very brief periods of the year (Burch 1972; Sturdy 1975; David 1973). One possible implication of this is that the massive kills so abundantly recorded in the Late Pleistocene represent only occasional or at best seasonal events in the lives of hunters who otherwise relied on a more general assortment of game. It may be that such spectacular kills, which would have attracted fairly large and semipermanent group settlement, are both better preserved and more often recovered than more typical but less noticeable instances of solitary hunting (Burch 1972). A related possibility is that seasonal reindeer hunting might have alternated with coastal exploitation of fish and marine mammals, as occurs among many populations today in or near the Arctic Circle where no other resources are available (Sturdy 1975). As Sturdy notes, the evidence of antlers at many Upper Paleolithic sites confirms the strictly seasonal nature of the occupations.

If in fact human populations attempted to rely on reindeer for any significant portion of the year, they would have been forced to adopt one of two basic patterns (Burch 1972, Sturdy 1975). First, they could have attempted to kill enough animals in the course of an occasional encounter to last them through some substantial portion of the year. It is tempting to visualize a large group of people banding together periodically for an organized drive which would result in massive and indiscriminate slaughter of the herd (Soergel 1922). But, such an exploitation pattern would involve great problems. For one thing it would imply a scale of social organization only rarely attested among earlier groups of human beings (Freeman 1973; S. R. Binford 1968). It is interesting to speculate with Hammond (1974) that such large-scale cooperation is likely to have emerged only as people perceived that the resources available to them as individual hunting units were becoming increasingly insufficient.

Another major problem encountered in this type of massive seasonal exploitation would be storage. In order for the kill to have value for more than a few days or weeks, it would be necessary to preserve it. In winter, of course, the availability of natural cold storage would have made this relatively easy, and in this sense the migration to cold climate may have eased man's hunting burden greatly. If the same resources were used in the summer, however, then intensive efforts at preparing and drying meat would have been necessary. It is difficult to imagine this type of advanced labor investment being made unless there was considerable concern that fresh meat might not otherwise be available reliably and in quantity in coming seasons.

The other alternative would be for the human group to adapt its own movements to those of the herd in an attempt to have the resource continuously available. This would demand an alternate splitting and joining of the human groups paralleling the concentration and dispersal of the reindeer (David 1973), and it would involve fairly rapid movement. Burch (1972) has argued that the reindeer's great capacity for sustained movement would mean that people could not keep up and follow on a continuous basis but would instead have to rely on a more sporadic pattern involving periodic interception of the herd as it moved. As various authors have pointed out, however (Burch 1972; Sturdy 1975; David 1973), this kind of strict dependence on reindeer could not have supported a large human population; the high density of Upper Paleolithic population would seem to imply that such dependence was not the rule.

A number of scholars (most recently Higgs and Jarman 1969, 1972; Sturdy 1975; Barker 1975; Coles and Higgs 1969; J. G. D. Clark 1967) have suggested that the close relationships between man and his specialized prey in the Late Pleistocene may have involved a kind of herd management or protection, control of herd movements, or even incipient domestication. Barker (1973, 1975) has noted, for example, that Upper Paleolithic and even Mousterian sites often display a distribution that points to a kind of loose herding (similar to that described ethnographically among the Lapps) in which the seasonal movements of the prey are manipulated by the herders. The assumption of incipient domestication in the Paleolithic has been challenged by others (see, for example, Smolla 1960), and is, in any case, highly speculative. One obvious objection is that if Upper Paleolithic man were in fact managing these herds, the animals might be expected to have had a better survival rate when climate change began to interrupt their natural grazing areas at the end of the Pleistocene.

What is important, however, is that migratory exploitation of reindeer, whether by hunting or by herd management, would seem to involve very significant labor costs. Neither such migratory specialization on a particular herd nor the pattern of occasional interception of the herds and storage of the meat seems an attractive strategy, and it is plausible to argue that this pattern would not have been adopted except in the absence of other reliable prey.

A number of lines of archaeological evidence support the argument that specialization was a strategy adopted out of necessity rather than by choice—that it was in fact a particular adaptation to population pressure in northern latitudes. In the first place, the continued northward expansion of the hunting territory belies the assumption that meat was plentiful. And the increased reliance on coastal resources and small trapped game suggests that large mammals were not as readily available as we have been led to believe. Similarly, as Wilkinson (1975) has pointed out, the exploitation of the musk ox, which became increasingly evident in the later stages of the Upper Paleolithic, implies that other resources were scarce. Musk oxen typically are hunted today only as emergency resources in periods of shortage or when human groups are forced to occupy regions poor in other resources.

In addition, Klein (1969a,b) has pointed out that there is a good deal of stability in the specialized hunting patterns observed in any region. For example, in locations where certain specialized hunting patterns are observed among Mousterian populations, specialization on the same prey appears to continue with some regularity into the Upper Paleolithic. As Klein argues, this consistency suggests that specialized hunting patterns were a function of actual environmental limits on hunting choices rather than simple expressions of cultural preferences.

A pattern observed in Greece (Higgs et al. 1967) and in parts of Italy (Barker 1973, 1975) is also of interest. In both regions Middle Paleolithic hunters seem to have practiced relatively broad-ranging, generalized hunting patterns but to have confined themselves to a limited range of ecological niches, noticeably avoiding highland regions. In both cases the emergence of specialized hunting economies is coupled with an expansion of the pattern of seasonal movements in which hunters began to follow certain selected migratory species into highland grazing areas. Barker suggests that it was more economical for Middle Paleolithic groups to remain in lowland areas and exploit the range of fauna available.

The selective pattern, he argues, emerged only after the disappearance of many of the large nonmigratory fauna. According to his reconstruction, the increasing cold and aridity of the last glaciation favored large herbivores with the ability to move overland between seasonal pastures. The selection pressures favoring the survival of migratory species, and the elimination of their less mobile neighbors, may have forced human hunters to adopt a specialized economy and a more migratory life-style. The increased mobility of Upper Paleolithic populations has in fact been cited by Smith (1972a) as evidence of population pressure on resources. He argues that the proliferation of artificial shelters in the Upper Paleolithic is indicative of the fact that hunters were traveling further in the pursuit of game and were using hunting territories at greater distances from natural shelters.

Finally, there is evidence that fire as a manipulative tool became more important in the Upper Paleolithic, which may be indicative of human efforts to increase productivity. Van der Hammen (1957a,b) and Narr (1961, 1963) have called attention to the importance of widespread burning in Alleröd horizons correlated with the Tongiran or late Magdalenian culture in Belgium and the Netherlands. The fires are assumed to have been artificially set for the purpose of increasing the productivity of either game or vegetable resources.

There are thus a number of lines of evidence that Late Pleistocene populations in Europe were already undertaking a variety of strategies to alleviate population pressure on resources. That these strategies took a somewhat different form from those encountered in Africa or in other parts of the world (to be discussed below) would appear to be largely a function of the glacial environment.

Mesolithic Cultures in Europe

After the end of the Pleistocene in Europe a number of economic changes occurred. Early Holocene (Mesolithic) cultures in Europe are problematic and a number of controversies exist concerning their importance. We can proceed best, perhaps, by stating at the beginning what is clearly known about these groups and then going on to consider some of the points of contention.

First, it is fairly clear that the specialized dependence on large herds of gregarious animals such as the horse, reindeer, and mammoth, which had characterized the Late Pleistocene tundra dwellers of Europe, declined and that these animals became scarce or totally extinct. Mesolithic industries seem to represent a period of readaptation to the expanding forest biomes of Europe, involving the hunting of solitary forest-dwelling animals along with the rapid expansion of the scope of the food quest to include a wide range of animal species. There is a particularly heavy dependence on small game, avifauna, and the resources of the sea, rivers, and lakes as sources of calories and protein. These economic shifts appear to have entailed increasingly sedentary behavior on the part of many populations. Newell (1973) suggests that there was a gradual increase in sedentism among European Mesolithic populations (see also Whitehouse 1968; Jacobi 1973), and Vallois (1961) has pointed out that cemeteries containing multiple burials (including members of more than a single family) are typically a post-Pleistocene phenomenon.

The change in economy presumably also included increased use of plant material (Newell 1973). In contrast to other areas to be considered, the postglacial adaptation in Europe does not seem to have involved significant processing of grain or other ground vegetable material. References to grindstones occur (cf. Petersen 1973) but, as usual in the material on Europe, these references are scarce and the use of the grindstones is ambiguous. There are a number of references to plant materials recovered from archaeological sites, but preservation is too spotty for any serious evaluation of the economic importance of the species identified as foods. The most economically significant of the Mesolithic plant species appears to have been the hazelnut, which is recorded in large quantities from a number of different locations (J. G. D. Clark 1972; Newell 1973; Petersen 1973; Price 1973; J. Renfrew 1973). There is some speculation that the postglacial dispersal of the hazelnuts as evidenced in pollen profiles may indicate human interference (Iverson 1967; Waterbolk 1968). Similarly, Waterbolk (1962) has noted that increases in the frequency of other plants such as *Chenopodium* in the pollen profiles of early Holocene Europe might indicate the use of these plants by man or even their incipient cultivation. He notes, however, that these plants might simply be weedy volunteers which thrived in environments altered by man although they were neither encouraged nor exploited by human populations. A number of other plant species have been identified in the refuse of the period, including water nuts, water lily seeds, raspberries, strawberries, acorns, cherries, and pears; but there is no evidence that any of them were of particular economic significance (J. G. D. Clark 1972; Newell 1973; Renfrew 1973).

The minor role which plant foods appear to have played in the European Mesolithic may reflect the peculiar geography of Europe. Because of its very extensive coastline, Europe may have been able to absorb a good deal of population pressure by the exploitation of aquatic resources. Thus the same pressure which in the central Sahara forced man to rely on intensive exploitation of wild plants may have found a somewhat different outlet in Europe. This in turn may be a significant factor in accounting for the failure of Europe to contribute significantly to the development of agriculture in the Old World. As an alternative, it is possible that the vegetable foods, such as hazelnuts, which were extensively exploited by Mesolithic man in Europe, simply were not amenable to domestication.

To the extent that any particular order can be discerned in the economic shifts that characterized the Mesolithic economies of Europe, it would appear that the Late Pleistocene pattern of dependency on herd animals first gave way gradually to a system characterized by a wider range of land-based hunting. For example, Maglemosian peoples of the tenth millennium B.P. were hunting a wide spectrum of animals, mostly typical forest dwellers such as elk, aurochs, red deer, roe deer, and wild pigs, in addition to a variety of smaller animals including a number of furbearing species. A variety of wild fowl of coastal, bog and freshwater environments were also being hunted, including wild ducks, swans, grebe, coot, heron, cormorants, gulls, the greak auk, razorbills, and gannets. It was in this context of solitary animal hunting, especially the hunting of smaller animals, that the bow and arrow and the other products of microlithic technology took on their expanded role in the chase. There is even evidence of the use of blunt-ended wooden arrows in the hunting of small mammals and birds (J. G. D. Clark 1952).

The emphasis on aquatic resources appears to have developed gradually and increased in significance throughout the Mesolithic period. It is particularly in the closing phases of the European Mesolithic that specifically coastal, lacustrine, and riverine cultures emerge, with heavy reliance on fish and sea mammals. The late appearance of coastal sites is in part a function of rising sea level (Newell 1973; Clark 1968). As already indicated, there is evidence that earlier coastal adaptations in many locations may have been destroyed by subsequent rises in sea level. However, since this trend seems also to be observable inland, it is presumably not due entirely to differential preservation. Moreover, there are a number of stratified sites recorded where the density of fishbone or shellfish increases markedly in upper layers (Barker 1975). Neither Newell nor Clark, who call attention to the importance of rising sea level, argues that the increased importance of aquatic resources is due entirely to the differential preservation of coastal sites. In fact, both suggest that increased reliance on these resources was a major adaptive trend.

It is worth stressing that Mesolithic use of aquatic resources was for the most part *not* based on new technology. Mesolithic man added one or two new techniques to the quest for aquatic foods, but for the most part he simply placed increasing emphasis on resources which had long been readily available but which had previously been ignored or exploited only minimally.

For example, the use of stranded whales is recorded among a number of Mesolithic groups in Europe including the Tardenoisian, Larnian, Obanian, and Ertebølle people (J. G. D. Clark 1952). Since such strandings occurred with great regularity well into the twentieth century, when more than 400 strandings were recorded between 1913 and 1926 (J. G. D. Clark 1952), and since a stranded whale must have represented rather easy prey for anyone who had ever hunted a large land mammal, the scarcity of evidence for the use of this resource before the end of the Pleistocene is rather striking.

In addition, shellfish, eaten in Europe since at least the Middle Pleistocene in gradually increasing numbers, assumed great importance in the diet only well after the end of the Pleistocene, at which time, however, their use became very widespread (J. G. D. Clark 1952; Savory 1968; Barker 1975; Tringham 1971, 1973; Whitehouse 1968, 1971; Jacobi 1973). Similarly, seal hunting, traceable back to the Late Pleistocene, assumed gradually greater importance in Europe during the Mesolithic, and this trend continued after the appearance of Neolithic farming economies (J. G. D. Clark 1952). The importance of fishing also increased gradually from the Late Pleistocene through the beginnings of the Neolithic period in Europe, but in this case the economic transition resulted at least in part from the development of new technology at the end of the Pleistocene. In particular, along with such freshwater species as pike, taken in lakes and rivers, and marine species caught from the shore, there is evidence in the Mesolithic period of the exploitation of species that must have been taken in relatively deep marine water with the use of boats. Thus, for example, Tardenoisian middens of south coastal Brittany have yielded bones of marine fish in some quantity, indicating that their importance as a source of food was second only to that of shellfish and greatly in excess of that of animals hunted on land. Similar middens of the Obanian people of Scotland as well as those of the Ertebølle people also contained bones of deep-sea fish (J. G. D. Clark 1952).

Incidentally, even if the archaeological record is richest for western coastal Europe or the literature for that region most accessible, these trends toward dependence on water-based resources are by no means limited to that area. Tringham (1971, 1973) describes the intensified interest in solitary forest animals, fish, and shellfish displayed by post-Pleistocene populations along the rivers and lakes of eastern and central Europe. Similarly, Sulimirski (1970:28-50) indicates that eastern Europe and Russia as far as the eastern side of the Urals contained a number of early post-Pleistocene populations whose adaptive changes paralleled those of western Europe, with evidence of economic dependence on small game, fishing, and the gathering of aquatic and terrestrial molluscs and edible plant foods. He indicates further that some of these populations achieved rather large sedentary clusters, as shown by the size of their cemeteries. He describes, for example: the Mesolithic Shigir culture of the eastern side of the Ural Mountains, where a settled existence based primarily on lakeshore fishing had developed during the eighth millennium B.P.; the Kunda culture complex of the south Baltic area east of the Vistula River, whose sites are found exclusively on ancient sea and lake beaches and river banks, and whose artifact repertoire consisted of harpoons and fishhooks as well as projectile points; and contemporary populations of fishermen occupying northern Finland and the shores of the White Sea and the Arctic Ocean.

It is fairly clear, therefore, that all of Europe and European Russia underwent a series of parallel ecological changes in the early millennia of the Holocene, changes involving an altered style of hunting and an increased reliance on aquatic resources. There are, however, two major points of controversy in the interpretation of the European Mesolithic: first, it is not clear what happened in terms of actual population dynamics at the end of the Pleistocene. Second, it is not clear what caused the ecological shifts which are so well documented.

One widespread assumption is that Europe underwent an economic crisis at the end of the Pleistocene. It is well established that the open tundra environment of Pleistocene Europe was gradually replaced by a more closed forest vegetation after the retreat of the last glaciation, and it is evident that this alteration of the environment would have resulted in the decline or migration of the large herds of reindeer and other species which had provided the subsistence base for Upper Paleolithic hunters. The forest fauna which gradually replaced the Pleistocene herds would primarily have been solitary species less susceptible to mass slaughter. It is widely assumed that the total biomass of huntable animals would have been greatly reduced and hence that this change was a disastrous one for the large and dense populations of the Upper Paleolithic dependent on the exploitation of the large herds. The Mesolithic cultures are frequently said to represent a period of economic crisis and concomitant major reduction in European population; and it is often assumed that during the Mesolithic period in Europe population was low until the region was recolonized by farmers from the east (Price 1973; Sonneville-Bordes 1963; Waterbolk 1968; Dolukhanov 1973; Bibikov 1950; Radmilli 1960; Whitehouse 1968).

This model is supported both by the apparent reduction in the number of sites inhabited during the Mesolithic and by the fairly drastic economic changes that some scholars have observed. Both Radmilli (1960) and Whitehouse (1968) argue, for example, that the replacement of big game hunting by shellfish collecting and the hunting of small game in the Italian Mesolithic is so marked as to deserve classification as a crisis.

Various aspects of this model are subject to question, however. In the first place, it is not entirely clear that the extinction of the gregarious fauna resulted entirely from changes in the natural environment, although it can hardly be denied that these changes were a contributing factor. Kurtén (1968) has argued that since comparable mass extinctions are not observed at the beginning of earlier interglacial periods, climate change alone is unlikely to have had such a profound effect at the end of the Pleistocene. There is some evidence suggesting that

Upper Paleolithic man himself could have been a significant factor. As discussed above, there is evidence of massive and unselective kills and of the apparently extensive use of fire by the hunters of the period. On the other hand (Kowalski 1967), the extinct species are all associated with the tundra environment which was disappearing; and, conversely, not all of the species can be shown to relate directly or even indirectly to known human hunting patterns. Hence, it is probably wrong to ascribe too important a role to "overkill" by European hunters. What does seem plausible is that extensive human predation in conjunction with changing climate may have resulted in extinctions somewhat more pronounced than would result from climate change alone (Vereschagin 1967).

A much more important question has been raised by Clark (J. G. D. Clark 1968; see also Newell 1973; Jacobi 1973). He argues primarily that the replacement of tundra by temperate forests, while it would certainly have eliminated the large gregarious herds of grazing animals, would not necessarily have resulted in an overall reduction of the available biomass of big game. He notes that by Bourlière's (1963) estimate, temperate forests are in fact a relatively rich environment. Clark argues that the more significant change may have been a rise in post-Pleistocene sea levels which significantly reduced the area available to hunters. He suggests that in response to this reduction in their hunting territory, Mesolithic populations would have had only two possible strategies: first, to intensify their food quest, particularly by fishing; and, second, to enlarge their occupied territory by migration through the colonization of areas previously unoccupied or covered by ice, such as the Scandinavian peninsula. Implicit in Clark's model, as Newell (1973) points out, is the assumption that Late Pleistocene populations had increased to the maximum the existing economy could support, such that further population increases or reductions in territory would seriously have threatened the standard of living unless some compensation had occurred. Clark's model also rests on the assumption that we are dealing not so much with an environmental crisis as with a gradual process exacerbated by environmental factors. If Clark is correct we might expect to see a gradual supplementation of hunting by other strategies and not an abrupt replacement of one by the other. In this light it is interesting, though hardly conclusive, that a number of recent studies have tended to reevaluate the changes in economic strategy which are evident in the archaeological refuse and to see a more gradual process of change. For example, research on the Italian Mesolithic (Whitehouse 1971; Taschini 1964; Barker 1975) has shown that the decline in large game resources is neither so total nor so abrupt as once pictured.

Whether or not there was actually a decline in post-Pleistocene population in Europe, however, is a difficult question. In many portions of Europe, at least, Mesolithic sites are considerably less frequent than either Upper Paleolithic or Neolithic sites. In particular, inland areas, as opposed to coastlines or lakeshores, seem often to have been abandoned at the end of the Pleistocene (Whitehouse 1968, 1971; Tringham 1971, 1973; Berciu 1967; Petersen 1973). Some areas, in contrast, appear to have enjoyed relatively dense—and growing—populations. Newell (1973) indicates that, based on the number, size, and permanence of sites from each phase, the population of the area of the Netherlands seems to have undergone a fairly steady increase. Similarly, Jacobi (1973) argues that growing population in Britain is evidenced both by the density of sites (in regions of good preservation) and by the colonization of parts of Ireland and Scotland for the first time at about 8000 B.P. Piggott (1965:28) similarly has estimated on the basis of site density and distribution that the population of the British Isles went *up* from approximately 1,000 to 2,000 people in the Late Pleistocene to approximately 10,000 by 9500 B.P. Brothwell (1972) also argues that population in Britain went up after the end of the Pleistocene.

The interpretation of post-Pleistocene population trends is complicated by the fact that a number of changes occurred in geological conditions controlling site preservation, and there is some reason to believe that Mesolithic sites are poorly preserved and poorly represented either because of changing geomorphology or because Mesolithic man selected habitats less suited to archaeological site preservation. Jacobi (1973) has suggested that subsequent alluviation may have resulted in significant destruction of Mesolithic sites in Britain. Tringham (1971) notes that in eastern Europe Mesolithic populations moved into areas of shallow soil and high acid content which are not favorable for site preservation. Butzer (1971:563ff.), however, has offered the most comprehensive critique of the apparent population decline. He argues first that, while the density of Mesolithic sites in western and central Europe dating between 10,000 and 7000 B.P. may be considerably less than that for the

Upper Paleolithic as a whole, it is doubtful whether the number is substantially less than that for any particular, equivalent period (i.e., 3,000 years) of the Late Pleistocene. He points out that settlement was continuous across the Pleistocene-Holocene boundary in a large number of caves and that in fact the degree of discontinuity in cave occupations is no greater for the Paleolithic-Mesolithic transition than it is for comparable intervals within the Upper Paleolithic itself. He admits that there is a decline in open-air sites but suggests that there are several factors in the changing depositional environment to account for differential preservation favoring earlier sites. First, streams were actively down-cutting during the Wurm-Holocene transition so that riverine sites formed at this time would be particularly likely to be lost under subsequent alluvial fill. Second, the rapid and continuing rise of sea level until about 5000 B.P. would have submerged the great bulk of sites associated with littoral and marine exploitation at this time (see above). Third, the absence of loess sedimentation or reworked loess colluvium on upland surfaces after the end of the Pleistocene would mean that there was an absence of good depositional medium of the type enjoyed by Upper Paleolithic sites.

My own inclination is to agree with those (Newell, Jacobi, Butzer) who suggest a fairly continuous growth of population. I argued in chapter 3 that certain changes in human exploitative patterns are likely to be more reliable indicators of population growth than changes in the relative density of sites, since the latter are more subject to sampling error. Given the occurrence of several of these economic changes, in conjunction with acceptable explanations of the selective destruction of Mesolithic sites and Clark's reevaluation of the temperate forest as a resource base, it seems probable that population growth of a fairly continuous sort is the best available explanation of the patterns observed. There is, however, an alternative explanation of these patterns which cannot be ignored. On the one hand, population pressure, at least as I have defined it, seems to be increasing in post-Pleistocene Europe. On the other hand, there is some evidence that the absolute population of the subcontinent may have declined. Such a combination is theoretically possible in a particular region over a limited span of time. Population pressure might have been created wholly or in part by population growing against a static or declining resource base; but it might also have been created even in the face of declining population if there had been a catastrophic decline in available resources resulting from changing climate.

In either case, however, the role of climate change or of environmental changes of any sort in dictating the nature of post-Pleistocene adaptations in Europe should not be overplayed; it is important to see these changes in their proper context. To define the European Mesolithic economy *simply* as an adaptation to the particular post-Pleistocene environment of Europe, as is often done, seems to me to be untenable. The economic changes represented are in most instances simply intensifications of processes which have their beginnings well before the end of the Pleistocene; moreover, these economic changes are parallel to those witnessed already in Africa and in other regions of the Old World yet to be discussed where no comparable climate change occurred. Whatever the relative importance of climate change and of population growth as factors bringing about these changes, it would appear that the primary effects of climate change were indirect, operating through the medium of population pressure. The main role of climate change seems to have been to accelerate the imbalance between population and resources.

It was in this context of already high population pressure that agriculture spread into Europe. As J. G. D. Clark points out (1968), it was the economic stress of the post-Pleistocene period that explains the Europeans' receptivity to the new economy.

There is fairly widespread agreement that agriculture did not originate in Europe but spread into the subcontinent from the Middle East by some combination of the intrusion of new populations and the diffusion of Neolithic traits to Mesolithic peoples. Parts of Europe, most notably the southeast including the Balkan Peninsula, may have developed an early farming economy on their own, since fairly early evidence of agriculture (8100 B.P.) as well as industries and economies thought to be transitional have been identified (Renfrew 1968; Rodden 1962). There is some evidence that domestication of certain animal species may have occurred independently in parts of Europe (Higgs and Jarman 1969, 1972; Bökönyi 1970), but there remains a fairly clear consensus that the new pattern of domestication is primarily attributable to the intrusion of new populations from the Middle East or to the diffusion of new economic habits from that region (J. G. D. Clark 1965; Dolukhanov 1973; Tringham 1971; J. Renfrew 1973; Murray 1970;

Whitehouse 1971). Certainly the distribution of early Neolithic carbon-14 dates in Europe is consistent with the pattern of an east-to-west spread (J. G. D. Clark 1965; Ammerman and Cavalli-Sforza 1973). Moreover, although many of the domestic animals which accompany the Neolithic, particularly cattle and pigs, could have been domesticated independently in Europe, the plants which provided the agricultural staples for early European farming are all essentially foreign to most of Europe but native to the Near East. In addition, the spread of these early cultigens was often accompanied by the diffusion of art and artifact styles traceable to an eastern origin.

Farming seems to have developed in or spread to southeastern Europe, including the Balkan Peninsula, the lower Danube, and the northwestern Pontic area, between 9000 and 7000 B.P. From there, the new economy seems to have spread along two major routes: along the Mediterranean coast to the southern part of western Europe, and across the loess plains and the Danube Valley of central Europe to the northwest. The extent to which the spread was accomplished by migration as opposed to diffusion is not always clear, however. In much of southeastern Europe, notably the Danube Valley and the upper Elbe, Vistula, and Rhine rivers, farming seems to have spread via the intrusions of new populations. The early farming sites of these regions share a whole complex of traits both of economy and artistic style which links them with one another and with the Middle East while distinguishing them from the local Mesolithic sites. Moreover, the early Neolithic expansion in this region seems to have centered on parts of the terrain which were separate from the centers of Mesolithic habitation (Tringham 1971; Dolukhanov 1973; Gramsch 1971). At the same time some Mesolithic cultures appear to have persisted with very little cultural change side by side with the new groups (Tringham 1971; Dolukhanov 1973).

In other areas, a much more complex picture emerges, involving various mixtures of indigenous and intrusive culture traits. In some regions, the evidence seems to indicate that indigenous Mesolithic populations borrowed stylistic elements from the intrusive populations while retaining their own traditional hunting and gathering economies. In other areas the indigenous populations seem to have adopted the new economy while retaining their separate ethnic identities. In still other instances, the intrusive populations seem to have abandoned farming in favor of the hunting and gathering economy practiced by their new neighbors (Tringham 1971; Savory 1968; Dolukhanov 1973; Whitehouse 1971; Barfield 1971; Barker 1975). At present it is not really possible to distinguish these processes from one another in many areas; therefore, more detailed statements about the actual dynamics of this spread and its relation to population variables on the local level will have to await fuller description.

The Middle East

The history of the Middle Eastern region of the Old World shows broad parallels to both Africa and Europe in the nature of its economic evolution and in the general time sequence followed. Here too the archaeological record displays gradual transition from economies geared heavily to the exploitation of big game to those exploiting an increasingly broad spectrum of foods including small animals, aquatic resources, and vegetables. These changes are accompanied by increases in the density of sites that predate the beginnings of agriculture. The buildup of population pressure in this region is thus well documented, and several authors have already called attention to population pressure as a contributing factor in the origins of agriculture in this area (Binford 1968; Flannery 1969, 1973; Smith and Young 1972; G. Wright 1971; Reed in press-a).

Interpretation of the early history of the Middle East involves many of the same problems discussed above and there is no need to consider this time span in great detail. As in the regions already discussed, the relatively homogenous Acheulian and Mousterian occupations of the Mid-East appear to represent an economy predominantly based on hunting, judging by the associated faunal remains. There is some indication that the Acheulian hunters concentrated on a limited range of large fauna and that this pattern was replaced by a broader Mousterian emphasis on a mixture of large and small game (Braidwood and Howe 1962). This altered hunting pattern may reflect a decline in the available big game fauna. The extinction of much of the earlier Pleistocene big game of the Near East, notably elephants, rhinoceros, and hippopotamus, seems to have occurred approximately midway in the period of Mousterian occupation (Howell 1959). Lower Paleolithic sites in the region, such as Ubeidiya in the Jordan Valley and Latamne in northern Syria, contain spheroids and anvils (Stekelis et al. 1969:182; J. D. Clark

1966:217), and this Kraybill (in press) considers to be evidence of vegetable processing. But, as elsewhere, a numerically significant accumulation of tools clearly representing plant processing does not occur until the Late Pleistocene and early post-Pleistocene. Similarly, there is no indication that the Lower or Middle Paleolithic economies of the region placed any particular emphasis on the exploitation of microfauna or aquatic resources.

Early Upper Paleolithic sites in the Near East continue to display a significant emphasis on the hunting of large game. Flannery (1965, 1969) estimates that in Antelian and Baradostian assemblages, which represent the early Upper Paleolithic in the Levant and the Zagros regions respectively (Howell 1959; Solecki 1964), the hunting of hoofed mammals accounts for approximately 90 percent of the animal bones recovered and probably for fully 99 percent of the total meat in the diet. There is still little evidence for the processing of plant foods or for the use of significant quantities of fish, invertebrates, or even small mammals. The nature and distribution of sites from the period also seem to suggest an organization mainly geared to large game exploitation. Hole and Flannery (1967) divide the early Upper Paleolithic sites in the Zagros into several types: seasonal base camps of several families; "transitory" stations apparently placed to provide a good vantage point for the stalking and killing of game; and butchering stations for the preparation of the game. As Flannery notes, similar patterns are observable in the Levant (Binford and Binford 1966; Ewing 1949; see also G. A. Wright 1971).

By Flannery's estimation (1969) it is midway through these early Upper Paleolithic assemblages, probably at about 20,000 B.P., that localized subsistence patterns and local stylistic differences in artifacts emerge. At the same time, the concentration on big game hunting begins to break down and a series of trends become evident which involve a considerable broadening of the subsistence base to include a variety of small fauna as well as increasing amounts of vegetable foods including wild cereals. At first, as Flannery points out, many of these trends were not quantitatively significant. Thus, despite the broadening of the range of supplemental foods, the meat of large ungulates often continues to provide a very high percentage of the diet, or at least of the animal portion. Moreover, some of these trends are highly irregular, with sites even in adjoining regions showing strikingly different patterns. Nonetheless, by 9000 to 10,000 B.P. the economic transition seems to be so marked that the actual appearance of domesticated crops at this later date represents nothing more than a fairly minor variation on an already well-established pattern.

One trend of importance is the increasing use of ground stone tools for the processing of vegetable foods. There are scattered reports of mortars and pestles at Late Pleistocene sites in the Middle East (see, for example, the report of a mortar and two pestles at Ein Guev in Israel dated to about 16,000 B.P. mentioned by Stekelis and Bar Josef 1965). But these implements do not become common until the very end of the Pleistocene. Mortars are reported at a number of Natufian sites between 11,000 and 9500 B.P., including Nahal Oren, El Khiam, Mugharet-el-Wad, Erq-el-Ahmar, Kebara, and Ain Mallaha (Garrod 1958; Perrot 1966, 1968). They are also reported from Mureybat in Syria, Zawi Chemi and Shanidar Cave in Iraq, and Ganj Dareh Tepe in Iran during approximately the same time span (van Loon 1968; Rose Solecki 1969).

Actual grinding slabs also become common only fairly late. Reed (in press-a) reports only one reference to a grinding slab in the Middle East dating before about 12,000 B.P., and that is to a "quern fragment" at the Zarzian site of Palegawra dated between 14,000 and 12,000 B.P. (Braidwood and Howe 1960). On the other hand, grinding slabs occur at a number of sites after about 11,000 B.P.—at Shanidar at a level dated to about 10,450 B.P. (Ralph Solecki 1957); at Ain Mallaha (Perrot 1966); at Ganj Dareh Tepe; and at Zawi Chemi, dated at about 10,900 B.P. (Rose Solecki 1969). Thus, although different types of tools occur in different proportions at the various sites, grinding equipment of one type or another is quite widespread by the time of the preagricultural Natufian and Karim Shahirian assemblages shortly after 11,000 B.P., and in some sites such implements are quite numerous (see also Garrod 1958; Braidwood and Howe 1960).

At the same time, stone implements interpreted as axes occur at a number of sites in the region. Smith and Young (1972) infer that these tools are related to forest clearance. They also note the occurrence of perforated stone balls at sites of the period, which they interpret as digging stick weights, suggesting either incipient cultivation or increased harvesting of wild plant foods.

Sickle blades show a somewhat similar distribution. They are recorded

at Ein Guev and at Nahal Oren in Israel, both dated in excess of 15,000
B.P. (Stekelis and Bar Josef 1965; Noy et al. 1973), but they too occur
widely only from about 11,000 years ago in the Levant, Syria, and Iraq
(Garrod 1958; Braidwood and Howe 1960; Reed in press-a). There is
some direct evidence of the harvesting of wild grains extending back
more than 15,000 years. Small quantities of grains including emmer
wheat and barley occur in Kebaran and later levels at Nahal Oren (Noy
et al. 1973). The wheat even shows some evidence of morphological
change from the wild state suggesting to the excavators that there was
some human interference in its growth, although they hesitate to consider this as evidence of domestication. Elsewhere of course there is no
evidence of domestication of grains or other crops for several thousand
years.

Another trend evident some time before 12,000 B.P. and increasing
with time is the fairly widespread utilization of underground pits for
storage, some plastered with mud for impermeability and some either
sterilized by fire or used for *in situ* roasting of their contents. Such pits are
known from Zarzian levels at Shanidar (Ralph Solecki 1964a,b); from
Zawi Chemi (Ralph Solecki 1964a,b); from Karim Shahir (Braidwood
and Howe 1960); from Mureybat (van Loon 1966); and from a variety
of Natufian sites (Perrot 1966; Garrod and Bate 1937). As Flannery
(1969) points out, these pits would have been more useful for the storage of invertebrates or vegetable foods than for any activity associated
with hunting.

There is one other line of evidence which also suggests the increasing
importance of vegetable foods in the diet. The teeth of skeletons associated with Natufian sites at Ain Mallaha and El-Wad show severe wear
and tooth loss indicative of a gritty diet with a preponderance of cereals
or other coarse vegetable matter (Dahlberg 1960; Smith quoted in
Butzer 1971:556).

The transition to a broader pattern of exploitation is also noticeable
in the pattern of fauna harvested. Although large ungulates continue to
play an important role in the economy and specialized hunting patterns
are observable at many sites, the supply of meat from these sources
seems increasingly to be supplemented by exploitation of a variety of
small fauna including a number of nonmammalian and even invertebrate
species: water fowl, fish, crustacea, small reptiles, and terrestrial and
aquatic molluscs (Flannery 1969). For example, at Palegawra in Iraq a

variety of birds, tortoise, crabs, clams, and land snails supplement the larger fauna (Braidwood and Howe 1960). At Shanidar, large quantities of land snails as well as tortoise, aquatic molluscs, and fish accompany the larger fauna (Perkins 1964); and large quantities of land snails are also reported at Ksar'Akil (Ewing 1949). Barbed spears or harpoons and fishhooks of bone are recorded from sites such as the Mt. Carmel Caves (Garrod 1958), Kebara, El-Wad, Erq-el-Ahmar, and Ain Mallaha (Perrot 1968); and fishing is assumed to have become an economically signifi-cant activity at least by the time of these Natufian occupations (Perrot 1962; G. A. Wright 1971; Reed in press-a).

There is also considerable evidence that in the Late Pleistocene and particularly by about 11,000 B.P. fire may have been a significant tool for human modification of the environment. Lewis (1972) suggests that fire was an integral part of the broad-spectrum revolution. He has sug-gested that prior to the beginning of agriculture in the Middle East, fire was being used to a significant extent to create conditions favorable to the growth of cereals and other crops and to the animals on which hu-man populations relied. Lewis notes that ash horizons are widespread in archaeological sites of the late pre-farming era and that these horizons often correspond to changes in the pollen spectrum in favor of the pol-len of the large-grained cereals. At Shanidar, for example, evidence of fire becomes more widespread at the end of the Pleistocene and ash horizons correlate with the appearance of cereal pollen (Leroi-Gourhan 1969). Similarly, there is good evidence of burning at Mureybat (van Loon 1966). Lewis suggests that the occurrence of wild Einkorn wheat at this latter site, which is well outside the presumed normal range for this species, may reflect the artificial extension of the species' preferred habitat by the elimination of competitive species through fire. Lewis argues that general patterns of vegetational change observable in the Middle East prior to the advent of agriculture (van Zeist 1967) suggest the maintenance of a fire climax system. In addition, van Zeist (1967: 311) argues that the forests of the region may have been modified and kept relatively open by human interference.

On the basis of this new pattern of broad-spectrum exploitation, a number of populations appear to have developed sedentary settlement patterns at least by about 11,000 B.P. As evidenced by stone founda-tions, a number of Natufian sites in Palestine seem to have been seden-tary occupations before any domestication of wild cereals took place

(Perrot 1966). At Ain Mallaha, stone-walled houses indicate a sedentary population of between 100 and 200 persons (Perrot 1968). Similarly, at Tell Mureybat in Syria, approximately 10,000 B.P., a sedentary village of clay-walled houses seems to have emerged on the basis of the exploitation of grain without any evidence that the grain was domesticated (van Loon 1966; van Zeist and Casparie 1968). In the region of the Zagros Mountains, M'lefaat (Braidwood and Howe 1960), Zawi Chemi (Solecki 1964a), and Ganj Dareh Tepe (Smith 1968) all show signs of permanent or semipermanent habitation without any evidence of the use of domestic crops.

These changes in subsistence and settlement are paralleled by independent evidence that the absolute population density of the Middle East was increasing at least from the Upper Pleistocene through the early post-Pleistocene period, long before effective farming is known to have been established (ca. 9000 B.P.). Perrot (1962:148ff), pointing to the Palestine sequence, notes that beginning as early as the Upper Paleolithic Aurignacian industries and extending through the period of Natufian occupation there is a buildup in the number of sites, particularly open-air sites, on the Mediterranean coastal plain, indicating a rise in population. The Natufians themselves are generally conceded to have continued to increase in numbers in the years preceding the emergence of agriculture (G. A. Wright 1971:467; Reed in press-a). As G. A. Wright points out (1971:468), this increase in population in the Levant was accompanied by the use of previously unused areas. Natufian populations appear to have continued the Upper Paleolithic occupations of the Mediterranean coast, the valleys opening on to the coastal plain, and the Judean desert. Of these areas, the coastal regions seem to have been the most densely populated, as judged by the sites of El Wad and Nahal Oren. The Jordan Valley, however, which today has the largest and densest stands of wild cereals, was not occupied in the Upper Paleolithic and was intensively settled only by Natufian populations. Thus, in Palestine at least, the settlement of areas with exploitable wild cereals was a relatively late phenomenon concomitant with evidence of population growth and of population pressure on other resources. G. A. Wright (1971) suspects that among Natufian populations themselves the coastal sites may be earlier than those of the Jordan Valley associated with wild cereals.

A similar buildup of population is attested in the eastern half of

Mesopotamia. In the Zagros Mountains, in the period between 12,000 and 9000 B.P., both the number of sites and the degree of permanence evident implies growing population. In fact, Wright (1971:469) notes that there are more archaeological sites from this time range (which is prior to the first evidence of agriculture) than for the entire preceding period of Paleolithic occupation.

Exactly how much population growth or what rate of growth is implied by these trends is not clear, since the absolute size of population cannot be calculated with precision for any particular period; but an example can perhaps suffice to indicate the general magnitude of the demographic trends envisioned. Reed (in press-a) has suggested that because sedentary living patterns occurred in the Near East well before the beginning of domestication, the rate of growth (.1 percent) commonly ascribed to the Neolithic may in fact have begun as early as the end of the Pleistocene. On this basis, he suggests that a Palestinian population estimated at 6,000 in 11,000 B.P. might have increased to approximately 16,000 by 10,000 B.P., and to 45,000 by 9000 B.P., approximately the time of the first appearance of cultivation. As Reed himself states, however, the significant point is not the numbers themselves but the realization that even a fairly slow rate of growth would have had profound effects on the total population of the region.

The Pattern of Early Middle Eastern Domestication

In the period between about 9500 B.P. and 8000 B.P., settled village farming communities making use of morphologically domesticated crop plants appear throughout the Middle East, from Khuzistan on the east to the Anatolian Plateau and the Levantine coast on the west (Renfrew 1969, 1973; Flannery 1973; G. A. Wright 1971; Singh 1974; Reed in press-a). The agriculture of these early sites seems to have been based primarily on the domestication of wild two-rowed barley (*Hordeum spontaneum*), emmer wheat, and einkorn wheat (*Triticum dicoccoides* and *T. boeoticum*) and the simultaneous, or nearly simultaneous, domestication of a variety of wild legumes (pulses), including the pea (*Pisum*), lentils (*Lens*), the broad bean, and the vetches (*Vicia* spp.) (Harlan in press; Zohary and Hopf 1973). As Zohary and Hopf (1973) have pointed out, the combination of cereals and legumes is an interesting one. As in the New World, where a similar combination of crops is encountered (maize [*Zea mays*] and beans [*Phaseolus*]), early Middle Eastern agriculture seems to have been based on a complementary pattern of storable starch and storable protein.

The early agricultural sites in the Near East provide a fairly clear picture of the independent development of domestication in a number of regions. Early cultivated plants are known at approximately the same time level from Iran, Iraqi Kurdistan, Syria, Palestine, and western Turkey. Thus the earliest centers of domestication are scattered widely through the Middle East and there is no evidence of a clustering of early dated sites within this region or of a gradient of radiocarbon dates, such as that encountered in Europe, which might suggest diffusion from a single source. This is despite the fact that trade, particularly for raw materials such as obsidian, was underway at least by about 9500 B.P. (G. A. Wright 1969).

In southwest Iran, the earliest domestic crops are known from the Bus Mordeh and Ali Kosh phases of the Ali Kosh site, Deh Luran, where cultivated emmer wheat occurred mixed with some einkorn wheat, hulled barley, naked six-row barley, wild flax, and a lentil. These are accompanied by a variety of wild plants, particularly wild legumes, wild grass seeds, and a small variety of fruits and nuts (*Prosopis,* capers, and pistachio). In the earlier of the two phases (Bus Mordeh, dated between 9500 and 8750 B.P.), two species of wheat and two species of barley already appear to have been domesticated (Helbaek 1969). To the north, in the Zagros foothills in Iraqi Kurdistan, early agriculture is attested at Jarmo, dated to about 8750 B.P. At this site both wild and domesticated forms of einkorn and emmer occur, but the bulk of the material consists of barley. Samples of rachis fragments which were unshattered suggest that the barley may have been cultivated. In addition, several of the pulses occur, including field peas, lentils, and blue vetchlings. Among the wild plants are grains of *Aegilops* (a close relative of wheat), acorns, and pistachio nuts (Helbaek 1960). At Tepe Guran, midway between Jarmo and Ali Kosh, wild and cultivated two-row barley has been found (Meldgaard et al. 1963).

In southern Syria, Tell Ramad has yielded remains of cultivated plants including barley, emmer and einkorn wheat, club wheat (a more highly evolved wheat form), and lentils dated to about 9000 B.P. (van Zeist and Bottema 1966). In Palestine, at the site of Jericho, Pre-Pottery A levels dated in excess of 9,000 years contain domesticated emmer wheat and barley. In overlying B levels, einkorn, emmer, and barley occur along with peas, lentils, and horsebeans (Hopf 1969). At Beidha in south Jordan, wild and cultivated emmer wheat and barley occur along with a number of wild pulse seeds and pistachio nuts (Helbaek 1966).

The early Aceramic settlement at Hacilar in west central Anatolia has also yielded evidence for early agriculture. Here wild einkorn and emmer wheat, barley, and lentils are identified beginning about 9000 B.P. (Helbaek 1970). At Cayönü, also in Turkey, einkorn and emmer wheat, both presumed cultivated, are known from levels dated between 9400 and 8800 B.P. (Braidwood et al. 1971; Reed in press-a).

Aside from their scattered distribution, one of the most interesting patterns encountered in these early sites is the continued use of a variety of wild, gathered seeds including the wild forms of the very species (wheat and barley) which had already come under domestication. In addition, the wide-ranging exploitation of fauna seems to have continued. Braidwood and Howe (1962) report, for example, that large numbers of terrestrial snails along with freshwater crabs and fish occurred at Jarmo in addition to larger hunted animals and domestic fauna. Similarly, at Ali Kosh small mammals, fish, waterfowl, and molluscs continued to provide a significant portion of the diet (Flannery 1969). The implication of this pattern is that farming did not revolutionize the economy of the Middle East so much as it provided what was at first only a supplement to the existing economy of eclectic food gathering. As Flannery points out, the early farmers in the area were still in the period of "broad-spectrum" subsistence.

This pattern becomes particularly striking if an actual quantitative assessment is made of the relative importance of domestic and wild foods in the diet as reconstructed from site refuse. Flannery (1969) provides such an assessment for the Bus Mordeh phase at Ali Kosh. At this site, cultivated cereals represent only about 3 percent of the carbonized seed recovered in the site refuse. Because the seeds of cultivated cereals are somewhat larger than the seeds of many of the wild plants that were gathered, however, the cereals constitute about one-third of the total seed material by weight. The seeds of small, wild legumes of clover-alfalfa type constitute about 94 percent of the seed, but because of their small size they too make up only about one-third of the total weight. The collected seeds of wild grasses constitute only about 1 percent of the sample but they provide about 15 percent of the total weight of seed. Similarly, domestic sheep and goats accounted for about one-third of the total meat weight and hunted ungulates still provided about 60 percent of the meat supply. The remainder of the diet was supplied by a variety of nuts and fruits and the range of small fauna listed above.

The Interpretation of the Sequence from the Middle East

The interpretation of the sequence of events in the Middle East depends in large part on obtaining accurate knowledge about the ecological habits of the main domestic species, particularly the cereals which provided the staple basis of the new agriculture.

Harlan and Zohary (1966) and Zohary (1969) have provided a detailed analysis of the ranges and ecological tolerances of the various wild cereals based on their contemporary distributions. Allowing for the possibility of recent climatic shifts or the recent redistribution of weedy races in habitats disturbed by man, they make the following observations. Barley, which they consider to be the most important of the cereals, was found in wild stands over a very wide range, from Libya and western Turkey on the Mediterranean coast through Iran and Afghanistan, and into Pakistan and the southern corner of the USSR. They suggest, however, that massive stands worthy of exploitation by primitive man and occurring in primary habitats (as distinguished from weedy races in habitats which might have been artificially created) were mainly confined to a broad arc surrounding the Fertile Crescent and extending from the Levant across Syria into southeastern Turkey, western Iran, and northern Iraq. Einkorn wheat was found to have occurred in primary or natural habitats in western and eastern Turkey, northern Iraq, and western Iran, while secondary or weedy occurrences extended westward as far as Greece and Yugoslavia, covered all of Turkey, and stretched eastward over much of Iran (Palestine was not included). They conclude that southeastern Turkey with its rich basaltic rubble on the lower slopes of volcanic mountains was the best source of large pure stands. Wild emmer was found to occur in primary habitats chiefly on the Levantine coast of Israel, Jordan, Lebanon, and Syria, as well as in Turkey, northern Iraq, western Iran, and the southwestern USSR. But they argue that the race found in Turkey, Iraq, Iran, and the USSR was always rare, occurring only in isolated patches, whereas the races in the

Jordan Valley occurred in massive stands. On the basis of these observations, Harlan and Zohary conclude that emmer was probably domesticated in or near the upper Jordan watershed area of the Levant; einkorn in southeastern Turkey; and barley almost anywhere in a broad arc bordering the Fertile Crescent.

Having noticed that the various wild cereals occurred in dense stands comparable in density and scope to cultivated fields, Harlan (1967) proceeded to test the potential of these wild stands for harvest by men equipped only with tools known to have been available to early Holocene populations in the Middle East. Working by hand with einkorn wheat, he found that he could harvest about 2 kilograms of grain and glumes in an hour, whereas using a primitive flint-bladed sickle he could harvest about 2.45 kilograms per hour. The actual grain content of the harvest, however, was only about 46 percent by weight, so that he was in fact harvesting about 1 kilogram of edible grain for each hour's labor. This suggested that in a very brief period, approximately three weeks of work, a family could easily harvest enough grain to feed itself for a year. The labor costs, thus measured, are small even by Bushman standards.

Harlan's experiment raises some interesting questions. If wild cereal grains could be harvested with such ease, and if, as Harlan suggests, this would provide a welcome alternative to people living by hunting and gathering, the question must then be raised why people did not shift to harvesting wild grasses long before the end of the Pleistocene. The problem does not appear to have been a lack of technology. Sickles are probably a fairly recent invention, but even they are known in several localities before about 15,000 B.P., which is somewhat earlier than cereal harvesting can be shown to have been a major economic activity in the Middle East. Moreover, Harlan himself has demonstrated that sickles were not necessary for the harvest, and in fact added only slightly greater efficiency to the harvesting process. Grinding stones or mortars and pestles, the other major tools necessary for grain processing, have been shown to be of considerably greater antiquity than is the intensified harvesting of cereals.

The delay also cannot be related to the evolution of the plants themselves. In the New World, many of the major crop plants, most notably maize, did not become economically significant until long after they were first used by man, because it is only under human control and selection that they became truly productive. In the Middle East, however, Harlan's experiments suggest that in the wild state the cereals were already substantial resources and were, in fact, very little different from their domestic counterparts.

One possible explanation for the delay in the exploitation of the cereals is that some sort of climate or environmental change was necessary to stimulate their use. As suggested in chapter 1, Childe's postulation of an economic crisis resulting from dessication has been largely dismissed as a result of repeated demonstrations that the fauna and flora of the Middle East display no changes of a magnitude and direction suggesting such a crisis (Braidwood and Howe 1960). There is another possibility, however, which receives considerable support from modern pollen studies. This is that climate change in the Middle East may have helped to create plant communities of the type used by early agriculture only shortly before domestication took place. Pollen studies of the region reflect the dominance of cool, dry steppe conditions, characterized by a preponderance of *Artemisia* and chenopods, during the Late Pleistocene. These relatively cool conditions may have resulted simply in the restriction of the oak forest flora to lower elevations within the region or they may have brought about the total exclusion of major stands of oak from much of the Middle East (H. Wright in press). It is only by about 11,000 B.P. that these steppe flora were replaced to a significant degree by a new open forest vegetation characterized by oak and pistachio and by occasional concentrated stands of wild cereals (H. E. Wright 1968, 1970; van Zeist 1967, 1969; van Zeist and Wright 1963). Wright (in press) hypothesizes that the wild cereals immigrated at this time into the hill lands and plateaus northeast of the Mediterranean, setting the stage for their domestication. Arguing along similar lines, Smith and Young (1972) have suggested that this climate change, bringing with it dense stands of wild cereals, was an essential triggering device for Middle Eastern development since it permitted and even encouraged sedentism and concentration on cereal resources.

But this explanation is not entirely adequate. It may well be that the fortuitous migration of appropriate wild cereals into the region, in conjunction with rising population pressure, helped to account for the existence of an early center of domestication in this particular locality; but the climate change hypothesis alone is insufficient to explain the economic shift that is observed. By itself, such a hypothesis does not account for the fact that agriculture in the Middle East, as elsewhere, occurred in conjunction with a specific constellation of other economic changes. Moreover, the climate change hypothesis does not really explain why these crops were not exploited earlier. Climate change may account for the expanded range of cereal grasses in the Middle East at this particular time, but it does not account for their origins. As Wright himself points out, it is necessary to assume that some refuge area existed for the oak-pistachio group during the Pleistocene from which it could recolonize the Middle East when warmer, moister conditions prevailed. Pollen profiles from Israel (Rossignol 1962, 1963), for example, indicate that cereal grasses were present there during the Late Pleistocene, even if their range was somewhat more limited than it was in the Holocene.

Thus, if the grasses were as readily accessible and as desirable as Harlan describes them, one would expect them to have been harvested in their previous locations long before the twelfth millennium B.P. It is possible that the earlier evidence of grass utilization and barley pollen in northwest Africa in the Late Pleistocene represents just such a period of prior exploitation in a neighboring region. But even this would extend the period of their use back only a few thousand years, while we would expect from Harlan's description that grass grains would have been regularly harvested much earlier in the Pleistocene. This should have been true as long as the wild cereal grasses formed dense stands in nature, even if prior to the end of the Pleistocene such stands were small, scarce, or isolated. Given the ecological sophistication and geographical knowledge displayed by hunting and gathering groups, we would expect man to have searched out and exploited even isolated stands from a very early date.

One plausible explanation of the failure of early man to make use of these grasses has been offered by Butzer (1971:565), although it is an explanation not acceptable to many ecologists. He argues that the dense stands of "wild" grass which Harlan showed could be easily exploited in quantity are in fact not truly wild in character at all but occur only in environments disturbed by man. Butzer assumes that the oak-pistachio forest would have been a fairly closed environment incapable of supporting grasses in large stands until clearings were artificially created. As discussed above, Lewis (1972) similarly argues for the importance of man, with fire, as a creator of the open forest conditions suitable for the wild cereals. If these authors are correct, the implication is that grass harvesting was possible only after human populations had gone to significant lengths to assist the growth of the grasses or the growth of some other plant or animal species that benefited from fire or from forest clearance. This explanation does have the advantage of accounting for the association of early grass harvesting with other indicators of population pressure.

Another possible explanation of this delay may be that grass seed is not really an easy commodity to harvest and process. Harlan's figures on labor costs may be slightly deceptive. Grasses, even in dense stands, are easily harvested; but once gathered they are notoriously difficult to process into food, since threshing and possibly a number of additional steps are still involved. Harlan points out (1967:199) that the material harvested would be unfit for food since the grain is tightly bound in glumes and must be threshed out using a mortar and pestle and possibly toasted or roasted as well. And the wild forms of these cereals are even more difficult to process than are their domestic counterparts. In addition, dependence on wild grasses beyond their season of ripening involves significant labor costs (and crop losses) in storage. Storage pits or vessels would have to be constructed and probably renewed and disinfected by fire annually, and the grain would have to be parched to guard against premature germination (Murray 1970:41). Storage also requires that people stay near their stored grass seeds or, in other words, become sedentary. As Harlan (1967:201) points out, harvested cereal grains are bulky and would be extremely difficult to transport in large quantities. As indicated in the previous chapter, there are numerous reasons why I suspect that this type of enforced sedentism would not be attractive to a population given its choice.

It should also be noted that grass harvesting *must* take place during a particular season of the year. Not only is it *possible* to harvest enough for a year in three weeks of work; it is *necessary* to do so, because of the very brief time period between the ripening of the seed and the shattering of the heads. Flannery (1973:279-80) has suggested that under ideal conditions a particular patch of wild cereal might be harvestable for as much as three weeks, but that under poor conditions it might be necessary to harvest the whole stand within two or three days. This would mean that a human group harvesting grasses would be required to work very intensively for a period of several days or more, although much of the return on their labor would not be realized until well in the future. This in turn implies a future orientation very foreign to hunting and gathering groups and their ideologies, as has been discussed in chapter 2.

The most probable explanation, however is that grasses simply are not foods that most people prefer (see chapter 2). The grasses, in fact, are particularly striking examples of the low priority most populations place on vegetable foods in general and on agricultural foods in particular. As Yudkin (1969:551) has pointed out, people worldwide eat meat and various fruits when they can (when they have enough food or enough wealth) and eat cereals and tubers only when they must. Moreover, the evidence from chemical analyses suggests that these preferences are sensible. Cereals, although rich food sources compared to tubers, are not good sources of protein compared, for example, to meat or even to most nuts or wild leguminous plants also available to prehistoric Middle Eastern gatherers. Yudkin (1969:549) has suggested, moreover, that human beings are very poorly adapted to digest cereals, since allergies to the different varieties are fairly widespread even in modern populations and since all cereals contain phytate, which interferes with the proper assimilation of calcium by the body and is hence conducive to rickets.

The grasses, then, are rather undesirable as foods. In fact, they have only three things in their favor: first, they represent consumption at the lowest trophic level and therefore support relatively dense populations; second, among organisms at that trophic level the cereals are a type that forms dense stands readily harvestable in bulk (or at least a type that will form such dense stands in response to human interference); third, the cereals provide calories in readily storable form. Similar observations can be made about the legumes, which formed the other half of the early domestication complex. The latter provide large quantities of storable protein, but not in the form perferred by most human populations. These qualities suggest that both cereals and legumes would be utilized out of necessity, not out of choice.

It therefore seems to me very unlikely that a human population would have settled down near a field of wild wheat for the dubious pleasure of harvesting, threshing, and grinding the grain and eating the gruel year-round, thus forgoing some of the other portions of their diet. Even Harlan (1967:198) agrees that cereal pottage would be pretty dull fare. It seems more plausible to assume that two things happened. First, people became accustomed to harvesting wild cereal grasses as part of their annual cycle. This probably occurred because, with growing population and increasing competition, other foods they were accustomed to eating during the season when the grass ripened gradually became less available. Thus people began to harvest grass when it was ripe in default of other resources. The second step in the process took place when populations began to outgrow or exhaust the resources usually eaten at other times of the year or when they found that portions of their accustomed seasonal round were no longer available because other people had taken them over. The fact that other resources in other seasons had become increasingly scarce, combined with the realization (which must have been apparent to anyone harvesting grasses) that cereal grain could be stored if kept dry, forced people to make maximum use of the potential of the grass harvest and to store the extra harvest for use during later seasons. Grass use and seed storage thus probably emerged as behaviors practiced by people under population pressure when the need for stored calories outweighed the costs involved. It is therefore not surprising that the Middle Eastern stands of wild cereals were among the last resources to be heavily exploited in the region.

Harlan's study suggests a second major problem, this one concerning the actual domestication of the wild cereals. Apparently wild wheat could be harvested with great ease in quantities comparable to the yields of prepared fields, at least in certain limited locations. Moreover, as Harlan has pointed out (1967:198), wild wheat is richer in food value than domestic wheat, comparing favorably to premium modern wheat in its protein content. Why then should anyone bother to domesticate it? Domestication would simply appear to add to the labor costs of the cereal crop without increasing either the quantity or the quality of the product.

A possible answer is that domestication might in some manner make the work easier. One of the immediate consequences of domestication (and in fact the morphological criterion on which identification of domesticated forms is usually based) is the emergence of tough-rachised, nonshattering heads. As one harvests wild wheat there is a tendency for the head to shatter and grain to be lost, particularly if the timing of the harvest is poor. Nonshattering varieties of the various cereals appeared about 9000 B.P., apparently in response to human selection, and these varieties would presumably have increased the efficiency of sickle harvesting. Harlan's experiment in harvesting wild wheat indicates, however, that the importance of this change in terms of conservation of labor is not nearly so great as has been assumed in the past, since the harvesting process itself, even with wild wheat, is the most efficient operation in the whole preparation process. Possibly more important would be the evolution under domestication of naked (free-threshing) grains, for with these varieties some of the labor costs involved in removing the grain from the glumes would be eliminated, and the costs of preparation of the cereal after it had been harvested cut down. Early domesticated forms of barley did in fact develop this free-threshing characteristic. Both einkorn and emmer wheat, however, retain their enclosing glumes in the domestic state and would have continued to require extensive preparation after they were harvested. Free-threshing polyploid wheats are not encountered archaeologically until about 7500 B.P., long after agriculture was established (Harlan in press). It seems unlikely, therefore, that domestication and cultivation, which add several steps to the food procurement process in the form of planting and tending the crop, would have been undertaken in order to cut labor costs.

It would appear, instead, that the primary reason to domesticate wheat or barley would have been to obtain more total grain, either by causing the grain to grow in areas outside of its normal range of distribution or by enlarging the existing stands. (Alternatively, it might be domesticated in order to maintain the status quo in areas where either climate change or human depredation was tending to eliminate the natural stands.) This type of reasoning led Harlan and Zohary (1966: 1079) to argue that domestication of the cereals probably did not occur in the regions of maximum abundance but rather at the margins of these areas. Various authors, notably Flannery (1965) and Reed (in press-a), have pointed out that it would have been primarily in transporting the grain beyond its normal self-seeding range or in attempting to extend that range that man would have introduced selection pressure for the tough-rachised domestic forms.

If these theoretical arguments prove correct, I believe they provide further evidence of the importance of population pressure in stimulating the domestication process. Why should the range of the wild cereals be extended? An individual human group living near a stand of wild cereals might move out of the range of easy access to that resource; but, since the cereals were unlikely to have been considered a choice resource, they probably would not take the cereal with them to plant unless they expected to encounter no other, more desirable resources in their new home. An occasional group might move out of the range of the cereal and bring it along simply because they had grown to enjoy it in their diet. The latter behavior, however, would result only in occasional redefinition of the distribution of the crop; it would not cause a general spread of cereal cultivation. Such a general spread of cultivated cereals outward from their wild centers would only happen if people from these regions were systematically and continuously expanding out of the area, not just moving about haphazardly. Or this pattern might result if people in surrounding regions became increasingly short of more desirable foods and found that the wild grasses could be transplanted to their regions to alleviate their problems.

The archaeological record, as it is known at present, does in fact suggest that domestication occurred primarily in the attempt to extend the range of the major cereals. In the first place, there is a delay (2,000 years or more) between the use of the wild cereals and the emergence of the first domesticates. This delay is out of all proportion to the time required for the fairly minor genetic changes necessary to produce the domestic forms, but it might reflect the time span necessary for population to build up in the region to a level necessitating more widespread use of the wild cereals. More important, the spatial distribution of early centers of domestication appears to conform to the model suggested by Harlan and Zohary. The earliest settled villages in Mesopotamia are associated with wild cereals or with regions where wild cereals are assumed to have been available (Smith and Young 1972; Flannery 1969; Perrot 1966; G. A. Wright 1971; Reed in press-a; van Loon 1966). However the distribution of the earliest *agricultural* settlements appears to indicate that these are peripheral to the primary areas of wild cereal harvesting, as has already been pointed out by a number of authorities. Both Reed (in press-a) and G. A. Wright (1971), for example, show that the earliest sites displaying morphological domestication of cereal crops in the Levant are marginal to the known distribution of the wild cereals. The sites mentioned are Beidha in southern Jordan, Jericho in a desert oasis in Jordan, and Tell Ramad in Syria. G. A. Wright (1971:468) suggests that all of these are sites where early farmers domesticated their crops in an attempt to preserve the status quo in the face of growing population, expansion to regions less suitable to the cereals, or shrinking wild distribution of the cereals. He also discusses (1971:468) the "wild" einkorn wheat recovered at the site of Tell Mureybat in Syria (dated between 10,200 and 9500 B.P.) in an ecological situation where the plant is unlikely to have occurred in the wild. This, he says, may represent a very early stage of domestication when the plant had been transplanted from its natural ecological zone but before any morphological changes had occurred. Similarly, Flannery (1969) and G. A. Wright (1971:469) have pointed out that the earliest agricultural site in the eastern portion of Mesopotamia (Ali Kosh in Iran, where the Bus Mordeh phase is dated between 9500 and 8750 B.P.) is also in an area where, according to Harlan and Zohary, dense stands of cereals could only have been maintained by cultivation.

Also along this line, we might expect that populations within the region of the natural cereal stands would have been reluctant to replace their natural resource with cultivated resources and would have done so only when they had either outgrown or begun to exhaust the natural density and distribution of the cereals. In this regard, Perrot's argument (1968) that in the Jordan Valley, where there were dense stands of wild grasses, agriculture as evidenced by morphological domestication did not occur at all until sometime after 6500 B.P. is of considerable significance. Here, in a region of fairly dense wild cereals, the origins of agriculture were delayed too long for there to be any possible suggestion that the "concept" of agriculture had not yet been invented or made available by diffusion.

G. A. Wright (1971:469) has noted that in the Zagros, in contrast to the Levant, domestication, although still occurring first outside the range of natural distribution of the wild cereals, nonetheless appears fairly rapidly within the zone of wild grasses, as evidenced by such sites as Jarmo in Iraq. Why should there be this difference? Wright points out, citing Harlan and Zohary's study, that the wild grass stands of the Zagros-Taurus arc were in fact much less dense than those of the Levant. Thus it might be argued that, in the face of growing populations, the relatively sparse grass resources of eastern Mesopotamia required artificial extension sooner than the relatively dense stands of the Levant.

Incidentally, the extension of the range of the cereals by domestication becomes even more difficult to understand (and the role of population pressure conversely more apparent) when we realize that as the shift to increasing dependence on agriculture occurred there would have been a concomitant loss of dietary quality and an increasing risk of crop failure. Extending the range of wild grass stands would have meant the progressive elimination of other wild food sources which both provided dietary supplements and helped buffer a population against starvation. Dependence on artificially maintained cereal stands in habitats where they did not naturally occur would presumably have been more precarious than dependence on the original wild stands. Hole and Flannery (1967) note that dry-farming attempts on the Deh Luran Plain meet with failure two or three years out of every five. Similarly, in Iranian Kurdistan modern wheat farmers lead a precarious existence; Watson (1966) reports that they lost their entire crop in the years 1958, 1959, and 1960. Such profound difficulties in agriculture suggest that the shift would only have been made when the risks and high costs were outweighed by the need to obtain more food per unit of space. Flannery (1969:86) has expressed it admirably:

> There is no reason to believe that the early "food producers" were significantly better nourished than their "food collecting" ancestors. Nor was their subsistence base necessarily more "reliable"; . . . The one real advantage of cereal cultivation is that it increases carrying capacity of the land in terms of kilograms per hectare.

A Brief Overview of Late Pleistocene and Early Holocene Cultures in Eastern Asia

At the eastern end of the Eurasian landmass the data are somewhat scantier, but a picture emerges which is in general very similar to that observed in Africa, Europe, and the Middle East. In almost all areas studied, the end of the Pleistocene and the early Holocene prior to the emergence of agriculture is characterized by the emergence of broad-spectrum economies showing the characteristic signs of increasing population pressure.

Most of northeast Asia appears to have been occupied for the first time only fairly late in the Upper Pleistocene. There are a number of scattered and poorly documented or described finds which represent possible Levallois-Mousterian occupation of the region, presumed to be contemporary with the early portion of the Wurm glaciation. It is not until the late glacial period, however, that the first substantial evidence of occupation is found. In Siberia, paleolithic cultures, which are well defined after about 20,000 B.P., seem to have been based on the hunting of big game centered on the loess steppe environments. Mammoth, rhinoceros, reindeer, antelope, horse, sheep, and aurochs or bison are among the species recorded from Paleolithic sites in the region, although smaller game including ptarmigan, arctic fox, and hare are also recorded. Aquatic remains are noticeably absent from these early sites. Pleistocene populations appear strictly to have avoided the *taiga* or forested regions. It is only at the very latest Paleolithic sites, dated to about 12,500 B.P., that the first evidence of fishing is noted; and it is only in the early post-Pleistocene period that the expanding *taiga* vegetation was first penetrated for economic purposes, with concomitant shifts to the exploitation of solitary forest species. Subsequently fishing became a major source of food for some populations, with hunting persisting as the dominant portion of the economy only in tundra environments. Many of the fishing groups appear to have achieved a sedentary life-style based on runs of migratory fish, in a manner analogous to the populations of the northwestern coast of North America. Shell middens also occurred on the coast, and by about 3000 B.P. the first evidence of truly specialized maritime hunting of sea mammals is known. At some of the latter maritime sites, sea mammals seem to have composed as much as 90 percent of the food remains. Inland, Neolithic farming settlements with grindstones and milling stones, processing domestic millet, appear to have spread in southern portions of the region as early as about 4500 B.P. These farming communities seem to have relied heavily on fishing to supply their protein (Chard 1974).

In Japan, despite some tentative identifications of sites in the early Upper Pleistocene, there is no firm evidence of extensive occupation until late in the Wurm glaciation, sometime after about 20,000 B.P. The early sites after that date are predominantly inland occupations, although their apparent distribution may be skewed by Late Pleistocene changes in sea level. It is only after about 9000 B.P. that a tradition of sedentary hunting, collecting, and fishing populations using pottery emerges. The refuse of these early populations, which are collectively referred to as the Jomon tradition, reveals an economy geared to the

hunting of forest animals such as deer and pigs along with the exploitation of fresh, brackish, and saltwater molluscs, and fishing with hooks,
spears, and nets. In later phases a truly maritime exploitative pattern
emerged involving the use of boats for offshore fishing and true maritime hunting. The sites themselves, numbering about 75,000, are predominantly coastal shell middens, although a number of inland sites are
recorded. The middens contain grinding stones and mortars indicative
of the preparation of plant foods, and some plant remains, particularly
nuts, occur in the refuse. Sedentism is evidenced by the accumulations
of refuse themselves and by the foundations of permanent dwellings
which have been uncovered. Despite some fluctuations, there appears to
have been a general trend toward an increase in the number, size, and
permanence of the sites, indicating an overall growth in population.
Late in the Jomon sequence, cultivated plants including varieties of
millet, buckwheat, and beans appear in the middens. Then, sometime
after about 3000 B.P., the tradition was interrupted by the arrival of
new groups using rice agriculture (Groot 1951; Chard 1974).

In China, although the record of pre-agricultural man is poor, there
are also indications of a "Mesolithic" economy predating and helping
set the stage for the development or adoption of agriculture (Treistman
1972, 1975; Chang 1962, 1968, 1970, 1973; Ho in press). Upper Pleistocene exploitation patterns geared toward the hunting of large fauna
and a variety of smaller game seem to have given way at the end of the
Pleistocene to a variety of regional economies directed toward the use
of coastal and riverine resources and the processing of vegetable foods.
Treistman describes a picture of increasing sedentism after about 7000
B.P., focused on the intensive use of local resources. She compares this
to the model of "forest efficiency" used to describe the pattern of Archaic development in the eastern United States (see chap. 5). This pattern
both preceded and accompanied the early spread of farming in China.

In northern China, including Tungpei (Manchuria), post-Pleistocene
adaptations resemble those of Siberia, with food refuse displaying
heavy concentrations of fishbone and the shells of riverine molluscs.
Along the coast specialized economies geared to the exploitation of
littoral resources emerged, and a number of large settlements occurred
on bays and estuaries. In the north, also, the presence of grinding stones
indicates the importance of vegetable processing. Farther south, in the

vicinity of the Yangtze delta, semi- or fully sedentary settlements are known along streams. The economy reflects some hunting of game, but the primary food strategy appears to have been the intensive gathering of fish and shellfish.

According to both Chang (1970, 1973) and Ho (in press), the first full-fledged Chinese agriculture occurred among peoples of the Yangshao tradition, which has recently been dated to the early sixth or seventh millennium B.P. The earliest cultigens include two varieties of millet (*Setaria* and *Panicum*). Rice appears to have been added to the list of Chinese cultigens sometime before the end of the sixth millennium. Although the source of the farming in the Yangshao tradition is not entirely clear, Ho (in press) has presented a case in support of the assumption that Chinese agriculture was a development indigenous to China itself, and he traces the origins of the Yangshao farmers to sedentary hunting and fishing, plant collecting (and possibly proto-horticultural) populations of the seventh or eighth millennium B.P., which are represented at at least one site lacking agricultural produce but already containing pottery, stone spades, grindstones, and fishing equipment.

Farther south, in Indochina or Southeast Asia proper, the record is clearer thanks to recent work by a number of archaeologists (see Gorman 1969 and in press; Glover 1973, for summaries of recent work). Here again archaeological assemblages bear a striking temporal and economic similarity to their counterparts in Africa, Europe, and southwest Asia. In particular, during the terminal Pleistocene, beginning about 15,000 years ago, there emerged a widespread series of archaeological assemblages which Gorman (1969, in press) refers to as the Hoabinhian technocomplex. The complex demonstrates all of the symptoms of intensive broad-spectrum use of environment characteristic of other cultures of the same time period. The list of fauna exploited (and, to judge from the treatment of the bones, very clearly used as food by man) is impressive both for its range and for its concentration on small, nonmammalian, and aquatic organisms. The tabulated faunal remains include some large animals such as elephant, rhino, bovines, antelope, various deer, and suids; but they also include primates, a number of small carnivores, squirrels, murids, bats, and porcupines. In addition, turtles and other reptiles, birds, freshwater snails and bivalves, crustacea, and fish occur widely at inland sites; whereas at coastal sites marine mammals such as dugong are found along with marine molluscs, crustacea, and fish. Botanical remains from Spirit Cave, a Hoabinhian site in Thailand, included almonds, betel, broad beans, peas, gourds, water chestnuts, butternuts, candlenuts, and cucumber. The plant remains occur in levels dated between 8500 and 9000 B.P. Gorman has suggested that several of these plants, notably the gourd (*Lagenaria*), the cucumber (*Cucumis*), and the water chestnut (*Trapa*), *may* indicate some sort of incipient cultivation.

In this general context, agriculture, involving rice (*Oryza sativa*) and presumably taro (*Colocasia*) and yams (*Dioscorea*), was developed or introduced at least by the sixth millennium B.P. and possibly somewhat earlier. The exact date of the beginning of farming in this area is a matter of considerable debate, however (cf. Flannery 1973). There appears to be firm evidence of domestic rice at Non Nok Tha in Thailand in the sixth millennium B.P., and it is possible that the evidence from this site extends back into the seventh millennium. Implements typically associated with rice cultivation and processing occur much earlier, however, and may indicate that rice agriculture dates as far back as about 9000 B.P. in this region (Gorman in press). There is also speculation that the incipient cultivation of the root crops may extend back even before this date, although the empirical basis for such speculation is poor (Gorman in press).

It is worth noting, incidentally, that Gorman (in press) has called attention to the possible role of demographic stress as a factor in the development of agriculture in this region. He has suggested that a Late Pleistocene rise in sea level would have resulted in the severe reduction of available land space for Southeast Asian gatherers, and that the concomitant high population densities may help to explain in part the economic shifts observed.

Perhaps the most problematic area is India, largely because almost nothing is known about the late pre-agricultural economy of the region. As summarized by a number of scholars (Sankalia 1962, 1974; Allchin 1973a,b; Allchin and Allchin 1968; Wheeler 1968; Piggott 1950; Fairservis 1971), post-Pleistocene archaeological sites on the Indian subcontinent consist mostly of densely scattered clusters of microlithic tools, indicating fairly dense population and perhaps pointing to ecological adaptations that parallel those of other regions where microlithic tools abound. Most of the finds, unfortunately, are surface finds, and they are poorly dated; hence it is often difficult to distinguish pre-agricultural occupations from those of marginal populations of later date. A few tools have been found which may be indicative of plant harvesting or processing. Grindstones occur at a few locations where they are presumed to reflect the processing of wild vegetable resources, and doughnut stones, interpreted as digging-stick weights, are also reported. Actual samples of food refuse are almost nonexistent, but those that do occur include a range of large and small mammals including fish and small reptiles. Shellfish also occur in association with microliths at a few locations. There are a number of sites showing orientations toward rivers, lakes, swamps, or the coast, and some of these are in positions indicating that boats must have been available and that fishing must have been the major economic activity (Allchin and Allchin 1968); however, most of these sites apparently lack organic refuse. As far as I have been able to determine, there is little evidence of large shell-mounds and few other indications of the type of intensive exploitation of aquatic resources recorded in other regions. From this scattering of evidence, Fairservis ventures a guess that the Indian Mesolithic populations were similar to their European or African contemporaries in terms of displaying broad-spectrum use of local resources, but it must be admitted that in this instance the case for economic (or demographic) change is weak.

The actual origins of agriculture on the Indian subcontinent are equally poorly defined. Singh (1971) has suggested that large cereal pollen grains dated between 9500 and 5000 B.P. may indicate early agriculture in India. Thus it is possible that Indian agriculture dates back substantially as early as that of the Middle East. However, according to Vishnu-Mittre (in press; see also Sankalia 1974), there is no valid evidence of agriculture on the subcontinent until some time in the fifth millennium B.P., at which time the cultivated species include wheat, barley, rice, peas, and lentils. These are accompanied by a variety of weeds and forage plants in the earliest sites with good preservation, but the weeds are typically those associated with wheat cultivation. As Vishnu-Mittre concludes, the whole complex is fairly clearly derived by diffusion from other regions; there is virtually no evidence of early indigenous domestication.

5 THE NEW WORLD: NORTH AND MIDDLE AMERICA

Although there are a number of significant problems and controversies in the interpretation of New World prehistory, a case can be made for the gradual buildup of population and population pressure in the Americas which is very similar to that just offered for the Old World. The expansion of population south and east from Alaska appears to have been accompanied in both North and South America by a progressive expansion of population out of open game-rich environments into forests and desert regions and into specialized coastal and riverine habitats. In the north there was even a recolonization of arctic environments as the glaciers withdrew. This territorial expansion appears to have been accompanied by a gradual transition from economies specializing relatively heavily in the exploitation of big game to those of a more eclectic type. As new territory for expansion was exhausted and as much of the big game became extinct, rapid parallel evolution of economies of "Mesolithic" or "Archaic" type occurred. This economic transition was accompanied by an increase in the size and density of population as well as by a trend toward sedentism, both evidenced by increases in the numbers, size, and depth of archaeological sites. Within this context, experiments in the cultivation of various plants seem to have occurred in a number of regions.

It is not entirely clear, however, why population pressure took so little time to build up to the threshold of agriculture in the New World in comparison to the Old. In the New World, population densities necessitating "Mesolithic" economies and, subsequently, agriculture appear to have been achieved in as little as 20,000 to 30,000 years by some estimates (and as little as 3,000 to 5,000 years by others), while in the Old World a similar buildup in population density is measurable in hundreds of thousands or even millions of years.

Although the causes for this differential rate of buildup are not entirely known, several hints can be offered which help to explain the greater rapidity of events in the New World. The most important point to be made is that by far the greatest time differential between events in the Old and New Worlds concerns the time required to complete the process of territorial expansion in the two hemispheres. Once the phase of territorial expansion was completed in each hemisphere, the actual rates of economic change involved in the evolution of Mesolithic and Archaic economies were not markedly different. In the Old World, the expansion of population from Africa through Eurasia required millions of years; in the New World, colonization of North and South America is measurable in thousands. The territorial expansion in the New World occurred so rapidly, in fact, that both the eastern portions of North America and the southern tip of Tierra del Fuego were colonized at about the same time that Eurasian populations took their first steps into the northern portions of the British Isles and Scandinavia.

The geography of the New World and the pattern of human migration must have favored a much more rapid rate of population growth and expansion than occurred in the Old World. In the Old World, man's expansion forced him to move out of areas to which he was well suited biologically and to develop artificial adaptations to cold. However, since the bulk of America, both North and South, lies north of the equator or within the tropics, man's southward expansion in the New World would have taken him into progressively less hostile environments, at least insofar as temperature was concerned. This gradual reduction of cold would have caused less stress and readjustment than were involved in man's northern expansion through the Old World. The increasing warmth of southern climates may even have provided some positive incentive helping to offset man's reluctance to alter his adaptive strategy. This does not imply by any means that man simply "turned south" once across the Bering Strait and sought warmer, greener pastures. He certainly had no comprehension of the principle that greater warmth lay to the south. Thus he would have left the subarctic niche to which he was culturally adapted only if he were under some pressure to do so. But, we can assume that somewhat less pressure would have been required to force people to move in a direction back toward the equator or within the tropics than had been required to force man to leave the tropics in the first place. It is only as man approached the southern tip of South America or ascended to great heights in the mountains, reentering cold climates, that we must assume severe population pressure to account for his movements. If I am correct, as argued in chapter 2, that human populations regulate their rate of growth and tend to tolerate a rate of growth in direct proportion to the degree of difficulty they perceive in coping with larger numbers, then a strong case can be made that population growth would have been much faster in the New World than in the Old. Paul Martin (1973) has gone so far as to propose that the New World offered an open and favorable habitat comparable to historically recorded situations described by Birdsell (1958, 1968) in which human populations expanded at rates as great as 3.4 percent annually, doubling every 20 years. By Martin's calculations, population growth at this rate would have resulted in the saturation of the Western Hemisphere with hunters in as little as 350 years. He also offers a somewhat more conservative estimate of a 1.4 percent annual rate of growth (doubling of population every 50 years), at which rate the Western Hemisphere would have been filled within 800 years. He argues that man probably did arrive in Tierra del Fuego within little more than 1,000 years after crossing the Bering Strait. In a similar vein, Haynes (1966) has argued that Clovis hunters expanding into an open and game-rich environment might have populated all of North America within 500 years.

Martin's and Haynes's models represent an extreme position. They are not entirely justifiable on theoretical grounds since the New World certainly provided more ecological barriers to human movement and necessitated more restraint on the part of expanding human populations than they imply. Moreover, their reconstructions may not be entirely in keeping with the available archaeological evidence, which suggests the possibility that man has been in the New World for longer periods than they allow. Nevertheless, the general principle that population would have expanded quite rapidly in the relatively open and increasingly hospitable environment of the New World is valid and important.

Having completed his territorial expansion in the New World, man proceeded to move from an overall emphasis on big game hunting to a broader focus on more generalized resources somewhat more rapidly than in the Old World. The reason in this case appears to be that the preferred big game resources failed as a food staple much more rapidly in the New World than in the Old. On entering the New World man was, as far as we know, entirely modern in his biology and possessed of a Middle or Upper Paleolithic technology. He was thus a relatively efficient predator for the entire time span of his habitation in the New

World. In fact, if some of the questionable early dates (see below) for man in the New World are accepted, *modern* man may have inhabited the New World essentially as long as he inhabited the Old. His prey in the New World, however, would not have been very formidable adversaries. Unlike the Old World animals, who were fortunate to have a long period of time in which to accommodate themselves to man's evolving skill as a predator, New World animals were faced suddenly with a skillful hunter with whom they had no prior experience (Edwards 1967, Jelinek 1967), and they would thus have been fairly easy prey. Moreover, as discussed below, there is reason to believe that the New World animals were rather poorly adapted to their own environment, exclusive of man, and thus might easily have been hunted to extinction. An additional possibility is that environmental changes at the close of the Pleistocene may have accelerated the buildup of population pressure in the New World in the same manner that they did in Europe, but on a much broader and more extensive scale. As will be discussed below, there is evidence of massive and widespread extinctions of game animals at the end of the Pleistocene, and there is reason to believe that, as in Europe, these resulted at least in part from climate changes. These changes may well have helped to create an imbalance between man and his preferred resources which both hastened his dispersal into secondary environments and necessitated his search for secondary resources.

A Brief Outline of the Prehistory of North America

The date of man's first entry into the New World is open to a good deal of speculation, although there seems to be little doubt now that the original entry occurred across the narrow gap between Siberia and Alaska sometime during the Upper Pleistocene. The most probable reconstruction is to assume that human populations crossed on dry land during one of numerous episodes in the Pleistocene when eustatic lowering of sea level corresponding to glacial advances created a dry land bridge. Hopkins (1959, 1967) has suggested that the land bridge between the continents was open prior to 35,000 B.P., closed between about 35,000 and 25,000 B.P., and open again between about 25,000 and 15,000 B.P. He indicates that in the terminal stage of the Wisconsin glaciation there were several additional brief periods when the land bridge would have been passable, including, for example, the period between 13,000 and 11,000 years ago (see also Flint 1957). Irving (1971) has suggested that man might also have traveled across ice at times when no land bridge per se was available; and it has even been argued that the Bering Sea might have been crossed by boat at times when it was spanned neither by ice nor by land. If these latter possibilities are accepted, man could have crossed at almost any time. However, neither of the latter possibilities seems to me to be plausible. I have argued that human populations do not alter their natural ranges for purposes of exploration or adventure but tend to do so only when such a move represents the sole, or at least the easiest, means of extending the old status quo. I cannot envision a boatload of sailors setting out for the New World, nor can I imagine a group of hunters chasing an occasional reindeer across the ice. Neither man nor the animals he pursued would have ventured en masse across the Bering Strait unless they were crossing on land at a time when the land was simply an extension of their existing niche or the best approximation of the old niche then available. It is of considerable importance that the Beringian environment as reconstructed (Colvinaux 1967, Hopkins 1967) would have been essentially similar to that of northeast Siberia, supporting an abundance of grazing animals. (It is worth noting, incidentally, that viewed in polar projection with corrections made for low sea level the Beringian land bridge represents quite a respectable extension of the Siberian landmass and that it is not nearly as tight a bottleneck as it appears in the map projections to which we are most accustomed.)

The most likely model for the expansion across the strait would be to assume that human hunting populations in Siberia were involved in random movements in pursuit of game and that the competition of other populations continuously helped to select in favor of hunting choices and strategies which tended to colonize new territories for which there was no competition. Hunting groups might thus gradually have expanded across the Beringian landmass with no conception of a significant alteration in their adaptive strategy. They would have moved east only at the rate at which competition made the colonization of new territory attractive.

Incidentally, there is no reason to perceive this colonization as a single event or even as a series of discrete events, although a number of separate crossings could have occurred (Byers 1959; Haynes 1967; Willey 1966). If we assume the gradual westward expansion of culturally homogeneous populations, then the mechanism of population flux (chapter 2) along with the availability of an open frontier in an ecologically homogeneous region would result in a fairly continuous tendency for individuals and groups to move eastward in their hunting, since the east would continue to provide the least competition and the most attractive hunting prospects.

Campbell (1963) has suggested that the best opportunity for the crossing of the strait might have occurred just after the final Wisconsin maximum, about 20,000 to 18,000 years ago, because at this time the ecology of Alaska would have provided an optimum environment for hunters already adapted to Siberian conditions. A date of 20,000 B.P. or sometime thereafter seems probable also because it is only at this time that there is substantial evidence of human occupation of northeast Asia (see chapter 4). As discussed in the previous chapter, there is tentative evidence of earlier Upper Pleistocene populations in Siberia who might have provided the ancestry for the first American immigrants; but this evidence is sparse and questionable. Even if it is valid, however, it does not present a picture of Siberia as sufficiently crowded with hunters to stimulate the territorial expansion necessary for the colonization of Beringia.

Even if the Alaskan environment itself was fairly attractive to Siberian hunters and thus the eastward expansion from Siberia relatively easy to explain, the southward expansion of human populations from Alaska presents something more of a problem in archaeological interpretation and reconstruction. It is possible that the early populations moved down the Pacific Coast of North America (Jennings, various) or across the Arctic and down the Atlantic Coast. Bryan (1969) and Butzer (1971:492) have objected to the coastal routes, noting that on both sides of the continent the coastline is interrupted by deep fjords or by steep cliffs where the glaciers calve directly into the sea, effectively preventing continuous migration. A further objection to this model is that there is little evidence that coastal adaptations had yet begun either in Asia or in North America. Although, as discussed in the previous chapter, some of the evidence may have been obscured by rising sea levels, there is no evidence that East Asian populations of 15,000 years ago or more were yet adapted to the use of aquatic mammals, fish, or shellfish either along coasts or along inland waterways. It is therefore highly unlikely that a littoral economy would have been adopted even if such an existence would have permitted movement along a coastal shelf.

The only other alternative route for a southward expansion seems to have been along the ice-free corridor which existed between the margins of the Cordilleran and Laurentide ice sheets where they merged just east of the Rocky Mountains. As reconstructed by Prest (1969), the corridor would have extended along the eastern edge of the Yukon and the western edge of the Northwest Territories, cut across the northeast corner of British Columbia, and extended southeast across Alberta, opening out below the ice sheets in the vicinity of Montana. Most important, the corridor would have provided early arrivals in North America with a southward extension of the open, game-rich environment to which they were accustomed. It is easy to envision both the initial colonization of this pathway and a more or less continuous movement along it by population flux, resulting in a rapid and fairly steady flow southward as long as the corridor was open.

There is considerable debate, however, about when and for how long this corridor was open. Haynes (1964) argues that the corridor was closed between about 25,000 B.P. and 12,000 B.P., and on this basis he suggests that the arrival of man in the New World is a very recent event, postdating the opening of this passage within the last 12,000 years. In a more recent article (1969), recognizing the possibility that certain sites in the Americas may predate this time, he has conceded that the initial southern migration could have occurred prior to closure, between 28,000 and 25,000 years ago; and he also seems to admit that the opening and closing of the barrier may have been more irregular than first assumed. Bryan (1969) has argued that the corridor was closed between 25,000 B.P. and 8000 to 9000 B.P. Since his date for the opening of the corridor is clearly later than a number of archaeological sites in North America accepted even by the most conservative archaeologists as valid evidence of man's presence, he argues that man must have moved south prior to the closing of the corridor at least 25,000 years ago. Since the Bering land bridge was inundated between 25,000 and 35,000 years ago, Bryan argues that North America must have been penetrated at least 35,000 years ago by men carrying a pre-Upper Paleolithic, Mousterian tool kit. However, other geologists and archaeologists are somewhat more cautious in discussing the opening and closing of the corridor, and it remains possible that it was open periodically between 25,000 and 12,000 years ago or that it opened earlier than Haynes and Bryan contend (see, for example, Ackerman, comments in Bryan 1969; Prest 1969; Butzer 1971).

Unfortunately, the earliest dated archaeological sites in the New World do not provide much aid in the resolution of this problem since there is a good deal of controversy among scholars as to which early dates are to be accepted. In the first place, there is no dated series of sites in Alaska or Siberia which is widely accepted as ancestral to the early North American populations. Thus we are forced to extrapolate backward from known sites farther south, and unfortunately many of these are only poorly understood. This chapter will concern itself only with the controversy surrounding some of the early North American sites. The South American sites, despite their obvious theoretical significance, provide problems of their own and do little to clear up the dispute. They will be dealt with in the following chapter.

Much of the controversy about early American tool horizons concerns the interpretation of a number of "typologically primitive" stone tool assemblages, generally lacking bifacially flaked projectile points, which have been found scattered through North (and South) America. Many archaeologists (most notably Krieger 1964; Chard 1959; and Bryan 1969) view the assemblages as indicators of the arrival of Middle Paleolithic Asian populations who crossed the Bering Strait prior to the Upper Paleolithic evolution of fine blade technology and carefully controlled bifacial flaking. Krieger (1964) has listed a number of assemblages which he lumps together into a "Pre-Projectile Point" stage in the occupation of the New World. The same assemblages are classified by Jennings (1974) somewhat less enthusiastically as "the Chopper-Scraper Problem."

Most of the assemblages which they list have been found on the surface or in very poorly defined geological contexts where it is difficult to obtain reliable dates or to establish valid faunal or geological associations. Often it is the typology itself or the advanced degree of weathering or patina ("varnish") on the tools that suggests great antiquity. Most of the assemblages also lack associated refuse or functionally diagnostic tool types which could provide clues to their economies. There are, however, a few sites attributed to the pre-projectile point stage which, although they are still problematic, do provide evidence that is more amenable to critical analysis. There are several locations where hearths or stone tools (with the noticeable exception of bifacial projectile points) have been found in presumed association with extinct Pleistocene fauna or with Pleistocene geological features or where these artifacts have been attributed radiocarbon ages of between 15,000 and 40,000 years.

For example, at Lewisville, Texas, bones of a variety of animals including bison, peccary, turtle, giant tortoise, glyptodon, horse, camel, and mammoth have been found associated with hearths and in apparent association with a Clovis point, the latter probably accidental. Radiocarbon dates for the whole assemblage are in excess of 38,000 years (Crook and Harris 1957; Krieger 1957). At Friesenhahn Cave, also in Texas, fossil fauna including more than 30 genera now extinct and thought to represent the Wisconsin glaciation have been found in association with a number of flint artifacts and cut bone (Kennerly 1956; Evans 1961). At Tule Springs, Nevada, radiocarbon readings in excess of 23,000 years have been obtained on a hearth and stone artifacts associated with split and burned bones of bison, horse, mammoth, camel, and sloth (Harrington and Simpson 1961). At American Falls in Idaho the bones of large bison, apparently perforated and cut, have been dated between 30,000 and 40,000 years (Hopkins and Butler 1961). On the coast of California at Santa Rosa Island a number of hearths and mammoth bones have been found with radiocarbon dates ranging from 10,000 to 29,000 years (Orr 1956, 1960, 1962). At the Scripps Campus site at La Jolla, California, fragments of charred bone and shell have been dated to 21,500 B.P. (Carter 1957). At Tequixquiac in Mexico, Pleistocene vertebrate fauna have been found associated with stone scrapers, splintered and sharpened long-bone fragments, and one bone bearing man-made carvings (deTerra et al. 1949; Maldonado-Koerdell and Aveleyra 1949; Aveleyra 1950). At Valsequillo in Mexico, artifacts occur with bones of extinct fauna and with radiocarbon dates in excess of 20,000 years. The fauna include horse, camel, mastodon, antelope, mammoth, dire wolf, smilodon, tapir, and glyptodon (Irwin-Williams 1963, 1967, 1969). Finally, at Tlapacoya, between Mexico City and the state of Puebla, bones of Pleistocene mammals occur near a hearth dated to 24,000 years (Mirambell 1967).

Beginning about 15,000 B.P., the residue of human occupation in North America begins to show evidence of fine bifacial flaking techniques and stone projectile points. Wilson Butte Cave in Idaho has produced bones of extinct camel and other mammals along with a small leaf-shaped biface, a blade, and a burin—an assemblage dated by carbon-14 between 14,500 and 15,000 B.P. (Gruhn 1961, 1965). At Fort Rock

Cave, Oregon, lanceolate projectile points have been radiocarbon dated in excess of 13,200 B.P. (Cressman and Bedwell 1968 cited in Bryan 1969; Cressman cited in Haynes 1969). At Marmes Rockshelter in eastern Washington human bones and projectile points have been found that are dated between 10,800 and 13,000 B.P.

All of these early North American assemblages, however, are subject to some degree of dispute. In general, the greater the antiquity claimed, the more widespread the criticism. In some cases the human manufacture of the early "stone tools" has been called into question. In others the association between tools and extinct fauna has been found unconvincing or the dating methods shown to be faulty. In still other cases the "stylistically ancient" tools have been discovered to be partial samples of more recent assemblages or unfinished forms of more modern artifacts (see Ascher and Ascher 1965; Lorenzo 1967; Sharrock 1966; Irwin 1971; Heizer 1964; Shutler 1968; Ferguson and Libby 1964; Haynes 1969; Wormington 1971; Martin and Plog 1973). In sum, most of the pre-projectile point assemblages are not held in wide esteem. Even Jennings (1974), once a firm believer in the existence of pre-projectile point man, concedes that most of the evidence so far put forward is faulty; and he admits that he becomes more skeptical as more years go by without firm evidence of early human occupation in North American being uncovered.

The most widely accepted evidence for man in North America prior to about 15,000 B.P. appears to be the assemblages from Valsequillo and Tlapacoya in Mexico. Willey (1966, 1971) accepts them as evidence of early man, and others such as Butzer (1971) and Haynes (1969), while more skeptical about some aspects of the sites and their dates, consider that they have the best claim to recognition. More recent locations such as the Wilson Butte site, the Marmes Shelter, and Fort Rock Cave appear to be more generally accepted (Haynes 1969; Bryan 1975; Lynch 1976), although Haynes raises some questions about the dating of the Wilson Butte site. Some authorities (Graham and Heizer 1968; Martin 1973) appear to question whether any of these early dates are acceptable.

There is thus considerable controversy about the existence or extent of human activity in North America prior to about 12,000 B.P. The controversy is of some importance not only because it affects the chronological framework against which the evolution of economic systems must be measured but also because it has a significant bearing on the interpretation of the history and distribution of subsequent, well-established archaeological groupings. Unfortunately, at this point, the argument is moot. My own inclination, as suggested above, is to presume a relatively late (post-20,000 B.P.) date for human arrival in the New World because I assume that population pressure played a significant role in man's expansion and because I see no evidence that a significant Siberian population had built up before 20,000 B.P. I am also inclined to accept the relatively late arrival of man in North America for another reason. As Martin (1973) points out, even by the most ambitious claims, human cultural remains prior to 12,000 B.P. in North America are remarkably scarce compared to amounts of debris left by contemporary Eurasian populations. Taken at face value, the evidence would appear to suggest a fairly widespread but extremely small population. This sparse record is not in keeping with the degree of success that contemporary populations were experiencing in similar environments elsewhere nor is it in keeping with known models of population expansion. As Martin concludes, if man had been in the New World at all by 30,000 or 40,000 years ago, his remains should be much more numerous than they are.

Aside from the question of antiquity per se, however, there is another reason for discussing these suspected pre-projectile point horizons in the New World and that is to call attention to the ecological and economic patterns which these groups display. The evidence is admittedly sparse and the conclusions tentative. But the point should be made that although many of these early groups may be classified as "pre-projectile point" cultures, they are in no sense to be considered pre-hunting groups or nonhunting groups, any more than were the pre-projectile point populations of the Old World. For one thing, there are good theoretical reasons to assume that the first Americans, whatever their actual antiquity, would have placed a heavy emphasis on the hunt. They were descended from Asian populations who were already adapted to a relatively narrowly focused hunting economy; moreover, they had passed through an arctic tundra environment which would, if anything, have strengthened their dependence on the hunt (Chard 1974:43; Jelinek 1967). Chard, who believes in the early arrival of man in the New World, has suggested that the Siberian environment would have been an effective cultural filter, favoring groups adapted to big game hunting. Moreover, as I suggested above, the emigration from Siberia appears to have occurred before the adoption of fishing (the only other major mode of arctic survival) in Siberia itself. The early arrivals must, therefore, have been accustomed to a hunting economy, and we would expect them to retain this emphasis in their new homeland until required to change.

The direct economic evidence from these early sites, such as it is, also supports the assumption that these early populations were primarily hunters. As Jennings (1974:81) points out, most of the evidence of pre-projectile point man appears to represent the takers of big game. Where economic refuse is reported, the early "sites" primarily contain, or are associated with, large mammalian fauna. Small mammals and small nonmammalian fauna (including some shellfish) occur, but it is the larger forms that predominate. There is little evidence of significant exploitation of the smaller vertebrates or of invertebrate fauna, and there is no evidence of major exploitation of aquatic resources. There is also no evidence of clearly defined vegetable-processing tools of the types (manos and milling stones, mortars and pestles) which become increasingly widespread after 10,000 B.P. There is thus some justification for thinking that these early populations (to the extent that any or all of the early assemblages are valid) reflect a relatively greater reliance on hunting than is evidenced in later post-Pleistocene sites.

After 12,000 B.P. the evidence of human occupation in North America is unequivocal, although, as we shall see, the controversies surrounding its interpretation are no less significant. The earliest archaeological evidence in North America which appears to be accepted by essentially all New World prehistorians is the Clovis horizon, which is widespread in North America beginning in the twelfth millennium B.P. It is characterized by large, bifacially flaked lanceolate projectile points fluted at the base. Typically the points are found scattered on the surface, and they have been reported very widely and in large numbers in the continental United States. The distribution of these points or close cultural affiliates in the plains, the southwest, and the eastern United States is well established. There is some debate (Meighan and Haynes 1970; Warren and Ranere 1968; Willey 1966; Tuohy 1968; Krieger 1964; Jennings 1974) as to whether or to what extent the area west of the Rockies and particularly the far west had been penetrated by Clovis populations. Clovis points are reported sporadically in the west: Campbell (1949) reports them from Nevada; Wormington (1957) from Lake Mohave in southern California; and Meighan (1959) from Borax Lake in northern California. But such reports are surprisingly isolated and in any case some of the identifications are questionable.

Dates on Clovis sites in the plains regions and the southwest display a remarkable uniformity. In Arizona, Clovis points have been dated at the Lehner site to 11,500 B.P. (Haury et al. 1959); to 11,200 B.P. at the Murray Springs site (Haynes and Hemmings 1968); and to 11,200 B.P. at the Escapule site (Hemmings and Haynes 1969). At Blackwater Draw, the Clovis type-site in New Mexico, the points are dated to 11,200 B.P. and at the Dent site in Colorado a similar date on Clovis material has been obtained (Agogino and Rovner 1964). A possible Clovis point (but one which is subject to varying stylistic identification) at the base of Ventana Cave in Arizona has been dated to 11,300 B.P. (Haury 1950). At the Union Pacific Mammoth Kill in Wyoming a date of 11,300 B.P. has also been obtained on a mammoth kill presumed to relate to the Clovis horizon, although here diagnostic Clovis points were not found (Irwin et al. 1962).

For those who deny the validity of the earlier archaeological assemblages (cf. Haynes various, Martin 1973), the Clovis assemblages represent the initial rapid expansion of an early hunting population which emerged out of the ice-free corridor on to the American Great Plains just prior to 12,000 B.P. and from there expanded through the rest of North America in a period of a few hundred years. To others, like Bryan (1969), the fluted point tradition represents the indigenous American development of fine bifacial flaking techniques out of the preexisting chopper-scraper tradition. The primacy of the western plains area as the epicenter of Clovis distribution seems to be fairly widely accepted (Willey 1966; Fitting 1968, 1970; Spencer and Jennings 1965; Jennings 1974; Haynes various; Irwin-Williams MS; Butzer 1971). It is assumed that from the plains the culture spread south (and west?) in a period when moister conditions created an extension of the plains environment (Irwin-Williams MS; Mehringer 1967; Martin and Plog 1973) and eastward into the forested regions east of the Mississippi at a time when much of the northeast was still a relatively open environment along the margins of the retreating ice (Fitzhugh 1972; Funk et al. 1970; Fitting 1968, 1970; Borns 1971).

The location of the ice-free corridor south from Alaska does seem to demand a northern plains origin for the Clovis tradition if it is assumed that this tradition is derived more or less directly from the Old World. Of course, if the Clovis point is a New World innovation from older New World traditions, the priority of the plains need not be assumed. The known early carbon dates on western Clovis sites (above) are older by at least a millennium than the earliest dates on fluted points (or on other well-established cultural material) in the east. Here the oldest dates occur at the Bull Brook site in Massachusetts (Byers 1954), about 9000 B.P., and at the Debert site in Nova Scotia (McDonald 1968), about 10,600 B.P. But most of the eastern Clovis materials are undatable surface finds, and there is some controversy as to whether either the Debert or the Bull Brook materials are typologically equivalent to true Clovis and whether either represents the earliest penetration of fluted point hunters in the east. It has been argued (R. J. Mason 1962; Byers 1959; Witthoft 1954; Griffen 1967) that the density of surface finds and the stylistic variety of large fluted points found in the east argue for an eastern origin for the Clovis horizon, but this position does not seem to be widely accepted.

Since the Clovis horizon represents the earliest, or at least one of the earliest, well-established culture traditions in North America, the inter-pretation of the economic patterns associated with this group is of considerable significance for an understanding of patterns of economic evolution on the continent. Like the Acheulian populations of the Old World, the Clovis people—as well as the more diversified and more numerous populations represented by the Folsom and Plano industries, which occur on the North American plains between 11,000 and 8000 B.P.—have traditionally been characterized as big game hunters. But, as with the Acheulian, a number of recent studies (Flannery 1966, 1967; Wilmsen 1968, 1970; Fitting 1968, 1970) have cast some doubts on the applicability of this description.

Certainly, there is ample indication that big game, most notably mammoth and bison, was a significant part of the economy of the Clovis group and its derivatives on the plains, although as evidence accumulates, the old-fashioned, simple equation of Clovis with mammoth hunting and Folsom with bison becomes more and more blurred (Martin and Plog 1973).

At the Lehner site in Arizona (Haury et al. 1959), Clovis points are associated with the bones of mammoth, horse, bison, and tapir; at the

Naco site in Arizona (Haury et al. 1953), they are associated with mammoth and bison; and at Murray Springs (Haynes and Hemmings 1968) with mammoth, bison, horse, camel, and wolf. At Blackwater Draw (Sellards 1952; Hester 1972) Clovis artifacts are associated with mammoth, camel, horse, and bison; and at the Dent site they are found with mammoth. Similarly, Folsom and Plano sites usually yield a variety of large fauna. At the Lindenmeier site (Roberts, various), from about 10,800 B.P., Folsom artifacts are associated with bison, camel, deer, mammoth, antelope, and some smaller fauna such as fox and wolf. At Blackwater Draw, Folsom and Midland points are found with bison. At the Plainview site (Krieger 1947) Plainview points are associated with mass slaughter of bison by means of the fall or jump technique. Similarly, among Plano populations at the Scottsbluff site in Nebraska (Barbour and Schultz 1932; Schultz and Eiseley 1935) and at the Olsen-Chubbuck site in eastern Colorado (Wheat 1972) large-scale bison kills are also in evidence.

The distribution of the sites of these groups also suggests an emphasis on hunting. In a survey of Paleo-Indian sites in the Rio Grande area, Judge and Dawson (1972) have noted that the sites typically are to be found overlooking grazing areas adjacent to water; and they argue that the most important determinant of site location was the presence of water not for the human population itself but for the prey species. Similar observations have been made by a number of other authors with reference to both western and eastern sites (cf. Wendorf and Hester 1962; MacDonald 1968; Wendorf 1961).

On the other hand, Flannery (1966, 1967) has noted that in some locations, such as the Tehuacan caves, a mixture of large and small game occurs which is not markedly different from that at later sites. There is the implication that where a reasonably broad sample of the Paleo-Indian diet is preserved small game may be seen to have played a larger part in the diet than is usually presumed. Small animals do occur at a number of sites of the so-called "big game hunters." Bird remains are found, for example, at two Paleo-Indian sites at least: in Folsom levels at Blackwater Draw, New Mexico, and in Plano levels at the Horner site in Wyoming. Similarly there are sporadic reports of turtle and rabbit and other small mammals from Paleo-Indian sites (Wendorf and Hester 1962). At the Lindenmeier site (10,780 B.P.), in addition to large fauna such as bison, deer, and antelope, small animals including jackrabbits, rodents, and small carnivores have been found. Nonetheless, the preponderance of large game at most known Paleo-Indian sites and the locations of those sites suggest that the large game, not the small, was the major focus of activity (Wendorf et al. 1961; Wendorf and Hester 1962; Irwin-Williams MS; Irwin-Williams and Haynes 1970; Judge and Dawson 1972).

If the balance of large and small game at Paleo-Indian sites is open to dispute, the preponderance of land mammals over other fauna is unquestionable. Although fish and shellfish are reported sporadically (and most often from late, Plano sites), there is a noticeable absence of significant accumulations of these resources or of other invertebrates. Nor is there any evidence that coastal, riverine, or lacustrine areas were a significant focus of occupation in the early stages except insofar as these zones provided an attractive habitat for big game. There is also very little direct evidence of significant exploitation of vegetable foods by the Paleo-Indian groups. Hackberry seeds have been reported from a few locations, but they are predominantly late (Plano) sites. There is no indication of the use of plants as staples.

Analysis of the Paleo-Indian tool kit may provide evidence of a somewhat broader diet, but here, too, I believe the evidence suggests an adaptation with a relatively heavy focus on hunting. The tool kit of the Paleo-Indian groups is composed overwhelmingly (though not entirely) of chipped stones traditionally described as a variety of projectile points, scrapers, perforators, knives, and so forth. Wilmsen (1968, 1970) has recently undertaken an analysis of tool function based on edge characteristics and a study of site function based on tool-type distributions. He recognizes a variety of functional activities, including hunting and butchering, woodworking and plant processing, and he argues on this basis that the Paleo-Indian diet was somewhat broader than previously assumed. Most of his inferences about tool use, however, are unsupported by direct evidence, since except for the bones of butchered animals the processed materials are largely absent. Wilmsen speculates about the diet at the Quad, Shoop, and Williamson sites of the Eastern Paleo-Indian tradition; but, unfortunately, none of these three has produced significant organic remains. The Levi site in Texas (Alexander 1963), which Wilmsen also discusses, has produced a comparatively broad range of organic remains including both hackberries and shellfish; but this site is a late (Plano) assemblage dated between 9300 and

7400 B.P., and is not necessarily a valid sample of the Paleo-Indian tradition. In fact several manos and a milling stone also occur, and the site has cultural affiliations to other sites of the Western Archaic or Desert Culture. Jennings (1974) views it as transitional from Plano to Archaic. In any case, there is no convincing relationship between the organic remains and the chipped stone tool types which permits extrapolation of the dietary pattern to other sites on the basis of the chipped stone tools alone. It seems to me that while functional analysis of chipped stone tools and assemblages of such tools may allow us to infer directions of use and the toughness of the worked material, specific dietary conclusions on this basis in the absence of confirmatory organic remains must be highly speculative. Similarly, Fitting's (1968, 1970) argument that the relative scarcity of projectile points among other stone tool types in Eastern Paleo-Indian contexts indicates a somewhat reduced reliance on hunting is suggestive but not conclusive.

Somewhat more convincing evidence of the use of vegetable resources among Paleo-Indian groups is the occasional recovery of grinding tools. However, such reports are rare, the numbers encountered are small, and the identifications and associations are often questionable. One grindstone was recovered in association with a Clovis (?) point at the bottom of Ventana Cave, Arizona, in a layer dated to 11,300 B.P. (Haury 1950). Grindstones are reported in association with Folsom materials at the Midland site in Texas, but the association is poor; "handstones" have been found in conjunction with Folsom materials at the Lindenmeier site, although they are associated with ochre and the functional identification of these tools as applied to plant processing is questionable. An object variously reported as a hammerstone and a mano has been reported from the Clovis site at Blackwater Draw; and at the Vernon site in Arizona a mano has been found in association with Folsom materials although this is a badly disturbed surface site (Wendorf and Hester 1962; Martin and Plog 1973).

In general, however, grinding equipment does not appear to have been a regular or significant part of the Paleo-Indian kit (Krieger 1964; Irwin and Wormington 1971). It is only among the later Plano sites that grinding equipment is usually found in the inventory, and various authors (cf. Jennings 1974) consider the arrival of such equipment to be the hallmark of the later Plano phases and of the plains "Archaic."

One other type of artifactual evidence which might provide clear, if indirect, evidence of broad-spectrum exploitation is also noticeably lacking. As far as I have been able to discern, there are no reports from Paleo-Indian sites of fishhooks, harpoons, sinkers, or other items specifically related to fishing or the taking of aquatic resources, except in the latest Plano and "Plains Archaic" phases.

Taken at face value, therefore, the evidence of economic activity from sites of the Paleo-Indian groups does tend to support the assumption of a rather marked emphasis on land-based hunting, even if this focus was not as single-minded as the term "big game hunters" implies. Even in the east, where Fitting argues that the Paleo-Indians would rapidly have assumed a broad-spectrum economy in a forest environment, there is no direct evidence of such an economic transition until somewhat later.

The picture is undoubtedly skewed to some extent by preservation. As indicated above, many Paleo-Indian sites have no preserved organic remains whatever, and this is particularly true in the moister eastern region where Fitting and Wilmsen have challenged the "big game hunting" interpretation. Moreover, it is clear that where preservation occurs at all, it will select markedly in favor of large faunal remains. We might expect, therefore, that the economy of the times was somewhat more eclectic than the direct evidence implies; but there appears to be no basis for arguing that this pattern was comparable to the well-attested broad-spectrum economies of later periods. It should be pointed out, moreover, that differential preservation is not as serious a problem in comparing Paleo-Indian economies with those of their descendants as it was in drawing comparisons between Acheulian and later populations in the Old World. The small time-differential between Paleo-Indian populations of 10,000 years ago and Archaic populations of 7,000 years ago is simply not sufficient to account for the enormous and widespread disparity in preserved organic remains.

On the other hand, the fact that the site sample from the Paleo-Indian period is skewed may be a significant problem. Wilmsen's classification of functional types among Paleo-Indian sites as well as similar studies by other authorities have cast into sharp relief a problem which has long been generally if vaguely recognized: that is, that most of our knowledge of the Paleo-Indians has come from the excavation of specialized "kill" sites which because of their spectacular nature have attracted a disproportionate amount of attention. Our picture of Paleo-

Indian economics is probably skewed to some degree in favor of the big game so prominently displayed at the kills. But this problem is being remedied somewhat as multipurpose sites and campsites become known, and it is significant that the picture of Paleo-Indian subsistence has not changed as radically as some scholars expected. At Hell Gap in Wyoming, a sequence of campsites has been described which spans the period between about 11,000 and 7500 B.P. Even at these sites, ground stone implements for plant processing occur only at the very end of the sequence, just before 8000 B.P., and a broadened spectrum of economic activity similarly is noticeable only in the later phases. Shellfish, for example, occur only in the Frederick occupation between 8400 and 8000 B.P. (Irwin-Williams et al. 1973).

Thus, even in the light of the fuller appreciation of the big game hunting Paleo-Indian economy which has resulted from recent studies, the conclusion that these early populations concentrated relatively heavily on large fauna as a source of food still appears justified. And, to the extent that these Paleo-Indian groups represent the earliest occupants of most of North America, we are justified in looking on hunting as the primary occupation of the early inhabitants of the continent.

The Relationship between Paleo-Indian Hunters and Other North American Culture Traditions

Unfortunately it is not entirely clear to what extent the Paleo-Indian hunters represent an early occupation from which other, local traditions are derived, and to what extent the plains hunters were simply one local variant among a number of separate cultural entities of equal antiquity. It is my impression, on the basis of the existing evidence, that the Paleo-Indian hunting tradition, at least in its earliest (Clovis) manifestation, does in fact antedate essentially all of the other well-defined regional traditions of North America. The only exception to this at present appears to be the artifact complex represented by Wilson Butte and some other sites in the northwest which are also considered to represent groups concentrating primarily on hunting (see below). It appears that the Clovis horizon represents a widespread and uniform hunting group which spread over most of North America as then defined by retreating ice. The various specialized regional economies then developed from this basis, contemporary with the Late Folsom and Plano phases on the plains. In areas where the Clovis hunters did not go, such as parts of the far west, the earliest sites still seem to represent a post-Clovis expansion of human occupation. The basic premises for this conclusion are four: first, Clovis points and closely related fluted forms occur very widely in North America; second, such forms, where dated, have consistently the earliest available regional dates; third, where stratified, Clovis artifacts appear fairly uniformly to represent the beginning of the local sequence; fourth, in areas where Clovis artifacts are lacking there is generally an almost complete absence of archaeological assemblages which are reliably dated to the same period (although various regional assemblages have been *estimated* to be of equal antiquity). Except in the Pacific northwest, well-dated archaeological assemblages lacking fluted points which can match the 11,000–12,000 year dating of the Clovis points are virtually nonexistent, and even assemblages which can match the 10,000–10,800 year antiquity of the Folsom sites are exceedingly rare.

Admittedly, however, neither the series of comparative dates available nor the number of stratified sequences known is sufficient to establish conclusively the temporal priority of the Paleo-Indian hunters, and a number of alternative interpretations of the culture sequence must be considered. It is possible that some of the other regional traditions may have evolved independently, deriving from preexisting chopper-flake groups, and that they will prove to be of antiquity equal to that of the Paleo-Indians; and it is also possible that some of these traditions may stem from separate ancient migrations from the Old World. It has been argued, for example, that the Eastern Archaic tradition with its broader gathering spectrum of exploitation may be a parallel development independent of the dispersal of the Paleo-Indians (Witthoft 1956; Fowler 1959; Byers 1959). Most eastern archaeologists, however, seem to accept the derivation of the Archaic tradition in the east from Paleo-Indian hunters of Clovis or closely related types (Fitting 1968, 1970; Fitzhugh 1972; Griffen 1964, 1967; Willey 1966; Ritchie 1969; Jennings 1974). This interpretation is supported by the stratigraphic superposition of Archaic sequences on top of Paleo-Indian layers at a number of locations: at Bull Brook, Massachusetts (Byers in Fowler 1959); at Flint Creek, Alabama (Cambron and Waters 1959); possibly also at the Quad site, Alabama (Cambron and Hulse 1960); and at Silver Springs in Florida (Neill 1958). At other locations, such as Modoc Rock Shelter in Illinois (Fowler 1959), where projectile points of Plains type appear in the midst of an Archaic sequence, they are fairly clearly late Plains styles.

Even in the absence of additional stratified sequences, it should be noted that dates on Eastern Archaic sites never reach the antiquity of the Clovis horizon. Stylistically distinctive Archaic points can be assigned considerable antiquity in some locations such as the St. Albans site in West Virginia and the Doershuck site in North Carolina (Coe 1964; Broyles 1966, 1971), and at Modoc Rock Shelter they occur at dates of 8000 to 10,000 B.P. But these dates do not rival those of the Clovis horizon, and, as discussed below, such early "Archaic" levels in any case lack the economic specializations which later characterize the Eastern Archaic. Grindstones and large accumulations of fish and shellfish all occur at later dates.

Similarly, Willey (1966) and Krieger (1964) have argued that the Old Cordilleran tradition of the western United States (Butler 1961), defined by projectile points, bolas stones, burins, choppers, and a variety of other stylistic elements distinct from those of the Plains hunters, may represent a separate evolutionary development paralleling the Paleo-Indian groups. The dating is questionable, however, and other authorities such as Jennings (1974), Butler (1969), and Browman and Munsell (1969) visualize the Old Cordilleran as a relatively late tradition in the northwest, coeval at its earliest with the Plano groups on the plains. But the argument is largely one of definition. If some of the earliest Northwest Plateau sites are included in the tradition then its antiquity is established; but specialists in that region such as Browman and Munsell tend to define the tradition more narrowly. Most important, the heavy fishing emphasis which Willey saw as one of the significant economic characteristics of the group does not emerge until about 7700 B.P., and even then there is only very occasional evidence for extensive fishing. As will be discussed below, while early northwestern sites are stylistically different from the Paleo-Indian tradition they are nonetheless not markedly different economically.

The most problematic argument concerns the origin of the Desert Culture(s) of the Great Basin and the southwestern and western United States. Jennings (various publications) and Spencer and Jennings (1965) trace this cultural tradition directly back to the chopper-flake traditions and see its evolution as virtually coeval with the development of the

Paleo-Indian hunters. Since the Desert Culture, like the Eastern Archaic, implies an economy which is far more diversified than the hunting tradition, one that relies to a considerable extent on the grinding of small seeds, the acceptance of this theory would affect the interpretation of economic evolution in the New World radically. However, a number of archaeologists (Willey 1966; Krieger 1964; Swanson 1966; Martin and Plog 1973; Kelley 1959; McGregor 1965; Haynes various) argue for the clear temporal priority of the hunting groups and there are several pieces of evidence to support this reasoning. In the first place, chipped stone pieces interpreted as Paleo-Indian artifacts (or as Old Cordilleran) underlie what are otherwise clearly Desert Culture or "Western Archaic" levels in a number of relatively well-defined situations: Ventana Cave (Haury 1950); Danger Cave (Jennings 1957; Krieger 1964); Burnet Cave, New Mexico (Howard 1935); Manzano Cave (Hibben 1941); the Lindenmeier site (Roberts 1940); Leonhard Shelter (Heizer 1951); Promontory Cave II (Steward 1937); and Fort Rock Cave (Hester 1973; but see Fagan 1975). In addition, even without such stratified occurrences the carbon-14 dates available on the two traditions indicate the temporal priority of the Paleo-Indian hunters over any full-fledged Desert Culture by something more than a full millennium. Even Spencer and Jennings (1965) seem to admit that no convincingly early dates are yet available on the Desert Culture. In comparison to Clovis dates in excess of 11,000 years, the lowest levels at Danger Cave which contain milling stones are dated to about 9700 B.P. (Jennings 1957). Levels at Fort Rock Cave in eastern Oregon containing what may be an early Desert Culture manifestation, including several manos and one possible metate, date to about the same period. In the southwest, the Double Adobe site, the type-site for the earliest (Sulphur Springs) phase of the Cochise culture (where grindstones are abundant), is dated between about 9300 B.P. and 8200 B.P. (Sayles and Antevs 1941; Irwin-Williams MS; Haury 1960). Similarly the San Dieguito culture, another early "Desert" manifestation in the southwest, has its earliest firm date at the C. W. Harris site in California, where it is dated between 8500 and 9000 B.P. Moreover, grindstones are not yet present in this phase (Rogers 1958; Warren and True 1961; Warren 1967; Irwin-Williams, MS).

These dates do not prove, of course, that the Desert tradition, the Old Cordilleran tradition, or the Eastern Archaic did not exist earlier as separate cultures paralleling the expansion of the Clovis hunters. The matter remains undecided, largely owing to confusion about the cultural classification of certain sites and tools. There is considerable disagreement, for example, as to what cultural tradition is represented by the bottom levels of Danger Cave and Ventana Cave, and it is possible to interpret these early levels as stylistic antecedents of later Desert Culture layers. The important point, however, is that even if such local traditions as the Eastern Archaic or the Desert tradition are traceable *stylistically* to an early date, it is fairly clear that none of them existed as a well-developed or widespread *economic* entity at the time of the Clovis hunters. Neither the Eastern Archaic nor the Desert Culture had yet developed the economic specializations which later characterized them.

It may be useful temporarily to ignore the cultural designations altogether and look simply at the evolution of tool functions and tool frequencies. Although some notice needs to be taken of the different interpretations of the cultural history of North America, arguments about stylistic culture groups and their derivation from one another should not be allowed to obscure the central theme of economic evolution. If we ignore the controversies concerning stylistic horizons and typologies, a pattern of economic evolution can be demonstrated which lends support to the theory developed in this book. The focus should not be on whether Paleo-Indian materials or Old Cordilleran materials underlie the Desert tradition at Danger Cave; the important point is that layers with grindstones overlie layers without them. Similarly, at Modoc Rock Shelter in Illinois stylistic identification of projectile points in the lowest layers dating to about 10,000 B.P. is less significant than the fact that grindstones occur only in the upper layers after 8000 B.P. At Ventana Cave there are numerous controversies about the stylistic affiliations of the various layers; the important fact, however, is that Haury (1950) sees an irregular trend from a predominance of animal foods in the diet (60 percent) in lower levels to a predominance of vegetable foods in upper levels (70 percent). On a larger scale it is important to note that whatever the relative antiquity of the Paleo-Indians, the Desert tradition, and the Eastern Archaic, there is a very clear tendency for both the Archaic and the Desert Culture to expand geographically at the expense of cultures of hunting type, and for direct evidence of broad-spectrum exploitation to increase and intensify (cf. Irwin-Williams MS; Fitting 1970). Similarly, for present purposes the cultural relationships between the Paleo-Indian tradition and the Old Cordilleran tradition are less significant than the fact that specialized fishing sites occur only fairly late even in the northwest and that fishing seems to become increasingly important through time.

Briefly: the archaeological sites in North America that are actually datable to the eleventh and twelfth millennia B.P. or earlier or which are attributable to this time span for good stylistic reasons all represent hunting groups; virtually all are related to the fluted point tradition, although a few such as Wilson Butte and Marmes Shelter may conceivably be stylistically ancestral to a different tradition of Old Cordilleran type. Only occasional, and often questionable, pieces of grinding equipment occur with these sites; actual plant remains are almost never found; and fish, shellfish, and fishing equipment are scarce or absent. There is thus a clear picture of relatively heavy concentration on meat as a food source.

Sites with numerous milling stones, suggesting a significant concentration on vegetable foods, occur beginning only in the tenth millennium B.P., and at this time such sites are quite rare and confined to a limited part of the desert regions of the western United States. It is only in the period after 9000 B.P. that sites of the Desert type become widespread, while sites of the Paleo-Indian hunting type are increasingly confined to the plains. It is also at this time that grinding equipment for processing plants begins to become significant on the plains themselves. Only after 9000 B.P. or later do the trends toward broad-spectrum exploitation which characterized the Eastern Archaic begin to disrupt the more specialized hunting foci of the Paleo-Indians or early Archaic groups in the east. It is also at this time that cultures of Desert type spread westward into California and the trends toward aquatic exploitation begin in the far west and the northwest.

The reasons for this economic transition are not entirely clear. The explanation appears in part to be climatic. Increasing dessication probably played a significant role in the expansion of desert culture groups (Irwin-Williams and Haynes 1970), and in the east, as will be discussed, the development of Archaic culture occurred against the backdrop of the northward migration of climatic zones (Fitting 1968, 1970). On the other hand, since the trend toward broad-spectrum exploitation after 9000 B.P. is remarkably uniform despite the varied climatic zones of the subcontinent, climate variation cannot be the sole or primary cause of the observed changes. In part the transformation appears to be related to the extinction of much of the big game fauna which had provided a large part of the subsistence of the Paleo-Indian hunters. As R. J. Mason (1962) has pointed out, the relationship between the emergence of generalized Archaic subsistence bases and the extinction of much of the Pleistocene fauna can hardly be coincidental.

I would argue, of course, that the economic transitions in large part reflect simply the growth of the American Indian population beyond the carrying capacity of its primary resources, with the result that some of these resources were speeded toward extinction while the human resource base as a whole gradually broadened.

The Extinction of the American Pleistocene Fauna

At the end of the Pleistocene, during a period roughly corresponding to the emergence and disappearance of the early big game hunting traditions, North America was the scene of a very widespread wave of animal extinctions (Martin and Wright [eds.] 1967). At least 200 genera of Pleistocene animals, the vast majority of which were mammals, became extinct or at least disappeared over much of their former range. A disproportionate number of these appear to have been the larger mammals, and many but by no means all of these groups are known to have been among those used for food by early human populations. Birds, the only other animal group significantly represented among the extinctions of the period, were only a minor component in the process (Martin 1967b; Martin and Guilday 1967). The extinct groups included mastodon, mammoth, tapir, various equids, pigs, camelids, some species of bison, cervids, mountain goat, musk ox, antelope, shrub oxen, yak, giant beaver, glyptodon, giant armadillo, ground sloth, a variety of wolves, coyote, cats including the saber-toothed tiger, bear, and a variety of rodents.

The exact dates of the period of extinction and the extent to which the extinctions were concentrated into one or more clearly defined episodes is a matter of some debate. As Martin (1967b) points out, the majority of the extinct genera have not been critically dated by radiocarbon methods, and in any case attempts to identify the "last" or most recent occurrence of an animal species or genus is fraught with problems. The major wave of extinctions has been variously placed at approximately 8000 B.P. (Martin 1958; Hester 1960, 1967; Slaughter 1967) and at approximately 11,000 B.P. (Martin 1967b; Lundelius 1967; Mehringer 1967). It should be pointed out, however, that while precise dating is of considerable significance in interpreting the cause of the extinctions, it is less important in considering their effects. Whichever date is accepted as marking the major wave of extinctions, it should be clear that during the early Paleo-Indian occupation many species were approaching extinction and becoming increasingly unreliable or unavailable as food sources.

The cause of these Late Pleistocene extinctions is open to dispute. One group (see particularly Martin 1967b, 1973; Edwards 1967) has suggested that man himself as a hunter is primarily responsible. Their general argument is that the extinction was an anomalous and unique event without precedent in world history, lacking correlated "natural" phenomena which might provide an explanation. Martin (1967b) points out, for example, that the end of the Pleistocene brought with it a sudden rapid increase in rates of extinction which correlates in North America and several other continents with the first arrival of human migrants but in no case seems to precede that arrival. He argues that the wave of extinctions cannot be explained entirely by climatic events correlated with the end of the Wisconsin glaciation since the record of extinction is not matched or even approached at the beginning of preceding interglacials. In fact, the number of extinctions recorded for the terminal Pleistocene exceeds the number recorded for the whole remainder of the Pleistocene period.

Martin also argues that the pattern of extinction at the end of the Pleistocene is unique for its selective effects on large terrestrial herbivores and the carnivores and scavengers which were ecologically dependent on them. Thus the whole complex of extinctions can be seen to be related directly or indirectly to human hunting patterns, which appear to have focused on the terrestrial herbivores. Conversely, Martin argues that the absence of evidence of the extinction of plant and invertebrate species, such as accompanied previous episodes of extinction and might be expected to result from a general ecological crisis, argues in favor of human hunting, with its selective effects, as an agent of extinction. He also notes that a surprisingly large number of species or genera appear to have become extinct without being replaced in their ecological niches; hence their extinction was not due to competitive displacement by immigrant or newly evolved species. (Guilday [1967], incidentally, challenges this observation, pointing out that although many of these niches are unoccupied today it is not clear that they have always been so and it is possible that many of these "open" niches are of recent origin.) There is also the problem that aside from their possible relationship to human hunting, the extinct fauna seem to share no common ecological pattern. Thus, unlike the contemporary extinctions in Europe, which appear to have focused on the disappearance of the tundra and the recolonization of Europe by woodland environments, the extinct fauna of North America seem to be related to a variety of habitats and no single natural phenomenon is likely to have affected them all (Mehringer 1967). As reconstructed by Martin, the extinctions at the end of the Pleistocene in the New World present a pattern for which only human agency can account.

The major critique of human hunting as an agency in the extinctions has been provided by Hester (1967) and Guilday (1967). One major focus in their critique has been to point out that man seems to have appeared in North America *after* many of the major groups were already extinct, or at least after many of them were already greatly reduced in numbers. A similar argument has been put forward by Kurtén (1965), who notes that many of the extinct species show a progressive diminution in the size of individuals; this, he argues, is suggestive of ecological stress. To the extent that this diminution is contemporary with human hunting it might result from the negative selective effects of such hunting on large animals (Edwards 1967). But by Kurtén's reconstruction the trend is often evident well back into the Pleistocene, suggesting that stresses leading toward extinction had a long history prior to human arrival.

Hester also questions whether the Paleo-Indians would have been capable of making serious inroads on the populations of their prey species. In reply to Edwards (1967), who suggests that the hunters would have been highly efficient and the prey unwary, Hester argues that there is no evidence that human hunting reached numerically significant proportions. He points out that Paleo-Indian populations were small, their demands minimal, and their technology fairly primitive. Hester questions whether the Paleo-Indian diet was as heavily geared to big game as it appears in the archaeological record (see above) and points out that the known kill sites do not indicate numerically significant destruction of prey species. The sites are relatively few (a fact attributable partly but not totally to poor preservation). Moreover, at most sites the number of animals killed was fairly small, the prey usually numbering fewer than twelve individuals. Kill sites demonstrating mass slaughter by means of stampedes are comparatively rare, and for the most part such large-scale kills seem to relate to the later phases of the Paleo-Indian occupation after the major wave of extinction had occurred. Even more striking, as Hester points out, is the fact that of all the extinct species only mammoth and bison occur with any regularity or in significant numbers in Paleo-Indian sites; and many of the extinct species, including some potential prey, have no provable association with Paleo-Indian hunters at all. This latter point can be confirmed by references to the bestiary of Pleistocene extinctions provided by Martin and Guilday (1967).

Hester's most striking argument is the comparison he draws between Paleo-Indian and modern Indian hunting patterns. The more recent groups, who by Hester's estimate had population densities a thousand times that of the Paleo-Indian groups, who appear to have worried little about conservation, who knew and used well-developed methods for mass slaughter, and who in many cases possessed firearms, did not succeed in destroying any prey species although their range of potential prey was by this time much more limited. What damage was done to the modern bison appears to have resulted mainly from commercial slaughter, the spread of farming, and the dissemination of new diseases. Viewed in this light, the ability of Paleo-Indians to bring about the extinction of the Pleistocene fauna unaided by other factors seems doubtful.

The most likely alternative explanation for the massive extinctions is to assume that the fauna succumbed to changes in climate, particularly in the distribution of moisture at the end of the Pleistocene. There seems to be no question that such changes occurred nor that they corresponded in a general way with the major wave or waves of extinction. The problems are encountered in reconstructing the pattern of climatic events accurately enough to show how, where, and when changes could have resulted in extinction of the various fauna. As suggested above, climatic explanations also face a problem in attempting to account for an extinction pattern which was geographically dispersed and ecologically generalized, but highly selective of relatively large mammalian fauna. The latter, under most circumstances, should be the species most tolerant of environmental changes (Edwards 1967; Mehringer 1967). It is well established, for example, that the presence or absence of large mammals is among the least accurate of paleo-climatic indicators. Paleo-environmental reconstrucfions have come more and more to focus on plants and microfauna as sensitive indicators of paleo-temperatures or humidity; yet relatively few of these life forms seem to have become extinct at the end of the Pleistocene.

Among others, Hester (1967), Slaughter (1967), and Guilday (1967) have all argued for the significance of post-Pleistocene environmental changes as a key ingredient in the extinction of the megafauna. They postulate a general deterioration of the North American environment related primarily to the decline in effective moisture; this would restrict both the availability of surface water and the extent of edible vegetation. They argue that this particular type of climate change could have acted selectively against large mammals, particularly herbivores. Climate changes might therefore have results paralleling the chain of related extinctions of herbivores followed by their dependent carnivores and scavengers which Martin attributes to hunting. Slaughter argues that there is a relationship which cannot be coincidental between climate change, patterns of extinction, and modern remnant ranges of the surviving species. He argues that the selective killing of the larger mammalian fauna results from the fact that the climate changes would have had the most significant impact on animals with long gestation periods. Size is not the important selective factor; it merely correlates roughly with the length of gestation. Guilday similarly argues that increasing dessication at the end of the Pleistocene would have decreased the number of habitats suitable to grazing animals, increased competition, and at the same time tended to create barriers to migration, which otherwise might have provided a means of adjustment for animal populations. Citing Taylor (1965), he suggests that the climate of the Wisconsin glaciation may have been the severest of the Pleistocene, with the implication that either the rigor of that climate or the relatively more pronounced shift to interglacial conditions which followed might have resulted in a more intense period of extinction than that previously encountered. Guilday argues also that in a deteriorating environment of the type envisioned for postglacial North America large herbivores might have been the animals most severely affected because of their greater demand for space and food. The larger animals, he argues, would be least able to make use of remnant microhabitats which would persist in conditions of deteriorating climate and protect smaller fauna.

Both Mehringer (1967) and Martin (1967b) suggest, however, that the onset of postglacial conditions would have been generally beneficial for the megafauna, providing improved natural conditions. Mehringer argues that the decline of the ice would tend to enhance rather than to eliminate the megafaunal habitats and that, in general, more habitats would be available in post-Pleistocene times.

In fact, the pattern of post-Pleistocene climatic change must have been considerably more complicated than is revealed if one simply ascribes to it a general tendency toward the amelioration or deterioration of the habitats of big game. Climate changes are zonal phenomena (see below, Fitting's [1970] correlation of culture phases and the northward migration of climate zones in eastern North America), and their patterns are complex. Therefore I find it difficult to envision climate changes as resulting in general and widespread extinctions. The pattern of regional variation which climate changes imply must have been overlaid by some phenomenon like hunting which applied a parallel selective force in the variety of heterogeneous environments.

One other major factor may have been involved. Edwards (1967) suggests the possibility that disease was a major factor in the reduction of the New World fauna. The opening of the Bering land bridge would have reestablished contact between previously isolated faunal groups exposing the fauna in both hemispheres to new disease organisms. The transmission of these organisms, which would occur most easily among gregarious and mobile species, might have worked selectively against the large herbivores; and disease would probably have acted selectively against large (and therefore slow-reproducing) species. On the other hand, the introduction of new diseases is likely to have been a regular phenomenon associated with lowering of sea levels, and it seems improbable that this would have resulted in such a marked pattern of extinction only at the end of the Pleistocene. Moreover, movement of animals and disease organisms would be primarily a function of low sea levels and of glacial periods rather than interglacials. It would therefore be difficult to associate this phenomenon with the postglacial timing of the extinctions. Finally, as Edwards himself points out, infectious diseases rarely cause extinction, because transmission between individuals statistically becomes prohibitively difficult before extinction occurs.

The most interesting argument from my perspective concerning the possibility of Pleistocene overkill is that relating to the balance or equilibrium between man as predator and his prey species. As Edwards (1967), Lundelius (1967), and Guilday (1967) have all pointed out, normal predator-prey relationships involve equilibrium mechanisms in which predators are regulated in numbers by social mechanisms or by the process of diminishing returns. Hence they do not, in most cases, overextend their consumption or destroy their prey species. The over-kill model (whether human hunting is seen as acting alone or only in conjunction with climate or disease) implies that no such limiting mechanisms operated to restrain human hunters. Edwards has suggested that this may simply reflect the fact that human hunters reached or perceived the point of diminishing returns only after the critical minimum density of the prey had been reached. Guilday suggests that because man was not dependent on particular resources but could supplement his diet, his own population could continue to grow despite the increasing scarcity of big game and the diminishing returns of the hunt. Although diminishing returns act as a homeostatic mechanism controlling the populations of most predators, no such control operated in the case of man. One might extend Guilday's case slightly. It could also be argued that because of their *perception* of the availability of other resources and because of their ability to regulate their behavior consciously the Paleo-Indians, in typical human fashion, were not restricted by any of the types of "automatic" behavioral controls that appear to regulate the numbers and densities of most animal groups.

In any case, it is clear that the terminal Pleistocene extinctions present a complex problem and one that has not yet been satisfactorily resolved. At present the best explanation, as set forth by a number of scholars (Hester 1967; Edwards 1967; Guilday 1967), is to presume a fairly complex interaction of variables involving both human and non-human factors. Guilday suggests, for example, that man may have played a part only in applying the final coup de grace to isolated remnant populations already doomed by a variety of other factors.

Whatever the causes of these extinctions, their occurrence is a well-established fact. The consequence was that American Indian populations in most parts of North America during the period between 10,000 and

8000 B.P. found themselves increasingly forced to do without the variety of large game which we can assume was their preferred food and which, to judge from earlier sites, had certainly been a major portion of their diet. The eclectic economies which emerge during this period and after appear in part to have been a response to this loss. On the other hand, the pattern of development of these economies for several millennia subsequent to the major waves of extinction suggests that extinction was not the only, or even the primary, stimulus to the economic evolution that ensued.

Post-Pleistocene Adaptations in North America

After the close of the Pleistocene and the extinction of much of the large mammalian game fauna, North America is characterized by a trend toward increasing regional differentiation of artifact styles. It now becomes necessary to treat each of the major regions as a separate unit with its own evolutionary pattern. At the same time, there is sufficient parallelism among the regions that the concept of a continent-wide "Archaic stage," which was originally used in a context of a simpler conception of American prehistory, has never quite been abandoned. As suggested by a number of scholars (see Swanson 1966; Meighan 1959a; Jennings 1974 for recent formulations), the common element which binds the various local stylistic traditions together is the tendency toward increasing efficiency in the use of space and the more complete exploitation of the available resources. The notion of "forest efficiency," first defined by Caldwell (1958) as a specific description of Eastern Archaic patterns in a forested environment, has been generalized as a concept applicable to human behavior in a wide range of environments.

Throughout much of the eastern United States, post-Pleistocene cultural evolution occurred in an area of deciduous forest. This environment, although possibly poorer in big game than the plains, nonetheless provided a rich and varied resource base for human populations once a broader spectrum of resources was subsumed within the cultural definition of acceptable foods. Caldwell's formulation of "forest efficiency" refers to a growth in the awareness of these varied resources and the development of technology for their use. The rich range of wild resources in the east seems to have absorbed a good deal of population growth and to have permitted the development of relatively large and stable populations. These eventually took on many of the characteristics of sedentary agricultural groups, while at the same time proving remarkably resistant to the diffusion of agricultural technology itself.

The general picture of post-Pleistocene evolution in the east as described by a number of scholars (Griffin 1964, 1967; Byers 1959; Sears 1964; Ritchie 1969; Caldwell 1958, 1962; Fitzhugh 1972; Fitting 1968, 1970) is one of gradually growing population densities accompanied by increasingly broad-spectrum and intense use of localized resources, particularly vegetables and aquatic foods, as supplements for hunted game. The pattern is somewhat irregular since these trends were played out against a pattern of varied and changing environments. For example, alterations in the courses and flow patterns of rivers seem to have influenced the local availability of species of fish and shellfish (Fitzhugh 1972). Thus the general trend toward increased use of these resources is by no means uniform or continuous in all locations. Fitting (1970) argues that high lake levels and periods of slow river flow permit the survival of relatively dense populations of slow-current adapted fish species, which in turn allows the development of human economies relying heavily on fish. He argues that in Michigan, where fishing became important only in the late Archaic periods after 5000 B.P., this economic transition reflects the availability of new resources resulting from altered patterns of stream flow rather than any human factor.

Similarly, as described by Fitting (1968, 1970) the interplay of deciduous and boreal forests and the northern tundra and parkland zones as these ecological zones moved north with the retreating glaciers had a significant effect on the distribution and evolution of Archaic economic strategies. Fitting in fact would argue as indicated above that the Paleo-Indians themselves were eclectic gatherers once they penetrated the deciduous forests of the southeast, and in his reconstruction of post-Pleistocene events in the east, the northward expansion of the deciduous forest with its rich and varied resources is the primary ingredient in the pattern of post-Pleistocene culture change.

However, if the effects of changing environmental zones cannot be ignored, neither, I believe, can they be given a primary role. The observable trends are similar in different environments. Even in the southeast, for example, where the deciduous forest environment was relatively constant, trends toward intensified exploitation occurred which parallel those at the northern margins of the migrating forest zone. Thus, throughout the eastern United States, as in other areas, local climatic variables appear only to have provided variations on a general theme of intensification.

The actual transition from Paleo-Indian to Archaic styles and economic patterns appears to have been in part a gradual cultural evolution and in part a replacement of old populations by new ones. In the southeast in particular the evolution seems to have been gradual, and the distinction between earlier and later phases is difficult and arbitrary (Griffin 1967; Williams and Stoltman 1965; Guthe 1967). In the north, where the retreat of the glaciers resulted in the northward contraction of the game-rich park and tundra environments and the expansion of the boreal and later the deciduous forest zones, it appears that many regions were abandoned by Paleo-Indian groups and later recolonized by northward-moving Archaic groups. In Michigan, Paleo-Indian hunting groups in the shrinking park-tundra environment seem even to have coexisted with Archaic groups in forested regions slightly to the south (Fitting 1970).

Among populations stylistically identifiable as Archaic, prior to about 7000 B.P. the changes from the Paleo-Indian hunting life-style were fairly minor. The early Archaic sites are distinguished from their predecessors by stylistic variations in chipped stone tools such as projectile points and by the increasing presence of grindstones and other ground stone forms, particularly atlatl weights and woodworking tools. Ground stone, however, is not yet as common as it becomes at sites of later periods and, as indicated above, many stylistically Archaic sites such as Modoc Rock Shelter still lack grinding equipment in their earliest layers. Some sites with good organic preservation—for example, Modoc Shelter—suggest a pattern of fairly broad spectrum hunting, the prey including large mammals such as deer and a variety of small mammals, fish, shellfish, and birds. In general, though, there is little direct economic evidence from this period. Most important, the known sites are still quite sparse and accumulations of debris typically small, indicating relatively brief occupations by small groups (Griffin 1967; Ritchie 1969; Fitzhugh 1972; Fitting 1968, 1970). Fitting has even argued that population may have been temporarily reduced in parts of the northeast at the end of the Paleo-Indian period. He notes that early Archaic sites are few in the northeast and suggests that this is because the boreal conifer forest which migrated north with the retreating glaciers would not have supported populations as great either as those surviving in the game-rich tundra to the north or those inhabiting the deciduous forests to the south. The relative scarcity of sites of the period and the lack of evidence of aquatic resources may, however, result at least in part from poor preservation. Fitzhugh (1972) suggests that early coastal sites may have been selectively destroyed by rising sea levels and riverine adaptations of the early periods lost in subsequent alteration of river profiles. I find it difficult, however, to attribute the lack of early Archaic evidence totally to such problems of preservation, given the variety of regional environments in which the same patterns are repeated.

It is not until the middle and later Archaic phases, after about 7000 B.P., that the trends toward broad-spectrum use of the environment really become evident. Shellmounds suggesting intensive exploitation of riverine resources date primarily from this period, grinding tools are now found in large numbers, and actual preserved remains of nuts and other vegetable foods become common. It is also at about this time that sites with large accumulations of refuse occur, suggesting increasing group size and either increased sedentism or carefully scheduled seasonal reoccupation of selected locations.

In the south, shellmounds become common in both coastal and riverine environments in Georgia, Florida, Alabama, Tennessee, and Kentucky by about 7000 B.P. Along the Tennessee River between 7000 and 3000 B.P., the refuse of freshwater mussels and fish occur in quantity at a number of sites along with the remains of varieties of large and small game (Lewis and Kneburg 1959). Similarly, the Lauderdale phase in Alabama (Webb and DeJarnette 1942, 1948) contains a number of deep shell midden sites which suggest long-term or repeated occupation. At the Eva site in Tennessee (Lewis and Lewis 1961) beginning about 7200 B.P., continuous occupation or reoccupation resulted in the accumulation of approximately 180 burials whose homogeneity of physical type suggests an isolated, inbred group. Here the diet seems to have been based primarily on the hunting of deer and certain smaller game supplemented by fish and shellfish, and, judging by the occurrence of a variety of pestles, anvils, and nutstones, a variety of vegetable foods as well. At Indian Knoll in Kentucky (Webb 1946; Lewis and Kneburg 1959) shellmounds approximately two acres in extent and reaching depths of eight feet testify to human occupation or cyclical reoccupation between about 6000 and 4000 B.P. In this case more than 800 skeletons have been excavated, and apparently many hundreds more were removed by pot-hunters prior to the main excavations. Here again there is evidence of the significance of deer, small mammals, and birds in the diet; but in this site mussel shells occur at all levels in such abundance that it is assumed that they provided a minor staple. Milling stones and nutstones occur, and the great quantity of walnuts, hickory nuts, and acorns suggests that these vegetable foods formed an important part of the diet.

Along the Atlantic coast, fishweirs such as the famous Boylston Street Weir in Boston (Johnston et al. 1942) appear to have been common after about 5000 B.P. As various authors have pointed out (cf. Byers 1959a), these weirs or traps, many of which are very large in scale, imply political manipulation of a relatively large work force and some degree of residential stability, since they involve significant labor investments which would not have been made for short-term exploitation. The labor investment also implies that fishing was not a casual activity, and it is a reasonable inference that it would not have been undertaken except in the case of the gradual failure or anticipation of failure of other, easier resources.

As already indicated, fishing becomes important in Michigan and much of the northeast only after about 5000 B.P., and in Michigan relatively dense permanent or semi-permanent seasonal occupations involving scheduled seasonal exploitation of fish, game, and vegetable resources date from this time (Fitting 1970). Although game, particularly deer, continues as a major constituent at many sites, depending on the season of occupation, some locations such as the Feeheley site display quantities of nuts—walnuts, butternuts, hickory nuts, and acorns—and in some places fishbone provides a large proportion of the faunal remains. Similarly, in New York (Ritchie 1969) a broad-spectrum economy becomes apparent only with the Lamoka site and contemporary occupations dated to about 5000 B.P. At the Lamoka site itself, although hunting, particularly of deer, is probably still the chief subsistence activity, fishhooks, gorges, and net sinkers are common and there is evidence of considerable reliance on fishing. Acorns are recovered in significant quantities and hickory nuts in somewhat smaller numbers. Grinding equipment, which Ritchie assumes was used primarily for the processing of acorn meal, occurs in considerable quantity.

In Illinois deep shell middens of the Riverton culture dated between 4000 and 3000 B.P. indicate significant reliance on mussels as well as hunted animals (Winters 1969). The Koster site, also in Illinois (Houart 1971; Asch et al. 1972), provides one of the most spectacular records of long-term occupation involving the intensive processing of vegetable foods. Here a thirty-foot accumulation of debris spans the period between Archaic occupations of about 7000 B.P. and Woodland and Mississippian occupations of the first millennium A.D. Asch et al. record the extensive use of hickory nuts and a variety of other nuts including black walnuts, pecans, hazelnuts, and acorns in Archaic levels. They note, incidentally, that nut protein is highly digestible and is of good quality, so that its dietary contribution parallels that of meat. Moreover, they point out that nuts with their high fat content resemble the meat of large mammals. They suggest that in view of the leanness of most freshwater fish, nuts would be a good and necessary complement to fish as a protein source. This suggests that the fish-nut combination which replaced or increasingly supplemented large mammals in the prehistoric diet of so many regions is an attempt to duplicate the dietary contribution of these animals once the latter were exhausted, extinct, or simply too scarce to provide a reliable food supply for a growing population.

At the Koster site Asch et al. also note the occurrence in small numbers of smaller edible seeds such as marsh elder (*Iva*), goosefoot (*Chenopodium*), and smartweed (*Polygonum*). None of these seeds occurs in Archaic levels in the site in sufficient quantity to establish with certainty that the plants were purposefully gathered or eaten, although the possibility is suggested. Asch et al. make the interesting observation that Archaic populations at the Koster site seem to have been fairly careful in their choice of vegetable foods, concentrating on hickory nuts rather than the less palatable and harder to process acorns, and tending to ignore the smaller seeds altogether. They note, however (see below), that in later sites in the same region acorns are extensively exploited along with small seeds such as *Chenopodium*. Making use of a population pressure model much like the one espoused here, these authors suggest that the shift from a selective focus on preferred nut resources to the use of acorns and smaller seeds by later populations represents the demands of an increasing population.

There is some evidence in other locations of the use of the smaller seeded plants by late Archaic populations. *Chenopodium* in particular appears to have been an important food as far back as about 3000 B.P. at such sites as Russel Cave in Alabama, the Higgs site in Tennessee, and the Cowan Creek site in Ohio (Yarnell in press).

In addition to these nuts and seeds which were apparently exploited in their wild forms, there is evidence for the indigenous domestication of at least two significant food crops, sumpweed (*Iva annua*) and the sunflower (*Helianthus annuus*), in the eastern United States by the beginning of the third millennium B.P. Moreover, according to present evidence, domestication of these two crops in Kentucky, Tennessee, and Missouri preceded the arrival of Mexican cultigens such as gourds, squash, and maize (Yarnell in press). A number of other plants which have been suggested as indigenous eastern domesticates, including *Chenopodium,* the Jerusalem artichoke (*Helianthus tuberosus*), apricot vines (*Passiflora incarnate*), maygrass (*Phalaris caroliniana*), and tobacco, are regarded by Yarnell as ruderal plants closely associated with man but lacking evidence of actual domestication. As reconstructed by Yarnell, there is a gradual increase in the importance of various small-seeded food plants such as *Chenopodium,* sunflower, sumpweed, and maygrass in the diet of parts of the eastern woodlands for a period of several hundred years prior to the domestication of sumpweed and sunflower and before the arrival of the Mexican crops by diffusion. All this suggests that by about 3000 B.P., in the later stages of the Archaic, the search for foods had become extremely broad spectrum and a good deal of effort had begun to go into the harvesting and processing of small-seeded plants. Not only had such plants been more or less ignored by the earlier big game hunters, but they were avoided or given only very minor attention even by vegetable gatherers, such as those at the Koster site, who were blessed with more palatable and accessible resources. Here, as in the Middle East, domestication appears to have involved, not choice resources, but low priority foods which were used extensively only after the range of the diet had been broadened to include a number of other items. Moreover, if Caldwell's characterization of primary forest efficiency is correct, this domestication took place only when man had essentially exhausted the variety offered by the eastern forests and could no longer expand his food base by turning to new foods. It is also important that the Mexican cultigens, which could have been received by diffusion at a much earlier date (Caldwell

1962), did not become part of the economy until a time when some stress had already forced man in the eastern United States to begin domestication of his own native resources.

It is worthy of note also that the arrival of Mexican cultigens during the third millennium B.P. does not appear to have had a major effect on the economy; instead these provided a fairly minor or local supplement in a broad-scale gathering pattern. It has long been assumed that the significant sociocultural achievements of the Hopewellian cultures were based on the use of maize agriculture; but the actual evidence for the importance of maize is minimal (Fitting 1970), and in many locations the Hopewell culture is thought to have been mainly centered on the gathering of local resources. Struever (1968), for example, describes the economy of the Hopewellian Middle Woodland occupation at Apple Creek in Illinois, where a settlement with storage pits and permanent dwellings seems to have been based on the hunting of deer, turkey, and migratory fowl and the extensive use of fish, hickory nuts, acorns, *Chenopodium, Polygonum,* and *Iva* seeds. The only cultigens recovered included the gourd, which presumably was not a food plant, and a few seeds of a squash (*Cucurbita pepo*). In addition, many contemporary populations outside the Hopewellian culture sphere continued to subsist primarily on wild foods (Griffin 1967; Fitting 1970). In Michigan, for example, variations on the seasonal exploitation pattern involving hunting, fishing, and wild nut and seed processing seem to have persisted through the early and middle Woodland periods (Fitting 1970), and in much of the northeast fishing and shellfishing appear to have provided the main basis for subsistence for sedentary populations as late as the middle Woodland period (Ritchie 1969; Cleland 1966). Griffin (1967) even argues that basic dependence on farming may have emerged in the eastern United States only with the Mississippian culture of 700–900 A.D.

In the central United States between the Mississippi River and the Rocky Mountains, a somewhat different pattern of post-Pleistocene culture history is evident. In contrast to the forested east, the central region is dominated by a plains environment at once richer in big game and poorer in the more varied resources which permitted dense populations to accumulate in the east. Thus, on the one hand, the hunting focused economy seems to have persisted longer on the plains than elsewhere; on the other hand, although this economy was supplemented increasingly with other resources, there is little evidence of a degree of population growth comparable to that observable in the east. In fact, to the extent that demographic patterns can be reconstructed, they seem to reflect an irregular development geared not only to changing economy but to climate shifts as well. Nonetheless, in a very general sense the pattern parallels that of other regions. Wedel (1964), for example, describes the plains culture sequence in terms of a succession of early big game hunting groups, followed by foraging groups of mixed hunting and gathering strategy, followed in turn by incipient horticulturalists.

As already suggested above, during the period between 11,000 and 7000 B.P., Clovis hunters were replaced on the plains by a number of derivative populations loosely characterized as the Folsom-Plainview-Plano tradition. These groups appear to have focused on big game as a resource, although with the later (Plano) groups grindstones and other indications of eclectic food gathering are increasingly found. The Plano sites, which are more numerous than those of their predecessors, also seem to indicate denser population. The evidence of mass kills, more typically associated with these later phases of the hunting culture, suggest both larger groups and better group organization. However, as suggested in the last chapter, the organized mass slaughter of whole herds, presumably accompanied by attempts at meat preservation, may also indicate a greater unreliability of big game resources that forced human groups to make the most of occasional herds. (See Wheat 1972 for a discussion of the social organization implicit in the Olsen-Chubbuck kill site.) The different Plano assemblages also suggest increasing regional stylistic variation indicating the break up of the loose population structure of the earlier periods. Finally, it is noteworthy that through the Llano-Plano progression the area occupied by the hunting groups became gradually narrower. Fringe areas, particularly deserts better watered during the period of the Clovis hunters, ceased to provide viable hunting territory, and the populations of these areas altered their economies accordingly.

After 7000 B.P., and particularly during the height of the Altithermal period, when it is probable that dry conditions severely affected the plains environment, the hunting tradition seems to have become even more locally restricted. Groups both stylistically and economically related to the eastern Archaic and the western Desert Cultures emerged on the eastern, southern, and western fringes of the plains.

Along the boundary between the plains environment and the eastern forest zone, a rather complex intertwining of Plano traits and eastern Archaic influences is evident, but the general trend seems to be in the direction of westward expansion of both Archaic artifact styles and eclectic economies of Archaic types. At Graham Cave in Missouri, for example (Logan 1952), basal layers dated between 9700 and 8000 B.P. contain chipped stone forms including lanceolate projectile points and knives of Plano type. In the next level, the projectile points suggest a mixture of Plano and Archaic forms, and in addition grooved axes occur. Most important, manos, milling stones, and nutstones are found in the upper level, suggesting increased reliance on the processing of vegetable foods. Similarly, the Rice site in Missouri displays an Archaic tool inventory typical of the Grove focus, which has a wide distribution along the south central United States (Bray 1956). The Grove focus (Bell and Baerreis 1951; Baerreis 1959) is better known from a number of other sites, primarily in Oklahoma, many of which display either long periods of use or frequent or systematic reoccupation. Notched projectile points of Archaic type as well as scrapers, drills, and choppers occur along with grooved axes, slab milling stones, manos, and mortars. Also in Oklahoma, the Fourche Maline focus (Baerreis 1951, 1959; Bell and Baerreis 1951) displays an Archaic tool inventory including blades and scrapers, celts, grooved axes, atlatl weights, hammerstones, and cupstones. These sites are middens containing significant amounts of shell as well as animal bone. In Texas, a number of regional variants of Archaic culture and economy have been described (Suhm et al. 1954; Crook and Harris 1952). Kelley (1959) combines a number of these local variants into a more general Texas Archaic or Balcones phase, defined generally by the presence of such items as mullers, manos, metates, and mortars in addition to chipped stone forms, and by the presence of plant food remains, fish scales, fishbone, mussels, snails, and saltwater molluscs in addition to animal bone. Stylistically, Balcones phase artifacts appear to resemble both eastern Archaic forms and western Desert styles, and Kelley argues that the Balcones phase is one link in a chain or cultural horizon linking the Archaic cultures of the eastern and western United States.

North and east of the regions just discussed, on the Great Plains proper, the prehistory of the period after 6000 B.P. is less well defined. Sites are scarce for a period of at least a thousand years after this date. As Wedel (1964) has pointed out, the period of scarcity corresponds with the Altithermal period, and he suggests that during this period of aridity human populations may have been forced to take refuge along the margins of the plains area. What sites there are suggest the influence of both eastern Archaic and western Desert Culture traditions. Willey (1966) seems inclined to view these plains "Archaic" cultures as indigenous developments permeated by outside influences, but the time gap between late Plano sites and well-defined Plains Archaic groups suggests the possibility of at least partial abandonment and replacement by new populations. On the eastern plains, in Nebraska, Kansas and Iowa, projectile points of eastern Archaic style are known primarily as surface finds. At one or two locations, better defined archaeological sites seem to indicate some continuation of older hunting-focused economies by populations with this new projectile point style. At Logan Creek, Nebraska, for example, bison bone and side-notched Archaic points occur dated between about 6700 and 7300 B.P. At the Simonson site in Iowa notched (Archaic type) spear points are found associated with bison and dated to about 8400 B.P. (Agogino and Frankforter 1960).

On the western plains beginning as late as about 5000 B.P. there are a number of sites scattered in Wyoming, Montana, Colorado, South Dakota, and western Nebraska which Wedel (1964) says point to a hunting and gathering stage, noting that they commonly possess milling stones and that the bone refuse typically includes higher percentages of small game such as rodents and reptiles than occurred in the sites of the earlier plains hunters. He suggests that the new economy, representing a fuller and more careful use of the environment, spread east from the Great Basin. At the McKean site in Wyoming (Mulloy 1954), where the occupation extends back from 3400 B.P. to perhaps 5000 B.P. or earlier, a variety of projectile points, knives, and scrapers have been recovered along with manos and milling stones. At Mummy Cave in Wyoming (Wedel et al. 1968) a stratified sequence has been found in which Plano points overlie what may be a Folsom point and are in turn overlain by notched points of Archaic type, while in levels dated to about 4400 B.P. grindstones occur. Similarly, at Pictograph Cave in Montana (Mulloy 1952, 1958) a stratified sequence extends from the early hunting groups through the Archaic to the historic period.

Wedel suggests, however, that not all of the western plains populations acquired the more eclectic dietary pattern. He notes that a number of sites occur at fairly late dates where grindstones are absent and where the nature of the chipped stone or the actual faunal debris indicates a fairly specialized hunting focus on bison. The assemblages at the Powers-Yonkee Bison Trap in Montana (Bentzen 1962) dated to about 4500 B.P. and the Signal Butte site in Nebraska (Strong 1935) dated between 3500 and 2000 B.P. seem to reflect heavy dependence on hunting, as does the Oxbow culture at Long Creek, Saskatchewan (Wettlaufer and Mayer-Oakes 1960). However, since some of these sites, notably Signal Butte, otherwise show strong stylistic affinities to the McKean assemblages (Mulloy 1954), it may be that we are dealing with special-purpose camps rather than discrete cultural groups with separate economies.

On the dry, western portion of the plains this hunting and gathering economy seems to have persisted relatively unchanged until the historic period and the arrival of the horse. In the eastern portion of the plains, however, where greater moisture supported lusher prairie grasses and gallery forests along streambeds, later influences penetrating from the east brought significant stylistic and economic changes. Pottery appears on the eastern plains just about 2,000 years ago, at which time it is fairly clearly derived from an eastern Woodland origin and is presumed to have been accompanied by the westward spread of maize agriculture. A few Hopewellian sites on the plains have in fact yielded maize dated between about 1,500 and 2,000 years ago (Wedel 1961, 1964), but actual evidence is rather sparse. It is not until the last thousand years that agriculture seems to have wrought a substantial change in the economy of the eastern plains. After about 800 A.D. a much more settled life-style is in evidence here, involving large and permanent villages based on agriculture and hunting.

West of the Rockies, in the arid environments which characterize the American southwest and the Great Basin area of Utah, Nevada, southern Idaho, eastern Oregon, and southwestern California, the scarcity of large game seems to have resulted in the emergence of eclectic economies of "Archaic" type at a relatively early date. At the same time, the paucity of food resources of all types appears to have resulted in relatively low population densities, although by all indications the pressure of populations on resources was quite high. In the southwest this pressure was relieved eventually by the adoption of Mexican cultigens, but farther north in the Great Basin, where the environment was apparently inhospitable to agriculture, populations remained small into the historic period.

The early post-Pleistocene cultures of this entire region have been grouped together by Jennings (1956, 1957, 1964; Jennings and Norbeck 1955) as the Desert Cultures. The group identity is based in part on common ecological patterns, particularly the adaptation to maximum exploitation of an arid and extremely limited environment, which required an intensive focus on the harvesting and processing of small seeds and the unselective gathering of varied wild game species, large and small. The use of the spatially limited resources of river and lakeshores is also part of the pattern (Baumhoff and Heizer 1965). Jennings suggests that exploitation of the desert environment would have required the maximization of seasonal differences in resources as these varied with altitude, and he argues that human populations would have had to make fairly rapid and widespread movements to make full use of what was available. The Desert Culture is also defined by a number of artifact types which occur commonly in the sites. Among the most important are frequent milling stones and mullers; typically small projectile points (for the taking of small game); and other chipped stone forms including knives, pulpers, choppers, and scrapers. The list also includes a number of perishable items including particularly basketry, woven sandals, fur cloth, vegetable quids, and more distinctive items such as deer-hoof rattles, medicine pouches, scapula grass-cutting tools, and digging sticks (Jennings 1956; Heizer 1956; Heizer and Krieger 1956; Day 1964). To the extent that the identification of Desert Culture sites is based on these perishable items, incidentally, it seems to me that the definition of the cultural unit is in some doubt, since the presence or absence of these items may reflect preservation in a dry environment rather than the actual cultural unity of the tradition. It may be more useful, particularly for our purposes, to think in terms of a series of generally similar economic adaptations rather than in terms of any well-defined cultural unit. Warren (1967) and Irwin-Williams (MS), among others, have already questioned the cultural unity of the Desert group.

As suggested above, sites representing this adaptive pattern begin to emerge in the western United States during the tenth millennium B.P. Probably the best known of the Desert Culture sites is Danger Cave, where strata bearing Desert Culture artifacts are first dated to 9700 B.P. Thereafter the cave appears to have been used on a continuous basis for about 7,000 years by people with an essentially unchanging economy of Desert type. The faunal remains include antelope, bison, sheep, jackrabbit, wood rat, bobcat, and desert fox; but the artifacts and the nature of the midden itself indicate the extreme importance of vegetable foods in the diet. Approximately 1,000 flat milling stones occur in addition to several hundred manos or handstones, numbers which compare strikingly to the relatively small number (2,000) of chipped stone artifacts recovered. Some 65 plant species were identifed from the cave fill, which was in fact largely vegetable in composition. In particular, large quantities of the chaff of pickleweed were encountered —a wild plant that seems to have been harvested for its seeds, which were parched, milled, and eaten. The importance of pickleweed in the diet is confirmed by its presence in quantity in human feces from the cave. Nearby, at Hogup Cave (Aikens 1970; Harper and Alder 1970), a similar Desert economy sequence has been excavated and dated over a period between about 8000 B.P. and 1470 A.D. Both the artifact inventory and the vegetal nature of the fill duplicate that at Danger Cave. During part of the sequence a more specifically aquatic focus is evident, and items such as aquatic birds, rushes, and the rhizomes of marsh plants are important in the diet. For most of the occupation, the economy appears generally to have been of Desert Archaic type, focusing heavily on vegetable foods, varied game animals, and aquatic birds. Interestingly enough, however, at about the time of Christ the Hogup people seem to have shifted away from this diet toward one of big game, and seed foods and milling stones become less important. In Nevada a number of additional Desert Culture assemblages have been found. At Gypsum Cave (Harrington 1933) millstones and atlatl weights occur along with digging sticks and "sickles," or harvesting tools of bone. Sloth dung from the cave has been dated to 10,500 B.P., but the association between dung and artifacts is questionable. Carbon-14 readings on the wooden artifacts themselves suggest that the occupation began about 3000 B.P. (Berger and Libby 1967). At Humboldt and Lovelock Caves near Humboldt Lake a more specifically lacustrine orientation is found. At Lovelock Cave (Loud and Harrington 1929; Heizer and Napton 1970, eds. 1970) the occupation dates from about 4500 B.P. through the historic period. The food remains indicate heavy reliance on lacustrine species such as fish, waterfowl, and halophytic plants as well as a range of small mammals. The artifactual assemblage is clearly related to the same pursuits, as it includes duck decoys, fishnets and hooks, horn and bone sickles, as well as hunting equipment. At

Humboldt Cave, where the occupation was roughly contemporaneous, a similar economy is evident. Surprisingly, neither site has milling stones or manos, but these items are common at nearby open-air sites and it is assumed (cf. Jennings 1974) that they represent part of the same economic cycle.

Similarly, in southeastern Oregon on the northern fringe of the Great Basin an economy of Desert type seems to have persisted over a long period. Reference has already been made to Desert Culture material or proto-Desert Culture material at Fort Rock Cave in the tenth millennium B.P., but typical Desert Culture assemblages are known from other locations such as Paisley Nos. 1 and 2 Caves, Roaring Springs Cave, and Catlow Cave in the same region (Cressman 1942), where dates range from about 7000 B.P. to the historic period.

In reference to the southwest, Irwin-Williams (MS) has recently summarized the development of Desert-Archaic cultures in terms of four regional traditions. The best defined of these traditions and the one that provides the earliest evidence of eclectic economy of Desert type, complete with grindstones, is the Cochise tradition of central and southeast Arizona and southwest New Mexico. The earliest, Sulphur Springs, phase of this tradition, mentioned above (Sayles and Antevs 1941), is known from six sites, one of which, the Double Adobe site, has provided a range of radiocarbon dates of between about 9300 and 8200 B.P. Manos and milling stones are common as well as chipped stone choppers, scrapers, and knives. The associated fauna represent a surprisingly big game oriented hunting pattern, but the association of the fauna has been called into question (Haury 1960).

In later Chiricahua and San Pedro phases of the Cochise tradition, after about 5500 B.P., the prominence of manos and milling stones indicates the continued importance of vegetable foods in the diet. At Tularosa Cave in Arizona (Martin et al. 1952), dated from about 2800 B.P., 39 floral species are identifed among the food remains, including yucca, a variety of cactus, walnuts, seeds of various grasses, sunflowers, and desert primrose. At Hay Hollow, also in Arizona (Martin and Plog 1973) and dated between 2300 B.P. and 1700 B.P., milling stones are common and a wide variety of wild plants occur including the seeds of various grasses, amaranths, chenopods, *chola,* and *Compositae.*

These later Cochise phases also include a number of domesticates. Primitive maize, presumably derived from Mexico, occurs in the southwest as early as about 5000 B.P. at Bat Cave (Dick 1965); squash and, somewhat later, beans are also added to the list of cultigens adopted from Mexico. These cultigens do not seem to have had a major impact on the economy; rather they appear simply to have been absorbed into the already eclectic pattern of food gathering (Whalen 1973; Reed 1964). Maize, beans, and squash all occur, for example, at Tularosa Cave; but they are minor elements in the refuse. Similarly, maize occurs at Hay Hollow but amidst a very wide variety of wild, gathered plants. The relative unimportance of maize in the diet is due in part to the primitive nature of the early maize itself. It is only after about 2750 B.P. that more modern maize forms occur in the southwest (Mangelsdorf and Lister 1956), but it is noteworthy that even after this date maize does not emerge as an economic staple for some time.

There is botanical evidence that at least one crop was domesticated in the southwest itself—Devil's Claw (*Proboscidea parviflora*)—and that a number of indigenous plants were tended or cultivated, including several species of tobacco, ground cherries (*Physalis* sp.), wild potatoes (*Solanum* sp.), and Rocky Mountain bee weed (*Cleome serrulata*). Sunflowers (*Helianthus annuus*), used for food in the southwest, spread from there to the eastern United States, but it is only in the latter region that clear evidence of morphological domestication occurs (Yarnell in press). The tending of these minor crops, which might seem unlikely after the arrival of more promising plants, suggests that the southwest had derived its own cultigens before it adopted those from Mexico, and thus may emerge as another independent hearth of domestication. However, at present there is no evidence that any of these domesticated or tended plants did precede the diffusion of the Mesoamerican cultigens into the region (Yarnell in press).

To the north of the Cochise area, in northern Arizona, southeast Utah, southwest Colorado, and northwestern New Mexico, Irwin-Williams (MS) describes a separate Oshara tradition of the Archaic or Desert complex. In this region the Desert economy is rather late, emerging only after the disappearance of the late Plano hunters sometime after about 8000 B.P. The early phases of this tradition, dated between about 7500 B.P. and 5300 B.P., lack grindstones. Irwin-Williams argues for a broader spectrum of economic activities than that characterizing the earlier Paleo-Indian hunting groups, largely on the basis of the tool kit and the patterns of settlement. It is only after 5000 B.P. that shallow-basined grinding slabs and manos appear, and these ground stone tools increase in frequency through the late phases of the tradition, dated between 3800 B.P. and 400 A.D. Domestic maize seems to have been added to the diet sometime after 3800 B.P., but here too it was probably a relatively minor increment in an eclectic gathering diet.

Irwin-Williams defines a third Desert-Archaic variant in southeast New Mexico and southwest Texas, known as the Hueco complex. Here too an Archaic economy seems to have appeared rather late following the abandonment of the area by Plano hunters; the earliest dates on the Hueco complex are on the order of about 4000 B.P. At Fresnal Shelter, whose beginnings are dated to at least 3600 B.P., a variety of projectile points, scrapers, and choppers occur with manos and metates. The organic remains consist of a variety of wild plant foods including cactus, sotol, grass seed, and wild squash, as well as domestic maize and beans.

In western Arizona and southeastern California, Irwin-Williams's fourth subarea, the definition of the Desert-Archaic economy and its relationship to putative early hunting groups is hardest to define. A number of chipped stone assemblages, mostly from unstratified surface deposits, have been identified from this region. As mentioned above, Heizer (1964) is dissatisfied with these surface assemblages and has reservations about their significance as proof of early occupation. According to Warren (1967) and Rogers (1939), the various materials can be lumped together as the San Dieguito complex, defined by a tool kit including various types of scrapers, knives, and projectile points. The dating of most of these materials is uncertain. Bennyhoff (1958) has suggested dates ranging back as far as 11,000 B.P., while others (Wallace 1964) give a more conservative estimate of about 9000 B.P. A radiocarbon date of 9640 B.P. on shell from natural shoreline deposits at Lake Mohave provides a maximum possible age for artifacts occurring on the surface of the beach, but there is no evidence that the artifacts are in fact this old (Heizer 1964). There are only two other reasonably well defined contexts in which such materials can be dated. The basal layer at Ventana Cave (Haury 1950) has been said to contain San Dieguito materials in conjunction with Clovis or Folsom items, which would give this material a date as early as 11,000 B.P., coeval with the earliest of the big game hunting traditions. Irwin-Williams questions the identification of the Ventana remains, however, and appears to accept the dates of 8500–9000 B.P. from the C. W. Harris site in California

(Warren and True 1961) as the earliest that are reliable for San Dieguito material.

The economy of these assemblages is also subject to question. Irwin-Williams includes San Dieguito in her grouping of broader spectrum Archaic economies, but other authorities such as Willey (1966), Warren (1967), and Jennings (1974) see the artifacts as pointing to a generalized hunting oriented economy. As stressed above, functional interpretation of the chipped stone artifacts is not entirely clear. But grindstones and related plant processing equipment are certainly absent from the C. W. Harris site, and they are absent or extremely rare in other surface assemblages of San Dieguito–Mohave type, which are presumed to represent the earliest occupation of southeastern California (Warren and True 1961). It is only in Pinto Basin assemblages (Campbell and Campbell 1935) derived from these early groups that grinding equipment becomes a regular and significant part of the assemblage. One Pinto Basin location, the Stahl site (Harrington 1957), has provided remains of seven circular structures which, along with storage pits and concentrated debris, seem to indicate long-term occupation. In addition to a variety of chipped stone forms, several types of manos and milling stones were recovered. The site, however, is undated.

The Pinto Basin assemblage with its more characteristic Desert-type tool assemblage and a population estimated to be greater than that of the antecedent San Dieguito phase (Wallace 1962) is difficult to date. It has been estimated to be as early as 9000 B.P. by Bennyhoff (1958), a date which seems too early in light of the dates from the C. W. Harris site. Meighan (1959b) has estimated the beginning of this phase at about 8000 B.P., which appears more plausible, while Wallace (1962) has argued for a date as late as 5000 B.P. In any case, subsequent cultural development in southeast California seems to have remained in the general Desert-Archaic economic pattern, with grindstones forming a significant part of the archaeological assemblages until the historic period (Wallace 1962).

The archaeological sequence of Baja California seems to have been derived from the Desert tradition described above. There is some possibility that the early San Dieguito or Lake Mohave groups penetrated the peninsula, but the first well-established archaeological assemblages (dated to about 7000 B.P.) are of Pinto Basin type, displaying milling stones and manos. Later occupants of the peninsula appear to have broadened the range of foods gathered. Extensive coastal shell middens are reported from the late prehistoric period (Massey 1961).

Farther north in California, the prehistoric sequence resembles the Archaic sequence of the eastern woodland and coastal regions. As in the east, a fairly rich resource base of wild vegetable and aquatic foods seems to have permitted the emergence of dense populations and large and sedentary human groups. At the same time it appears that the richness of the wild resources led to cultural resistance to the adoption of a farming economy, with the result that California Indians remained in an intensified hunting-gathering economy into the historic period (Kroeber 1917). Meighan (1959a) has divided this Archaic economy of California into three broad patterns. In parts of California, as indicated above, an economy of Desert type with primary dependence on plant seeds and small game occurs, while in the central valleys and mountainous areas of the state a different pattern, based on acorn processing, riverine fishing, and the hunting of deer and elk, was characteristic. Along the coast, a variety of economies emerged, focusing on shellfish, fish, or the hunting of sea mammals.

There seems to be fairly general agreement that this broad-spectrum gathering pattern, as evidenced by vegetable food processing tools along with shellfish and other aquatic remains, is of recent origin, dating after 7000 B.P. More controversy surrounds the origins of the late cultural traditions. One possibility is that the broad-spectrum economies of this time are derived from an ancient, indigenous hunting stratum represented by the very sporadic finds of Clovis points or chipped stone materials of San Dieguito type discussed above (Meighan 1959a). However, in most parts of California the evidence for this early occupation is sparse and usually derives from surface material of questionable affinities and date. Another approach to the prehistory of California has been to presume that much of the state was occupied only fairly late by groups pushed by population or other pressures out of adjoining regions. Heizer (1964), among others, has argued that the marked cultural and linguistic diversity of the California groups may reflect the repeated late migration of groups from other regions into available microhabitats.

The prehistory of California after 7000 B.P. has been divided into three broad chronological units (Heizer 1964): an early period extending to approximately 4000 B.P.; a middle period from 4000 B.P. to 2000 B.P.; and a late period from 2000 B.P. to historic times. The archaeology of these periods demonstrates fairly clear trends toward increased exploitation of vegetable and aquatic resources, greater geographical distribution of aboriginal populations, and growing size and density of population aggregates (Heizer 1964; Meighan 1959a,b; Wallace 1954, 1955; Beardsley 1948).

The early period is known primarily from the south central and south coastal regions and from the interior of central California. In southern California there are many dated sites belonging to the La Jolla or "Milling Stone horizon," dated between 7000 and 4000 B.P. The economy is presumed to have been one of seed grinding and hunting, with shellfish and other ocean resources as secondary foods (Wallace 1955; Heizer 1964). At the Little Sycamore site (Wallace 1954) a deep midden produced 116 metates along with refuse of shellfish and the bones of deer, sea otter, and waterfowl. Fish, however, was relatively scarce, and the use of water-based resources seems to have been less intensive than in later periods. On Santa Catalina and San Nicholas islands, beginning in the sixth millennium B.P., there is evidence of the hunting of sea mammals or at least of the use of stranded sea fauna.

In central California, contemporary populations of the Windmiller phase (Beardsley 1948) of the interior valleys seem to have developed a combined seed-processing, hunting, and fishing economy. Again, both metates and mortars are common, and bone fishhooks, gorges, and spears are found. The central California coast does not yet seem to have been exploited intensively, although some evidence of coastal economies may be hidden under the extensive shell middens of later periods. Farther north, both on the coasts and inland, the evidence for an early period seed-collecting culture is thin (Heizer 1964), and in general the early occupation is less well documented than that of later phases.

After 4000 B.P., in the middle and late periods, a number of changes are evident. In the first place, certain areas of California—notably the central coastal and northern coastal and interior regions—which had previously been only sparsely inhabited or not inhabited now provide evidence of substantial populations. In general, settlements throughout the state seem to have been larger and more permanent than those of the early period (Heizer 1964; Meighan 1959a), and there is evidence of increasing local stylistic diversity. Heizer (1964) notes that there are large numbers of burials which display skeletal trauma or imbedded projectile points indicative of increasing warfare.

Several economic changes are also evident. The characteristic hopper-mortar combining basketwork and ground stone, used historically in the processing of acorns, makes its first appearance during the middle period (Wallace 1955); and it may be that the extensive use of acorns, so characteristic of later California Indians, stems from this time. It is also during this period that the historic maritime exploitation pattern seems to have emerged (Meighan 1959b): barbed harpoons, net weights, and special blunt bone or antler projectile points for fishing all appear (Bennyhoff 1950; Heizer 1964), and populations with economies geared to the hunting of marine mammals appear to have established themselves on the offshore islands at this time as well. At the site of Little Harbor on Santa Catalina Island, a shell midden approximately one meter in depth, dated at its base to about 3900 B.P., consists almost entirely of shellfish and the bones of marine mammals including dolphin, porpoise, and seal, the hunting of which is presumed to have required the use of boats. The marine mammals compose fully 97 percent of the bone recovered from the site (Meighan 1959b). In addition, along much of the coast of central and northern California substantial shell middens appear for the first time during the middle and late periods, dating mostly after about 3000 B.P. (Heizer 1964; Pilling 1955).

Like California, the northwest coast and interior plateau of Washington and Oregon provided a relatively rich subsistence base for populations willing to use secondary resources such as riverine and vegetable foods. By the historic period, large, dense, and settled populations with many of the sociopolitical attributes of agricultural groups had emerged, basing their economy on the intensive exploitation of these resources— particularly the migratory salmon, which could be harvested in large numbers during their seasonal runs (Drucker 1955a,b). The prehistory of this region has been reviewed by a number of authorities (Cressman et al. 1960; Butler 1959; Daugherty 1962; Swanson 1962; Sanger 1967; Browman and Munsell 1969; Leonhardy and Rice 1970), and, although there is some confusion and difference of opinion about the precise dating and description of various stylistic groupings and phases and about the number of chronological divisions which are to be recognized, the broad economic trends are fairly clear. In general, as elsewhere, predominantly hunting cultures are early, while cultures with strong riverine orientation involving the intensive exploitation of fish and shellfish tend to be late (although the fishing life-style is firmly documented earlier here than in most other parts of the United States). Vegetable processing tools also occur primarily in the latter part of the sequence and increase in numbers and variety with time. In addition, most of the coast seems to have been occupied only in very recent prehistory. It is also evident that there is at least a general trend toward greater numbers of sites and increasing permanence of occupation.

Browman and Munsell (1969) divide the prehistory of the northwest into the following periods. Period 1 (15,000 to 10,500 B.P.) is defined on the basis of a small number of sites on the Columbia Plateau and in nearby regions of Idaho, which are among the earliest well-dated sites in North America. These sites demonstrate a subsistence economy based largely on the hunting of big game animals such as deer, antelope, mountain sheep, horses, and camelids, with a lesser degree of dependence on smaller forms such as grouse, rabbits, and some small carnivores. Wilson Butte in Idaho (Gruhn 1961, 1965), discussed above, suggests a subsistence base of this type as does Jaguar Cave in Idaho (Sadek-Kooros 1966), where levels dated between about 11,900 B.P. and 10,400 B.P. include the bones of a number of large and small mammals, with 80 percent of the bone coming from mountain sheep, deer, and antelope. At Marmes Rockshelter on the Columbia Plateau (Fryxell et al. 1965), in levels dating beginning 10,800 B.P. or earlier, some river mollusc shell was found, but the overall analysis of the faunal and tool assemblage suggested to the excavators that the economy was again predominantly a hunting one.

Browman and Munsell's period 2 (10,500 to 8000 B.P.) is again described by them as one in which the economy was predominantly oriented toward the hunting of big game such as bison, deer, elk, and antelope. There are hints, however, of the broadening of the spectrum of economic activities. The earliest levels of the Five Mile Rapids site at the Dalles Reservoir on the Columbia River, dated to about 9700 B.P., have provided evidence of a fairly varied hunting pattern, the remains including deer, beaver, a variety of small mammals, some birds, and even a seal. Some mollusc shells also occur, and a bone harpoon from this level may also indicate the beginning of aquatic hunting (Cressman et al. 1960). At the Lind Coulee site (Daugherty 1956), radiocarbon dated at approximately 8700 B.P., the remains of bison, deer, and a variety of smaller mammals along with chipped stone artifacts indicate a predominance of hunting; but here manos are present, suggesting some attention to the processing of vegetable foods. Similarly, at the Milliken site (Borden 1966), layers dated between about 9000 B.P. and 8200 B.P. yield a tool assemblage suggesting a predominantly hunting economy, although both anvil stones and burnt chokecherry pits point to some dependence on vegetable foods. Most striking, perhaps, the Millard Creek site (about 8300 B.P.) on Vancouver Island (Capes 1964) provides a midden fairly rich in shellfish and fishbone.

Period 3 (8000 to 6500 B.P.) is described by Browman and Munsell as the time of the florescence of the Old Cordilleran tradition. According to their reconstruction the high percentages of faunal remains and preponderance of hunting tools indicate that hunting was still the primary activity for most populations. On the other hand, manos and milling stones now begin to occur at sites in the region under study and edge-ground cobbles are found. Butler (1966) has argued that these tools may be the functional precursors of the mortar and pestle, and that, like the mano and metate complex, they may indicate the increasing importance of vegetable foods in the diet. There is also some direct evidence of vegetable foods: hazelnuts occur at Cascadia Cave (Newman 1966) and levels of the Milliken site (cited above) dating from this period display the gathering and storage of chokecherries.

At most sites of this period aquatic resources still represent only a minor ingredient of the diet. Fish and shellfish occur sporadically at sites on the Columbia River. There are two sites documenting more extensive use of fish and shellfish at this time, but Browman and Munsell emphasize their uniqueness. At the Marmes Rockshelter, levels dated to about 7600 B.P. (Fryxell et al. 1965) apparently contain fairly large quantities of shell, and at the Five Mile Rapids site levels dated to about 7700 B.P. display a very heavy dependence on salmon.

The major steps toward the broadening of the economic spectrum seem to have been taken during Browman and Munsell's periods 4 and 5 (not differentiated by them but representing the period between 6500 and 3500 B.P.). It is at this time that widespread and extensive exploitation of salmon and other fish becomes evident both from the quantity of bone in the refuse and the elaboration of fishing equipment. River mussels also become important in the diet of much of the region only during these periods. Mortars and pestles appear, as well as digging sticks and storage pits, all suggesting increased dependence on the processing of wild vegetable foods. Browman and Munsell suggest two explanations for this economic transition: first, that human cultures simply elaborated culturally to fill vacant niches; and, second, that the extinction of the big game fauna forced increasing reliance on new resources (Fryxell and Daugherty 1963). It seems to me that this transition is surprisingly late if it is to be related primarily to faunal extinctions, although the extinctions may have been one of the factors that, along with growing population, forced human populations to use new resources. In any case, as Browman and Munsell themselves point out, the expansion of riverine economic patterns seems to have been a matter of necessity rather than of choice.

In period 6 (3500 to 2000 B.P.) there is further expansion in the use of salmon, shellfish, and vegetable foods (the latter as evidenced in part by the spread of hopper-basket mortars). Browman and Munsell argue that it is at this time that the Northwestern Plateau economy achieved its historic form. Despite the early Vancouver Island finds, most coastal occupations in the northwest first appear only during this period.

The Major American Domesticates and the Mexican Sequence

The most important part of North America in the development of early agriculture is, of course, Mexico. Mexico, or the Mesoamerican region in general, has contributed by far the largest number of the native North American cultigens (see Mangelsdorf et al. 1964, for a fairly complete crop listing), and this area, on present evidence, appears to be the home of the three most important native food-crop plants: maize, beans, and squash. Mexico also has the longest archaeological history of domestication in North America, and it is the one portion of the continent where a clear case can be made for the independent, indigenous development of agricultural technology, a case reasonably unclouded by controversies about the diffusion of crops from other regions or even about the possibility of stimulus diffusion.

The archaeological sequence of domestication in Mexico is known primarily from the intensive study of three regions: the state of Tamaulipas in the northeast; the Tehuacán Valley in the south central region; and the Valley of Oaxaca also in the south. Two of the studies, those from Tamaulipas and Tehuacán, are primarily the work of Richard MacNeish and his associates (MacNeish 1958; Byers [ed.] 1967; MacNeish 1970; MacNeish 1972). The third and most recent study, that of the Valley of

Oaxaca, was conducted under the direction of Kent Flannery (Flannery et al. 1967; Kirkby 1973; Flannery 1973). In each of the three regions, dry highland caves have provided organic preservation that is among the best yet encountered by archaeologists anywhere in the world. As a result, in all three areas we have good, if sporadic, samples of organic refuse from which it is possible to piece together crude pictures of economic evolution during approximately the last 10,000 to 12,000 years. The pattern that emerges suggests that early experiments in cultivation began in a context of broad-spectrum exploitation approximately 7,000 to 8,000 years ago. But in each region agriculture appears to have developed very slowly as an economic strategy so that it is only by 4000 B.P. or later that we have evidence of sedentary populations making extensive use of domestic crops. The similarity of the three regions in this regard is striking and suggests that we have obtained a reliable sample of the pattern of economic evolution in highland Mexico as a whole.

Tracing the actual history of individual domesticates is still a complicated undertaking. One problem is that many of the major American domesticates have fairly complicated evolutionary histories. Unlike the major Old World cultigens, which can often be distinguished from well-documented wild ancestral forms on the basis of a small number of specific morphological changes, many of the Mexican cultigens are removed from their wild ancestry by long and complicated patterns of evolutionary change. In many cases, the wild ancestors of modern cultivated forms cannot even be identified with certainty, and as a result botanists often lack a knowledge of the morphology, habits, and distribution of the ancestral wild form from which to begin their interpretation of the domestication process. In addition, if the wild ancestry of a particular crop is not precisely known, collections of modern wild species are of limited value for comparative purposes. Under these circumstances the determination of the wild or domestic status of early archaeological specimens is extremely difficult.

Another related problem is that many of the changes that have occurred under domestication have been of a gradual, quantitative nature. Often even when the wild ancestry of a cultigen is known or presumed, the wild and primitive-cultivated forms can be distinguished from one another only on the basis of statistical analysis of reasonably large populations. Archaeological samples are often inadequate to allow positive discrimination. For all of these reasons, a good deal of controversy exists about the interpretation of many of the early Mexican archaeological specimens.

These problems are exacerbated by the uneven nature of the archaeological record itself. The high quality of organic preservation sporadically encountered in the three areas mentioned above highlights the degree to which our knowledge of the region as a whole is fragmentary, limited both by the irregularity of good organic preservation and by the lack of additional regional studies of equal intensity. There is the further problem that while the broad outline of economic change in these three regions is quite firmly established, many of the details of prehistory are obscure even in these well-studied sequences. The work from Oaxaca has not yet been published fully; and the studies of Tamaulipas and Tehuacán, while published in more detail, have raised numerous questions of interpretation and identification, many of which are not yet resolved or even reported in the literature. Flannery (1973) has recently undertaken the unenviable task of sorting out and then evaluating the variety of published and unpublished data and criticism from Mesoamerica. The brief historical outline of the major cultigens which follows is based largely on his synthesis.

The most important of the Mesoamerican cultigens, maize (*Zea mays*), is at the same time one of the most difficult to trace archaeologically since its botanical ancestry is a matter of some dispute. Until recently the most prominent theory concerning the origin of maize was that of Paul Mangelsdorf (1947), who postulated that modern maize was descended from a species of wild maize now long extinct and not from any known modern wild species. His hypothesis opened the door to a good deal of speculation about where in Mesoamerica (or in South America or even in the Old World) such a wild species might have existed and where and how many times it might have been domesticated. Recently, however, a growing number of botanists have been following the lead of W. C. Galinat (1971) and George Beadle (1972 and in press) in recognizing the wild grass teosinte (*Zea mexicana*) as the form from which domestic maize is derived. If they are correct, the historic distribution of teosinte in the semiarid, semitropical uplands between the state of Chihuahua in Mexico and Guatemala may provide some clue as to the approximate location of early centers of domestication (Flannery 1973). However, as Flannery warns, the present distribution of the grass may be misleading since modern forms of teosinte typically are weeds and may be heavily adapted in their distribution to human disturbances of the landscape (and possibly altered morphologically from their original wild state, as well, by artificial selection patterns or by crossbreeding with maize in recent prehistory).

The archaeological record of teosinte use by early human populations in Mexico is unfortunately very thin. Pollen believed to be that of teosinte occurs in archaeological layers dated to approximately 9000 B.P. in Guilá Naquitz Cave, Oaxaca (Schoenwetter cited in Flannery 1973), but apparently no seeds or other parts that might have provided more direct testimony of the nature and extent of human use of the plant were recovered. Teosinte seed has been found in archaeological layers dated to about 7000 B.P. at Tlapacoya in the Valley of Mexico (Lorenzo and Gonzales 1970); but, as Flannery points out, true maize is already known by this date. The earliest known maize cobs, also dated to approximately 7000 B.P., occur in the Tehuacán sequence. Mangelsdorf (Mangelsdorf et al. 1967) has argued on the basis of the small size and uniform nature of the cobs recovered that these specimens represent the wild maize he postulated, and it is his contention that the further development of maize, observable throughout the later phases of the Tehuacán sequence, demonstrates the evolution of the domestic form from its wild ancestor within the Tehuacán Valley itself. But if Beadle and Galinat are correct, as most botanists now believe, the early Tehuacán corn is already domestic. Since there is no evidence of teosinte at any time in the Tehuacán sequence, it would appear that maize arrived in Tehuacán from another region where it was already under cultivation. The diffusion of early cultivated maize at this period does not seem to have extended to the north however. Maize does not occur in the Tamaulipas sequence until approximately 2,000 years after its first appearance in Tehuacán (MacNeish 1958).

The archaeo-botanical history of the Mesoamerican squashes (*Cucurbita*) is equally difficult to unravel. Three domestic squash species (*Cucurbita pepo, C. mixta,* and *C. moschata*) can be ascribed considerable antiquity in Mexico, but apparently the interrelationships between the wild and cultivated cucurbits are not well enough known at present to allow early domestic forms to be reliably distinguished from potentially ancestral wild types in archaeological refuse, particularly if only the seeds are recovered. At present, *C. pepo* appears to have the longest history. A single seed considered to be "pepo-like" has been found in the Oaxaca sequence dated to about 10,000 B.P. The same sequence has produced a larger sample of seeds and peduncles dated between about 9400 and 8700 B.P. *Pepo* squash are also present in Tamaulipas by about 9000 B.P. Although discrimination is difficult, these early specimens are presumed to be wild. Specimens in Tamaulipas after 7000 B.P., however, appear to be domestic (MacNeish 1958). At Tehuacán, *pepo* squash does not appear until about 7000 B.P. and then only a single "wild" seed is encountered; this early specimen is questionable since the species is not found again in this region for approximately another 2,000 years.

In marked contrast to *pepo* squash, *C. mixta* is known earliest in Tehuacán and it is only the Tehuacán sequence that the species is known to have great antiquity. The earliest of such specimens date to about 7000 B.P., but the sample is small and hard to interpret. Samples from the same region dated to about 5000 B.P. are considered to provide more reliable evidence of early domestication. In Tamaulipas, the earliest *mixta* specimens date from the Christian Era (Cutler and Whitaker 1967; Whitaker et al. 1957).

Cucurbita moschata is the last of the three species to occur, appearing in Tehuacán possibly as early as 6500 B.P. and with more certainty between about 6000 and 5000 B.P. In Tamaulipas, *moschata* squash do not occur until after 4000 B.P. (Cutler and Whitaker 1961, 1967).

The archaeological record of beans (*Phaseolus* spp.) is somewhat easier to interpret because the wild ancestry of the domestic species is more clearly defined and criteria for distinguishing wild and domestic forms relatively well established (Kaplan 1965). Wild runner beans occur in Oaxaca in levels dated between 10,700 B.P. and 8700 B.P. and in Tamaulipas between 9000 and 7500 B.P. These early forms appear gradually to have been replaced by the cultivated form (*P. coccineus*), but apparently the date of this transition is not clear. Runner beans do not occur in Tehuacán until about 2000 B.P., at which time they are clearly domestic (Kaplan 1967). The common bean (*P. vulgaris*) is known archaeologically only in domestic form; it occurs in Tamaulipas between 6000 and 4300 B.P. and at approximately the same time in Tehuacán. Tepary beans (*P. acutifolius*) occur early only in the Tehuacán sequence, where they appear in domestic form about 5000 B.P. (Kaplan 1967).

A variety of lesser cultigens are also traceable archaeologically in the

Mexican sequences, only a few of which can be discussed. Avocados (*Persea americana*) occur possibly as early as 9200 B.P. in Tehuacán. A fairly good sample of avocado seeds presumed to represent a wild species is found in the same area between about 9000 and 7000 B.P. and the seeds continue through the Tehuacán sequence. The seeds show clear signs of morphological domestication by 3500 B.P., but it is not clear when in the course of the Tehuacán sequence the domestication was begun. Chili peppers (*Capsicum annuum*) are found in Tehuacán with dates perhaps as early as 8500 B.P. More firmly dated specimens occur in the same region between 8000 and 7000 B.P., but all of the early specimens appear to be wild. Domestication is first in evidence in Tehuacán at about 6000 B.P. (Smith 1967). Amaranth (*Amaranthus* sp.) is found by 6500 B.P. (and possibly earlier) in the Tehuacán sequence, but the dates of the beginnings of cultivation here are not clear.

One other interesting case concerns *Setaria*, a grass which appears to have supplemented teosinte/maize in the early phases of maize harvesting but gradually decreased in importance as cultivated maize strains slowly improved. Setaria occurs in Tehuacán possibly as early as 9000 B.P. and more clearly between 9000 and 7000 B.P. The grass seems to have had considerable economic significance in Tehuacán in the early portions of the sequence and may have been domesticated; in fact Smith (1967) suggests the possibility that domestication occurred by 8000 B.P. In Tamaulipas, where the grass was also harvested, Callen (1967) has noted an increase in the size of the grain, suggestive of domestication by about 5500 B.P.

From the partial review provided above, a very interesting pattern appears. Despite the marked temporal parallelism in the emergence and perfection of agricultural economies in the three regions, there is a noticeable lack of correspondence in the dates at which particular cultigens occur. Maize, *mixta* and *moschata* squash, and tepary beans are among the early cultigens in Tehuacán, each species occurring here far earlier than in the other sequences. Conversely, *pepo* squash and runner beans occur earlier in other sequences. Of the major domesticates, only common beans occur in the different regions at approximately the same time, and the dates of these early beans suggest that they were not among the earliest cultigens in any of the regions. The implication of this pattern is that the three sequences represent parallel economic adaptation rather than diffusion of crops or agricultural technology from region to region, and it is therefore reasonable to infer that these regions were undergoing similar pressures for economic change.

The source of these pressures however is debatable. Flannery, who espouses a population-density disequilibrium model for the development of agriculture in the Middle East (1969, 1973; see chapter 4), resists applying the same model to Mesoamerica on the grounds that the apparent population density and patterns of population growth are not sufficient to support a population pressure model. I believe, however, that the overall pattern of economic prehistory in Mexico and in particular the economic and demographic events in Tehuacán and Tamaulipas support the supposition that agriculture was only one part of a series of economic adjustments that point to population pressure.

The nature of the major Mexican cultigens is one of the key pieces of evidence in this argument. The three main Mexican cultigens—maize, beans, and squash—are all, in the wild state, what Flannery (1973:307) calls "third choice" (i.e., low priority) foods, used primarily in emergencies. Their primary value lies in their ability to be stored, their ready tolerance of habitats disturbed by man, and their rapid response to human intensification practices and artificial selection. Their main qualities are that they are expandable in quantity and density of production, not that they are desirable foods. Kaplan (1967:202) has suggested that the main economic significance of beans has probably always been that they provide vegetable protein in a readily storable form (dry seeds). Since for most human groups this is not a choice or palatable way to obtain protein, it is reasonable to assume that beans came into intensive use only when other protein sources were scarce or had been exhausted, or when a premium on storage had developed. The wild squashes (Flannery 1973:301) have flesh that is bitter or thick and dry, and so originally only the storable seeds would have been of value as food. But the squashes do well as weedy camp followers in that they tolerate soils disturbed by man. Similarly, teosinte is an unattractive plant, a species unlikely to have excited the interest of human populations while other more palatable foods were readily available. As Flannery has pointed out (1973:290) teosinte is used today only as a starvation food. The plant is difficult to harvest efficiently because of its brittle rachis and short period of peak maturation. Like the Old World cereals, it requires extensive processing once harvested to render it edible; but it is apparently even more difficult than the Old World cereals to process.

Thus the domestication of maize from teosinte amounts to a deliberate attempt to increase the availability of an emergency ration whose prime value was that it responded to such efforts to concentrate and improve it (Flannery 1973:296). Moreover since figures cited by Flannery suggest that teosinte in the wild does not have the naturally dense productivity of the Old World cereals and that domestic maize did not become a truly efficient staple for several millennia after its first appearance in the archaeological record (see also Kirkby 1973), it would appear that considerable human effort was expended over a long period of time to improve the productivity of this unfavored but responsive cultigen. The implication seems to be that the pressure for increased caloric productivity was both significant and fairly continuous.

There is other evidence of population pressure (though not necessarily of high absolute population density) in the Mexican archaeological sequence. The primary point is that the overall pattern of economic change is roughly similar to that in other regions in that it displays a number of trends that suggest a gradually broadening spectrum of exploitation. The earliest occupants of the region seem to have been "hunters"—at least the associated organic remains and the types of tools represented point to an emphasis on mammalian fauna, while the lack of preserved vegetables and the absence of grinding tools suggests that vegetable foods were not as important an element in the diet as they were later to become. Also, no fish or shellfish are found in the earliest locations and the absence of early dated coastal sites seems to reflect a lack of interest in aquatic resources. Such hunting oriented groups may be seen to have occurred as early as 20,000 B.P. or more if the finds from Tlapacoya and Valsequillo (discussed above) are considered. They are fairly clearly present between about 11,000 and 12,000 B.P., as evidenced by the distribution of Clovis points and of chipped stone assemblages with projectile points of other types (Weaver 1972; Willey 1966; Coe 1962, 1966).

In the dry highland areas of Mexico, beginning by about 10,000 B.P. or slightly earlier, this life-style seems increasingly to have been supplemented or displaced by new economic trends reminiscent of (and possibly culturally related to) the Desert tradition to the north (Weaver 1972; Willey 1966); both the increasing frequency of preserved vegetable remains and the appearance of grinding tools for plant processing indicate a general shift toward the greater use of plant foods in the diet.

Between 9000 and 10,000 B.P., Indians in Oaxaca were exploiting a wide variety of vegetable foods including acorns, pinyon nuts, mesquite beans (*Prosopis*), prickly pear (*Opuntia*), hackberries, maguey (*Agave*), and a variety of other species (Flannery et al. 1967). Comparable groups of wild plant species have been recovered from contemporary or slightly later deposits at both Tehuacán and Tamaulipas.

This does not imply, incidentally, that a hunting "stage" was followed by a gathering "stage" in any simple or absolute sense. The earliest dates in the Oaxaca sequence, for example, suggest that the broad-spectrum use of vegetable resources may have begun here earlier than in Tehuacán and Tamaulipas; hence varying economic styles may have overlapped in time. Moreover, as Flannery (1967) has pointed out, early "hunters" in Tehuacán already displayed a fairly varied strategy in the hunting and collecting of large and small animal species. The point here, as elsewhere, is simply that the nature of the early sites implies *relatively* heavy and widespread reliance on meat and probably on large game; the trend through time seems to have been one in which reliance on vegetable foods became both locally more intense and geographically more widespread. It is in the context of this general expansion of the vegetable portion of the diet that domestic plants began to play a significant part in the Mesoamerican economy, particularly after about 7000 B.P.

On present evidence, neither the Desert pattern nor the early development of domesticates was shared by coastal regions of Mesoamerica; but here a parallel, if slightly later, pattern of aquatic exploitation emerged. Coastal shell mounds are found beginning from about the sixth millennium B.P. and these become more widespread in late preceramic and more recent periods (Benson 1968; Lorenzo 1955, 1961; Mountjoy et al. 1972; Brush 1965). In some regions, such as Vera Cruz, coastal populations even seem to have achieved large and relatively permanent settlements before agricultural villages emerged inland. This parallel development of large and permanent population aggregates spanning two entirely different economic zones seems to suggest that a tendency toward larger and more stable groups was a general historical development in which the domestication of crop plants played only one part. It is also noteworthy that agriculture seems to have diffused to the coastal regions only after aquatic economies were well developed (cf. Coe and Flannery 1964, 1967).

The role of population pressure in stimulating the evolution of domestication in Mexico can be seen somewhat more fully in a study of
the archaeological sequences from Tehuacán and Tamaulipas. In both
regions, interdisciplinary teams have provided fairly detailed pictures of
the relationships between economic evolution and changes in demographic and settlement patterns over approximately the last 12,000
years. For the two areas, MacNeish has provided fairly complete quantitative analyses of the changes occurring in each of these spheres. Many
of his methods and conclusions are subject to question, and the quantitative estimates he offers are almost certainly more precise than his data
warrant; but viewed in very general terms his studies are still among the
best empirical descriptions of incipient cultivation we possess and the
trends he describes are of interest.

The two sequences have a number of significant points in common.
In both areas the earliest archaeological sites, dated prior to about 9000
B.P., are made up of a few small, temporary camps belonging to groups
whose economy MacNeish sees as mainly oriented to hunting. This
interpretation is based on the predominance and density of bone in the
refuse, the paucity of vegetable remains, and the absence of equipment
for food grinding, which occurs only in subsequent phases. In both sequences, later phases display increasing amounts of wild vegetable
refuse and grinding tools in growing numbers. By MacNeish's estimate,
in each region the role of wild vegetable consumption in the diet increases relative to hunting before agriculture begins, and even after the
first cultigens appear wild vegetable foods continue to grow in importance. It is thus evident in both sequences that the primary trend is not
the increased use of cultigens at the expense of wild foods; rather it is
the increased use of vegetable foods in general relative to meat. In
Tamaulipas, for example, MacNeish estimates that meat, which constitutes 50 to 70 percent of the diet before 9000 B.P., gradually declines
in importance until it accounts for only 10 to 15 percent by about
6000 B.P. Wild plant foods increase in importance over the same time
span until they make up as much as 70 to 80 percent of the diet. Domestic crops represent no more than 10 to 15 percent of the diet as late
as 5000 B.P., and it is only after this date that agriculture begins to expand significantly in proportion to the gathering of wild vegetable
foods. It is also only after this latter date that coastal oriented economies occur in Tamaulipas. Similarly, in Tehuacán, the economy of the

early populations (pre-9000 B.P.) is reconstructed as consisting predominantly (70 percent) of meat. This figure gradually diminishes in the sequence until the late phases when meat accounts for only 15 to 20 percent of the diet. Wild plant materials increase in importance until about 6000 B.P., at which time they constitute somewhat over half of the consumed food. Between 6000 and 4000 B.P., wild plant foods continue to provide about half of the total dietary intake while agriculture gradually expands at the expense of hunting. It is only after 4000 B.P. that agriculture displaces plant gathering as the major economic activity.

In both regions the slow economic transition is accompanied by gradual changes in demographic structure and settlement patterns. The number and size of archaeological sites identified increases gradually with time, suggesting a pattern of population growth which begins well before the emergence of agriculture as the major factor (or even a significant factor) in the subsistence base. At the same time both sequences display gradual trends toward larger archaeological sites and increased density of remains, which suggests that increased group size and a greater degree of sedentism developed gradually beginning well before agriculture became the basis of the economy. That these slow demographic changes began prior to the time when domestication assumed a significant role in the economy shows that they were not primarily dependent on the perfection of domestication technology. And the very pattern of economic evolution, its step-by-step movement first toward low trophic level consumption and later toward increasing reliance on domestic crops that provided abundant but low-priority foods, is good evidence of the role of population growth as a stimulus to economic change.

The population pressure model alone, however, does not explain why agriculture began so much earlier in Mexico than in other parts of North America. Although the reasons for this pattern are not entirely clear, I would suggest a two-part hypothesis. In the first place, if the population pressure model is generally correct we would expect agriculture to begin in the western desert regions of North America where neither rich game resources nor abundant wild vegetable foods were available to absorb significant population growth. Yet even within this zone the pattern of early agriculture is irregular. Therefore, in the second place, I think we must turn to Flannery's (1968) discussion of genetic plasticity in maize as a stimulus to agriculture and argue that domestication would have occurred earliest in regions such as Mexico where increasing population pressure overlapped the distribution of plant species whose genetic plasticity permitted relatively rapid and positive response to increasing human attention.

6 THE NEW WORLD: SOUTH AMERICA

The reconstruction of prehistoric cultural evolution in South America presents a number of problems. One of the most significant difficulties is that there is a good deal of regional variation in the quality and quantity of prehistoric research. Some portions of South America, most notably the deserts along the western coast, provide what is probably the best organic preservation encountered anywhere in the world; and parts of this western desert are among the world's most intensively surveyed regions. Other areas, such as the Amazon Basin, provide very poor organic preservation and, along with much of the eastern two-thirds of the continent, are only now beginning to undergo intensive archaeological study. As a result, the prehistory of many regions can be reconstructed only by generalizing from very small archaeological samples or by extrapolating from the patterns which have been observed in the more thoroughly studied areas.

These problems are exacerbated by two other limitations of the existing data. First, the overwhelming majority of the data now available comes from surface surveys; the number of well-excavated stratigraphic sequences is very small. Second, radiocarbon dates are only now becoming available in any quantity for most of South America. In the absence of good stratigraphic sequences and reliable radiocarbon dates, reconstructions of South American culture patterns have been forced to rely heavily on cultural "horizons" or stylistic correlations to establish chronological relationships between events in different regions (cf. Lanning and Hammel 1961; Lynch 1967; Lanning 1967a; Willey 1971). Unfortunately new radiocarbon dates have begun to show that such correlations are not always reliable.

The lack of stratigraphy and firm absolute dates has also necessitated heavy reliance on "seriation" as a means of establishing relative chronologies among the industries of individual regions. Too often, however, such seriations have not utilized careful analysis of the type described by Rowe (1961) but have relied instead on preconceptions about the relative antiquity of certain tool types. A number of stone tool assemblages have been attributed great age because of their apparently "primitive" character (cf. Krieger 1964; Lanning 1970; Lanning and Patterson 1967; Willey 1971), and yet often the actual antiquity of these assemblages cannot be firmly established. As a result there is still considerable controversy about the proper order of the archaeological sequence in many regions (cf. Lynch 1974; Lanning 1967a; Bryan 1973, 1975; Willey 1971).

For anyone attempting to reconstruct the early history of plant cultivation in South America there are two additional problems. First, we simply lack accurate botanical knowledge concerning the origin of most of the major South American cultigens; and, second, there is an almost total absence of overlap or correspondence between the botanical information which is available and the archaeological record. There is, for example, almost no archaeological evidence that can reasonably be interpreted as representing incipient cultivation. Most existing archaeological evidence of cultivated plants is relatively late, and most cultigens are first recorded in regions which are clearly well removed from their probable centers of origin. Moreover, the early dated archaeological specimens almost invariably represent fully domesticated forms. There are no sequences which convincingly display transitional stages in the domestication of any major crop and almost no archaeological sites where the use of ancestral wild species is clearly documented.

Pickersgill and Heiser (in press) have recently provided a review of known archaeological and botanical information about the major South American cultigens which highlights some of the problems involved. (See also Pickersgill 1969; Heiser 1965; Cohen 1975b; Flannery 1973; and, for a more complete if somewhat outdated review, Towle 1961.) They conclude that there is now reasonably good evidence that a number of the major South American cultigens were independently domesticated in South America itself. Moreover, there is some evidence to support Harlan's (1971) description of early domestication in South America as an essentially decentralized process involving the domestication of a variety of crops in a number of separate regions. However, the precise time and place of domestication can rarely be projected from the botanical data, much less identified archaeologically. We cannot even attempt as yet to describe in any detail the ecological, demographic, or social context in which incipient domestication took place.

Manioc (*Manihot esculenta*) provides what is probably the best case in point. The crop is today a major agricultural staple in much of South America, particularly in the lowland tropics, and is associated historically and in late prehistory with the tropical forest cultures of the Amazon Basin. Related wild forms are widely distributed at low altitudes in the tropical regions of South and Middle America. The crop could have been domesticated at one or more times at any place along this range since it is not yet known which wild species within the genus *Manihot* are ancestral. Interestingly enough, despite its historic association with the rain forest, the crop probably was not first domesticated there; its wild relatives and its habits and structure (the tuber is drought resistant but susceptible to rotting in poorly drained soils) all suggest, on the contrary, a savanna environment (Spath 1973; Sauer 1969). However, the earliest archaeological specimens of manioc, already fully domesticated, come from the coastal deserts of Peru at a date of about 3000 B.P. (Towle 1961; Cohen 1975b). This area is clearly outside the possible wild range of the species. The only other early evidence for the use of manioc consists of pottery griddle fragments assumed to be used for processing bitter manioc. These occur in archaeological contexts in Venezuela and Colombia dated to between 3000 and 4000 B.P. (and possibly earlier). The association of these griddles with manioc, while probable, is not certain since they are known to be used with other wild food crops (Sturtevant 1969). On the other hand, if the griddles are associated with manioc they probably already represent a fairly advanced stage in the development of the crop rather than incipient domestication.

Sweet potatoes (*Ipomoea batatas*) are equally problematic since their wild ancestry is still a matter of considerable dispute. Mexico, the eastern Andes, and the Amazon Basin have all been named as possible centers of domestication (Nishiyama 1971; O'Brien 1972; Yen 1971), with Pickersgill and Heiser leaning toward Yen's arguments in favor of a South American center of domestication. Again, however, the first archaeological documentation of what are *probably* sweet potatoes comes from the western desert coast of Peru in sites dated to approximately 4000 B.P. (Lanning 1967a; Patterson 1971a,b; Cohen 1975b).

Peanuts (*Arachis hypogaea*), another major South American domesticate, are also derived from unknown wild ancestry, although recent studies (Krapovickas 1969) favor derivation from the eastern side of the Andes at altitudes below 1800 meters in northwestern Argentina or Bolivia. Again, the earliest archaeological documentation occurs in coastal Peru beginning about 3800 B.P. (Cohen 1975b). The starchy tuber *achira* (*Canna edulis*) is another cultigen of unknown origin which is presumed to have originated on the eastern slopes of the Andes or in the Caribbean area (Pickersgill and Heiser in press). It too is first recorded archaeologically on the coast of Peru, at about 4300 B.P. (Towle 1961; Patterson 1971a,b; Cohen 1975b).

Potatoes (*Solanum* spp.), the most important of the series of root crops that provided starchy staples for high altitude populations in the Andes in the prehistoric period, are also of unknown origin. Related wild forms occur in both South America and Mesoamerica, but potatoes were not cultivated north of Colombia before the Spanish conquest. Within the Andes, however, several species were cultivated and the relationships among these species are complex and obscure. Hawkes (1967) has argued that the original cultivated form was probably developed in the vicinity of Lake Titicaca in the Andean highlands of southern Peru and Bolivia, and Pickersgill and Heiser also appear to see this region as a center of early domestication. Again, however, the archaeological record is very poor. MacNeish et al. (1975) claim possible evidence of domestic potatoes in Ayacucho in the Peruvian highlands by about 5000 B.P., and potatoes have been identified along with a variety of other tubers from layers at Tres Ventanas Cave in the upper Chilca Valley of the Peruvian coast dated to 10,000 B.P. (Engel 1970). In the latter case, however, the strata are almost certainly mixed (see discussion below) and the Tres Ventanas Cave specimens do not appear to be generally accepted as being of great antiquity (cf. Pickersgill and Heiser in press; Flannery 1973). Potatoes have been reported in archaeological contexts dated to about 3000 B.P. on the Central Coast of Peru (Cohen 1975b). Elsewhere on the coast, however, they are reported only within the last 1000 years. Tubers tentatively identified as potatoes have been excavated at Chiripa in the Bolivian highlands and dated to about 2400 B.P. (Towle 1961; Hawkes 1967).

The high Andes have also produced two grain crops which were and are of considerable local significance. Both are species of the genus *Chenopodium* grown primarily at altitudes of 3,000 meters or more. *Cañihua* (*C. pallidicaule*) is a semi-domesticate retaining many of its wild characteristics, including the ability to disperse its own seeds

(Gade 1970). The plant appears to have no significant archaeological record. The other species, quinoa (*C. quinoa*), is a full domesticate of much greater importance. Again the ancestry of the domestic form is obscure, potentially involving a whole range of wild species in both South and Middle America. Pickersgill and Heiser argue in favor of a South American origin for quinoa, independent of the domestication of other members of the genus *Chenopodium* in Mesoamerica. Domesticated quinoa is claimed by MacNeish (MacNeish et al. 1975) in archaeological layers dating before 6500 B.P. in Ayacucho in the Peruvian highlands, but this claim has been questioned by Pickersgill (personal communication and in Flannery 1973). Pickersgill and Heiser indicate that the earliest firm archaeological record occurs at about 2000 B.P. in northwestern Argentina (Hunziker and Planchuelo 1971). Otherwise the species is documented by 1600 B.P. in the Andean highlands of Bolivia (Towle 1961) and occurs only after about 1000 B.P. on the Peruvian coast.

The other major South American food crops are the complex of maize, beans, and squashes common to both South and Middle America. Maize, once believed to be native to South America (cf. Mangelsdorf and Reeves 1939; Grobman et al. 1961), is now generally assumed to derive from Mesoamerica, since teosinte, the postulated ancestral species, does not occur in South America. However, neither the route nor the time of this diffusion is clear. Maize is known on the coast of Peru from a number of late preceramic sites by about 4000 B.P. (Kelley and Bonavía 1963; Lanning 1967a), and it has been identified in Chihua phase deposits in Ayacucho five hundred to one thousand years earlier (MacNeish et al. 1975). There is indirect evidence of maize cultivation in coastal Ecuador before 4000 B.P. (Lathrap 1973), but so far as I can determine there is as yet no direct evidence of maize cultivation in either Ecuador or Colombia (along the postulated route of diffusion) which rivals the antiquity of the Peruvian specimens.

Of the squashes, two species (*Cucurbita ficifolia* and *C. moschata*) are known to have occurred prehistorically in both South America and Mesoamerica, but neither species can be traced reliably to its wild ancestry and it is not known whether these species represent independent parallel domestication in the two regions or diffusion from one region to the other. Both species are traceable back to dates of about 4500 to 5000 B.P. on the coast of Peru (Towle 1961; Lanning 1967a;

Cohen 1975b). One cultigen, *C. maxima*, is known only from South America, but it appears very late in the archaeological record, being found only within the last 1,500 years on the Peruvian coast (Towle 1961). Pickersgill and Heiser suggest that maxima squash may be derived from a wild form, *C. andreana*, which grows wild today in Argentina and Bolivia. Another possible ancestral form is *C. ecuadorensis*, which has recently been identified in the wild state of Ecuador (Cutler and Whitaker 1969). Both of these wild species are known from deposits on the coast of Peru dated to about 4500 B.P. in conjunction with the early domestic species (Lanning 1967a; Cutler and Whitaker 1969; Cohen 1975b); however, no record exists of the evolution of either species under domestication and the time gap between the known archaeological specimens of *C. andreana* and *C. ecuadorensis* and their putative descendant species, *C. maxima*, is puzzling. An unidentified species of squash has also been claimed by MacNeish (MacNeish et al. 1975) to have been under domestication in Ayacucho by about 6500 B.P., but this identification has been questioned by Pickersgill, and Pickersgill and Heiser do not include it in their summary.

Among the beans (*Phaseolus*) two species, common beans (*P. vulgaris*) and lima beans (*P. lunatus*), are both known to have been cultivated in prehistoric times in South America. Common beans are known in the wild state in both Mexico and South America, and it is not clear whether they were domesticated independently in each area or diffused from one region to the other. Pickersgill and Heiser suggest that they may have been domesticated separately in the two regions. Domestic common beans occur in the Callejón de Huaylas in the highlands of Peru by about 7600 B.P., the earliest record for any cultigen in South America; however, the specimens at this date are already fully evolved domesticates (Kaplan et al. 1973). Wild forms of *P. lunatus* are also known in Mesoamerica and in South America, but the size differential between the larger South American "lima" beans and the smaller Mesoamerican "sieva" beans suggests the independent domestication of distinct wild races in the two regions. Again, the earliest archeological record of domestic lima beans comes from the Callejón de Huaylas at about 7600 B.P., where the specimens recovered are already fully domestic (Kaplan et al. 1973). Since Kaplan (1965) presumes that lima beans originated as cultigens somewhere east of the Andes, it would appear that these specimens, despite their early date, are geographically removed from the original area of domestication.

Of the jack beans (*Canavalia* spp.) two species are known from pre-Columbian deposits in South America, but only one, *C. plagiosperma*, has a long history there (Sauer and Kaplan 1969). Little is known about the derivation of this cultigen. Related wild species tend to come from moist lowland areas such as seashores and river banks. Pickersgill and Heiser conclude that it is not yet possible to establish whether jack beans were domesticated in Mesoamerica or in South America nor whether they were domesticated east or west of the Andes. The earliest archaeological record comes from the Peruvian coast, where specimens have been recovered from sites beginning about 4000 B.P. (Sauer and Kaplan 1969), and from Ayacucho, where specimens occur in layers dated between 3700 and 4800 B.P. (MacNeish et al. 1970).

Given the limited archaeological data, the tentative and controversial nature of many of the existing syntheses of South American prehistory, and the discrepancies between botanical inferences about the origins of the major cultigens and the archaeological record of their occurrence, it is clear that a number of serious limits are placed on attempts to reconstruct the socioeconomic context of early agriculture on the continent. Controversies about the interpretation of local sequences cloud attempts to reconstruct the economic prehistory of many regions and other regions can be described only from very scanty data. The relative scarcity of sites in most areas also makes it difficult to quantify trends in site density, and hence the independent estimates of population size and density which are needed to support the hypothesis about population pressure cannot be obtained. Most important, we are effectively prevented from dealing with *incipient* cultivation.

We are, however, in a position to outline broadly the economic history of the continent as a whole by piecing together the available evidence. More important, although incipient cultivation cannot yet be evaluated, we can reconstruct, sometimes crudely and sometimes with great precision, the patterns of economic evolution that preceded the adoption of agriculture in secondary areas and the patterns of economic change that accompanied the transition from incipient to full-scale cultivation.

The partial reconstruction that is possible, although admittedly tentative, does suggest a number of significant patterns which recall developments already witnessed on other continents and lend support to the thesis developed in earlier chapters. A significant part of the pattern in South America as in other continents is the nature of the cultigens them-

selves. The qualities of the maize-beans-squash complex which suggest
the use of these crops only under conditions of population pressure
have already been discussed. Such arguments apply even more force-
fully to the highland and the lowland tuber crops which replaced
maize as the starchy staple in portions of South America where maize
could not be grown. The tubers, particularly manioc, are prodigious
producers of calories, but they are foods which are neither widely
prized for their palatability nor particularly valuable for the quality,
as opposed to the quantity, of their contribution to the human diet.
All are likely to have been cultivated only under conditions where the
demand for caloric productivity was increasing sufficiently to offset
human reluctance to use them.

A population pressure model is also supported by the pattern of
events which can be seen to precede and accompany the emergence of
full-scale agriculture in South America. Despite the numerous controversies about the correlation, classification, and dating of individual
sites and complexes, a reasonably clear pattern can be outlined. The
early inhabitants of the continent can be shown to have been relatively
selective in their choice of foods and habitats; both choices seem to be
centered on the pursuit of the larger game animals. In later periods, after
about 9000 B.P., there is evidence of an expansion into unused niches,
accompanied by an increasing diversity in the use of resources including
greater consumption of vegetable foods and invertebrate and aquatic
fauna. This pattern can be reasonably well documented in highland
areas of South America and in the majority of the coastal regions.
Perhaps most striking is the fact that, along the coast, agriculture is
almost invariably preceded or accompanied by the emergence of intensive coastal and/or riverine economies. Unfortunately, a similar
pattern of ecological expansion can only be postulated—largely on the
basis of negative evidence in combination with extrapolation from
highland and coastal patterns—for much of the interior lowlands, including the Amazon Basin.

The Archaeological Record of Early Man in South America

As in North America, the major debate concerning early man in
South America concerns the time of his arrival on the continent. One
school, most notably Willey (1971), Krieger (1964), Bryan (1965,

1973, 1975), MacNeish (MacNeish et al. 1975), Lanning, and Patterson
(Lanning 1967a, 1970; Patterson 1966; Lanning and Patterson 1967),
tends to accept a small series of early carbon-14 dates and a large number of apparently primitive industries, some correlated with dated geological features, as proof of man's arrival on the continent at dates
ranging from 12,000 B.P. back to 20,000 B.P. or more. Another school
(Haynes 1974; Lynch 1974, 1976; Martin 1973) has challenged the
validity of the early carbon-14 dates and tends to dismiss the typologically primitive industries and the geological correlations used to date
them. They note that most of these early industries have been obtained
from undated surface contexts or from poorly controlled stratigraphic
sections. In some cases they challenge the validity of the artifacts themselves and, in others, they argue that the "tools" are in fact quarry
blanks or other types of crude stonework which are of indeterminate
age but similar to artifacts occurring under well-controlled conditions
in a variety of relatively recent assemblages. These scholars argue that a
very high percentage of the sites dated before 12,000 B.P. are thus
questionable in one manner or another, and they argue that, as in North
America, well-dated, indisputable archaeological sites appear abruptly
in the archaeological record at about 12,000 B.P. They therefore
postulate a rapid and relatively recent colonization of South America
by Paleo-Indian populations, paralleling (and deriving from) the rapid
expansion of the Llano or Clovis hunters in North America.

As I am not in a position to evaluate either the stylistic comparisons
or the dating methods in question, for present purposes these issues
must remain unresolved. However, a review of the major early sites
and complexes in South America (pre-9000 B.P.) and some of the
controversies surrounding them will serve to provide a description of
economic activities among early inhabitants of the continent to which
later economic developments can be compared.

One of the most problematic archaeological sequences, and one which
is a key to much of the interpretation of early man in South America, is
the sequence of the Río Pedregal in Venezuela. From sites along this
river, Rouse and Cruxent (1963; see also Bryan 1973) have described a
series of stone industries associated with fragments of river terraces
presumed to be of varying ages. From the highest and presumably oldest terraces in the sequence they have described the Camare complex of
crudely worked chopping tools, knives, and scrapers. The complex

completely lacks projectile points or any evidence of fine bifacial work. On middle terraces, similar tools persist; but a new complex, Las Lagunas, is defined on the basis of the addition of smaller bifacial tools, still crudely chipped, which are interpreted to be large crude knives or points for thrusting spears. The third complex, isolated on lower terraces, is the El Jobo complex, characterized by the addition of large, relatively well-made lanceolate projectile points. The fourth or Las Casitas complex is characterized by the further addition of smaller triangular projectile points. Following the classic Old World system of dating artifacts by association with river terrraces and believing that the sequence conforms to the probable order of evolution of stone tool industries, Rouse and Cruxent consider the Río Pedregal sequence to provide evidence of early industrial evolution in Venezuela. Both Bryan (1973) and Willey (1971) accept the terrace sequence as evidence of such local evolutionary patterns, but the validity of the terrace identifications and associations, and the identity of the first two tool complexes have been challenged by a number of other authorities (cf. Lynch 1974).

Similar controversies surround the interpretation of the Manzanillo complex, also in Venezuela (Cruxent 1962). The industry, found near Lake Maracaibo, consists of chopping tools, scrapers, and crude bifacially flaked tools, including handaxe-like forms and large knives. Rouse and Cruxent (1963) compare this assemblage to Camare. On occasion a single carbon-14 date of 14,400 years from Rancho Peludo in Venezuela is tentatively applied to the Manzanillo complex (cf. Lanning 1970), but there seems to be general agreement that the relationship between the industry and the carbon sample is not reliable.

Neither Manzanillo nor the Río Pedregal sequence yields any direct evidence of the economic pattern of the populations represented nor is there any means of establishing the actual antiquity of the industries described, unless the questionable Manzanillo date is accepted. These problems may be solved in part by dates on archaeological associations at a series of nearby sites. The most famous of these is the Muaco site (Royo y Gómez 1960a,b; Bryan 1973; Rouse and Cruxent 1963; Cruxent 1961), where stone tools including an El Jobo point have been found in "association" with the bones of extinct Pleistocene animals (mastodon, sloth, glyptodon, and native horse as well as a variety of other large and small mammals). Radiocarbon dates of burned bone from the Muaco assemblage suggest an age of 14,000 to 16,000 B.P. The assemblage, however, also contains pieces of glass and pottery and it is generally conceded that the archaeological associations at the site are unreliable. On the other hand, the dates coming from cut and burned bone are accepted by some authorities (Willey 1971; Bryan 1973; Forbis 1974) as representing human hunting activites, even if these cannot be safely attributed to the El Jobo populations. Lynch (1976), who questions the early dates on bone, nevertheless suggests that the Muaco site with its concentration of bone debris in what would prehistorically have been a muddy mire could represent evidence of early communal hunting.

A similar site nearby may provide a more convincing, early dated association between El Joboid hunters and extinct fauna. The Taima Taima Water Hole about a mile from Muaco has produced a stone industry, including portions of three El Jobo points (Cruxent 1967, 1970; Bryan 1973), in association with extinct fauna similar to those at Muaco. One of the points was in position suggesting that it was actually embedded in the flesh of a mastodon. On the basis of a series of radiocarbon readings provided by Tamers (1966, 1969, 1970, 1971), Bryan (1973) has suggested that the whole assemblage is datable to about 12,000–14,000 B.P., and he argues that in this case the stratigraphy of the site precludes the possibility of disturbance. Both Haynes (1974) and Lynch (1974), however, raise questions about Bryan's use of some of these dates, although they seem to be willing to accept a date for the association between 10,000 and 12,000 years B.P. Lynch (1976) suggests that the concentration of bone at Taima Taima could be indicative of communal hunting.

At a third site, Cucuruchu, about three kilometers from Taima Taima, another probable association between El Joboid points and extinct fauna has been found (Cruxent 1970; Bryan 1973). The fauna are described as similar to those from Muaco. In this case there is evidence that some of the materials are not in their original position and that the association represents a secondary deposition; yet neither artifacts nor bones show signs of the type of wear indicative of "rolling" or significant displacement and it is assumed that the disturbance is minor. The available dates on the Cucuruchu assemblage (ca. 3000–6000 B.P.), however, are clearly anomalous and accepted as such by all concerned.

Two other unrelated, and somewhat less spectacular, finds from Venezuela should also be mentioned since they are potentially of great significance. Sanoja (1963) refers to surface finds of fishtail-shaped or "folsomoid" projectile points from two regions. These represent the northernmost finds within South America proper in a widespread and highly homogeneous series of projectile points which are known from surface collections and excavations all along the Andes and in the pampas of southeastern South America. The homogeneity of the fishtail group is commonly recognized and it is accepted as a true archaeological horizon (cf. Bird 1970; Lynch 1967, 1974, 1976; Mayer-Oakes 1966; Shobinger 1972; Willey 1971). (For a complete list of occurrences, including finds in Colombia, Ecuador, Peru, Chile, Chilean and Argentine Patagonia, the Argentine pampas, southern Brazil, and Uruguay, see Lynch 1974, 1976; Shobinger 1972; Willey 1971.) The similarity between South American fishtail points and the Clovis horizon of North America has been noticed and the derivation of the fishtail style from the North American horizon is widely accepted. Although most of the known finds, including those in Venezuela, are undated surface materials, fishtail points have been excavated in contexts (discussed below) similar to those of North American Clovis and Folsom hunting camps and kill sites. Dates available in other regions of South America clearly place this tradition within the period between 9000 and 12,000 B.P. Unfortunately there is at present no way of relating this early horizon to the Río Pedregal sequence or the dated kill sites in Venezuela, nor is its relationship to putatively early industries in other regions entirely clear.

In Colombia, in addition to surface finds of fluted or fishtail points at Manizales in the central Cordillera and at Restrepo in the western Cordillera (Reichel-Dolmatoff 1965), there are a small number of sites which can be dated with varying degrees of reliability to the period prior to 9000 B.P. The Tumba de Garzón site (Bürgl 1957; van der Hammen 1958) is reported to reveal an association of human artifacts and extinct Pleistocene fauna, including ground sloth and mastodon, although no date is available and the actual human manufacture of the "artifacts" is subject to some question (cf. Willey 1971). However, Lynch (1974), while apparently conceding that the pattern of chipping is unconvincing, does suggest that human activity may be represented by the presence of manuports or imported stone in combination with the fact that the chipping or fracture patterns are confined to the imported materials. He concludes that although the site is problematic it could represent a valid association of human industry with Late Pleistocene fauna.

Somewhat firmer evidence of early man comes from the El Abra Rock Shelters north of Bogotá (Hurt et al. 1972). The stone industry is fairly simple, consisting of nothing more than the retouching of the working edge of flakes. The tools are described primarily as varieties of scrapers, spokeshaves, and perforators. Projectile points are lacking, but the only evidence of economic activity, fragments of bone, suggests to the excavators that hunting constituted at least part of the economy. Three radiocarbon dates prior to 9000 B.P. are available from these shelters. The oldest is a reading of 12,400 B.P., but there are problems concerning both the stratigraphy and the possible contamination of this reading (Hurt 1971; Haynes 1974).

The only other reference to early material in Colombia concerns the Toquendana complex, defined by a series of straight and "incipient stemmed" projectile points. The original reference (Correal 1974) is unpublished though cited by MacNeish et al. (1975), who estimate that the assemblage dates between 12,000 and 9000 B.P. No further information is available.

The El Inga site in the highlands of Ecuador provides one of the richest known assemblages of fishtail points (Bell 1960, 1965; Mayer-Oakes and Bell 1960; Mayer-Oakes 1963, 1966). Unfortunately the site consists mostly of surface material and the assemblage is clearly mixed, including a variety of late projectile point styles and even pottery. The fishtail points, however, are considered sufficiently diagnostic to be segregated as representing a separate early occupation; and this distinction is confirmed to some extent by small-scale excavations, in which fishtail points predominate, at lower levels. The earliest carbon-14 date on the site is only about 9000 B.P.—surprisingly late if the site in fact represents the fishtail horizon, which is dated much earlier farther south. No direct evidence of food gathering activities is available.

Also in highland Ecuador, Chobshi Cave has produced radiocarbon dates ranging back from 7500 to about 10,000 B.P. on a stone industry associated with modern fauna. Deer (three species) predominate but there are also a variety of smaller mammals (Bell 1974; Wing cited in Lynch 1976). At Punín, also in Ecuador, a fossil stratum with extinct

Pleistocene fauna including mastodon and native horse is reported to have produced a human skull as well (Sullivan and Hellman 1925), but this association is now considered questionable or rejected entirely by most recent scholars (cf. Willey 1971).

Along the coast of Ecuador, on the Santa Elena Peninsula, Lanning and Patterson (Lanning 1970; Patterson 1968) have isolated a series of stone industries at surface locations. The sites have provided no direct evidence of economic activity nor are they directly datable, but Lanning and Patterson have defined three early industries on the basis of comparisons with Peruvian coastal material and on assumptions about the technological evolution of stone working. The Exacto complex is a flake and burin industry crossdated with Peruvian materials at approximately 12,000 B.P. The Manantial complex, dated between 10,000 and 11,000 B.P., consists of blades, scrapers, denticulates, and small crude bifaces which may be preforms for projectile points of El Jobo type (Lanning 1970). Bushnell (1971) and Lynch (1974), however, have challenged the seriation of these industries and the early age estimates which are given, preferring to see these industries as specialized quarry sites related to, or contemporary with, later and better defined assemblages. Lanning and Patterson's third complex, Carolina, is also problematic. It is defined by a variety of spokeshaves, notched pieces, denticulates, and cores; but it also includes one fragment of what appears to be a fluted fishtail point. Patterson compares the point fragment with El Inga specimens and suggests a date of approximately 9,000 years for the complex.

In Peru, both coast and highlands have produced a number of reportedly early assemblages, but many of the sites can be challenged on various grounds. The earliest dates come from the highlands around Ayacucho, where MacNeish (MacNeish 1969; MacNeish et al. 1970; MacNeish et al. 1975) has identified a series of industries radiocarbon dated to approximately 20,000 B.P. MacNeish's earliest phase, Paccaicasa from Pikimachay Cave, is described as consisting of some 71 artifacts and 100 flakes. The artifacts are primarily choppers and scrapers of various shapes including denticulate scrapers and spokeshaves. Two unifacial artifacts are tentatively labeled by MacNeish as projectile points. Most of the tools are made of tufa slabs from the cave walls themselves although others are of foreign materials. Accompanying the artifacts is a faunal assemblage consisting primarily of extinct ground sloth and horse, along with deer, a large carnivore, and a few rodent bones. Carbon-14 dates on the Paccaicasa assemblage suggest a time range of between 15,000 and 20,000 B.P. A number of other scholars, however, have questioned whether the Paccaicasa artifacts represent human activity (Cardich et al. 1973; Lynch 1974; Haynes 1974), so that the reliability of Paccaicasa as an indicator of the duration of human occupation in South America is suspect. MacNeish's second assemblage, the Ayacucho assemblage, also from Pikimachay Cave, is more generally accepted to represent human activity although the radiocarbon dates leading MacNeish to estimate an age of 14,000 years or more have been questioned (Haynes 1974; Lynch 1974). The tool assemblage consists largely of choppers; crude unifacial tools including scrapers, spokeshaves, knives, and burins; and several tools identified as unifacial projectile points. The material is associated with a large number of faunal remains, which consist predominantly of extinct sloth, but also include deer, native horse, and a type of extinct camelid. A variety of smaller mammal species including rodents are also represented.

A third phase at Ayacucho, the Huanta phase, has been identified from early levels of a different cave, Jaywamachay. The stone tool industry, which MacNeish estimates to date between 11,000 and 10,000 B.P., is characterized by more finely flaked projectile points, "rubbing and/or hammerstones," and a variety of scrapers, drills, and denticulate tools. The fauna are abundant and consist primarily of camelids and deer.

For the following, Puente phase, which is estimated to last until about 9000 B.P., MacNeish describes a chipped stone industry with a variety of projectile point forms along with a number of scraping and chopping tools. The bones of large mammals, primarily deer and camelids, are the primary evidence of economic activity, although MacNeish does suggest that some trapping and plant collecting supplemented hunting in the diet. Small mammals, particularly *Cavia* (guinea pig) were more important than in earlier phases. As discussed below, this trend continues in later phases of the Ayacucho sequence. MacNeish also notes that the Puente phase provides the first evidence of the use of a variety of ecological zones in the Ayacucho region and the earliest signs of seasonal scheduling of activities.

Excavation at several other sites in the Peruvian highlands also yields evidence of human occupation prior to 9000 B.P., although none of these other sites as yet displays an antiquity remotely approaching that of the Paccaicasa assemblage. At Guitarrero Cave in the Callejón de Huaylas (Lynch and Kennedy 1970) the lowest levels produced a variety of flake tools, choppers, and hammerstones along with one stemmed projectile point, microblades, and a small knife. The associated fauna include two species of deer, a variety of small mammals (including rodents, rabbit, and skunk), a small bird (the tinamou), and a number of snails. Published radiocarbon dates on the assemblage include readings of 12,600 B.P. and 9800 B.P., but Lynch (1974) indicates that other, unpublished readings fall somewhere in between the two, suggesting an actual age in the center of this range. A second assemblage at Guitarrero Cave, which dates from 10,500 B.P. to about 7500 B.P., includes a slightly broader range of small fauna and, in addition, a variety of vegetable remains. The latter have not yet been described, and it is not clear to what extent they represent food refuse nor to what extent the use of plants can be attributed to the earlier portion of this phase prior to 9000 B.P. A single grinding slab occurs in this phase, along with a variety of projectile point forms and scrapers, and a variety of perishable artifacts including cordage and textiles.

Lynch (1971) also calls attention to a series of hunting camps near the headwaters of the Río Santo at the south end of the Callejón de Huaylas. The sites are described as small shelters facing the water and are assumed to have been lookout points for hunters. Projectile points are very common in the refuse as is the bone of deer and camelids. At least one of these sites has a radiocarbon date of 11,600 B.P., but not all are presumed to be of equal antiquity.

At Lauricocha, in the central highlands of Peru, the lower levels of what appears to have been a hunting camp used over a long period of time provide radiocarbon dates suggesting that the initial occupation of the region began about 9500 B.P. (Cardich 1958). The site is described as providing abundant remains of various camelids and deer along with a tool kit consisting predominantly of projectile points, knives, and scrapers. "Hammerstones or perhaps incipient grinding stones" also occur, as do manos and metates. The latter, however, begin only with the second level of the occupation, estimated to have begun about 8,000 years ago (Cardich in Lynch 1972).

At Toquepala, in the southern highlands near Arequipa, a similar radiocarbon reading (9450 B.P.) has been provided on an assemblage with projectile points resembling those from Lauricocha, although no faunal remains are reported (Ravines 1967). And at Panalagua Cave, near Lake Junín in the central highlands, a hunting camp with considerable quantities of the bones of large mammals is reported (Matos cited in MacNeish et al. 1975). Although no dates are available, Mac-Neish compares the artifact assemblage to that of the Puente complex at Ayacucho and suggests a date of somewhat more than 9,000 years for the site.

The Peruvian coast has also provided a number of putatively early assemblages, many of which, however, are of questionable antiquity. At La Cumbre, in the Moche Valley on the north coast, Ossa and Moseley (1972) report a possible association between human artifacts and the bones of horse and mastodon, although they conclude that the association is not certain. Even more tentative is the superficial association between large projectile points and bones of mastodon and other extinct fauna at the Pampa de Los Fósiles on the north coast. As with the La Cumbre assemblage, the association between tools and fauna cannot be established conclusively. Potsherds also occur but these are assumed to be only accidentally associated (Bird 1948; Lanning and Hammel 1961). The Pampa de Paiján on the north coast has also produced a superficial assemblage of large projectile points which some authorities (Ossa 1974; Kornfield 1972, 1974; Bryan 1975) believe to be early; but since these assemblages are undated their presumed antiquity is based only on typological comparisons with more firmly dated sites. Ossa (in Lynch 1976) also records the surface find of a fishtail point on the north Peruvian coast; but again, except for its stylistic similarity to dated finds elsewhere, antiquity cannot be proved. Somewhat more firmly dated is the association between human artifacts and human skeletal remains with refuse including snail shells at Quirihuac Shelter (Ossa 1974). This assemblage, radiocarbon dated at about 10,700 B.P., provides one of the few firmly dated examples of Paleo-Indian use of molluscs.

On the central coast of Peru, Lanning and Patterson (Lanning 1967a, b, 1970; Patterson 1966; Patterson and Lanning 1964, 1967; MacNeish et al. 1975) have defined a series of early lithic industries based on the seriation of surface materials and the excavation of a single site, Cerro Chivateros near the mouth of the Chillón River. In their excavations at this site, Lanning and Patterson established a sequence of three superimposed tool industries. The uppermost, Chivateros II, is defined primarily by the presence of thick, heavy, bifacially flaked knives or choppers resembling paleolithic handaxes; large flake tools; large, heavy and crudely flaked projectile points interpreted as having been used on thrusting spears; and, in addition, a number of somewhat finer, narrower projectile points which Lanning considers suitable for use with spearthrowers. The underlying Chivateros I assemblage is distinguished primarily by the lack of the finer points. A single carbon-14 determination of 10,400 B.P. is available on the Chivateros I layer, and on this basis the Chivateros assemblages together are assumed to date roughly between 11,500 and 9000 B.P. (It should be noted also that it is this date along with the terrace sequence from the Río Pedregal in Venezuela that forms much of the basis for the presumed antiquity of the South American biface or handaxe assemblages.)

A third group of artifacts, the Red Zone assemblage, was found at a lower layer of the Cerro Chivateros site. The assemblage consists primarily of a variety of scraper forms, burins, and notched and pointed tools. Lanning and Patterson give this assemblage a date of more than 12,000 years, based on its position below a series of salitre layers which they believe to be correlated with phases of increased moisture, these being correlated in turn with phases of late glacial chronology in North America. A fourth assemblage in the Chillón Valley—the Oquendo assemblage of burins, denticulates, scrapers, and pointed tools—is estimated to date between 11,000 and 12,000 years B.P., based on seriation between Red Zone and Chivateros. In the adjoining Lurín Valley, the Tortuga and Achona complexes are judged to be similar to Red Zone and Oquendo and are given an equivalent date; and the Conchitas complex is compared to Chivateros and assigned an age of 9,000 to 11,000 years. If the antiquity of these assemblages can be upheld they may be of significance for the argument developed in this book, since both the Tortuga complex and the Conchitas complex contain the shells of marine molluscs and would indicate some degree of early prehistoric use of these resources. The Cerro Chivateros stratigraphy and the associated seriations, however, have been challenged. Fung has recently reexcavated the site and concluded (Fung Pineda et al. 1972) that the "strata" identified by Lanning and Patterson are in fact zones in a weathering profile, so that the segregation of industries by layer and the extrapolated dates for the lower layers of the site may have no significance. She has also challenged the association of the dated sample of "wood," suggesting that is was in fact a fragment of *Tillandsia*, a plant ubiquitous in the coastal desert and not necessarily associated with human activity. Lynch (1974), noting Fung's restudy and noting also the natural process of stratigraphic mixture that occurs at quarry locations, concludes that the stratigraphic sequence and the associated dates may have little meaning although he accepts the possibility that the carbon sample may in fact date a general Chivateros industry.

Two other highly problematic sites occur in the upper Chilca Valley of the central coast. Excavation at Tres Ventanas and Quiqche caves (Engel 1970) has produced essentially parallel records of occupation dating back to approximately 10,000 B.P. At Tres Ventanas, abundant projectile points along with the bones of deer, camelids, and a variety of small mammals suggest the importance of hunting throughout the sequence; but Engel indicates that fish and shellfish occur at all levels, as do grindstones and a variety of tubers including potatoes, sweet potatoes, and *jiquima* (*Pachyrrizus tuberosus*). Quiqche Cave is also supposed to have produced grindstones at all levels, but here there is no record of organic preservation. If the data are accurate these caves represent the only record of grindstones in South America prior to 9000 B.P., with the exception of the single stone in the broadly defined Guitarrero II unit, which may also be this early. Tres Ventanas would also represent by far the earliest archaeological record of the tubers listed. But there are strong indications that the cave sequences are badly mixed. First, at Tres Ventanas, Engel also reports the presence of *Opuntia ficus* at all levels, although this plant is thought to be a postcolonial introduction into South America. In addition, he reports the "reuse" of projectile points, noting that early projectile point styles recur in higher levels. Even if the stratigraphy is not mixed, however, there is reason to doubt the early date attributed to the lower levels. Engel admits that in both caves the lowest layers from which the early carbon dates were obtained contain an "additional" industry of heavy scrapers and crude bifacially worked objects. It seems probable that Engel has an early dated crude flake/biface industry underlying an assemblage of projectile points and grindstones of a type fairly common in western South America after about 8000 B.P. The bulk of the material that Engel describes is therefore probably considerably later than the earliest date he attributes to it.

Engel (1973) also provides two radiocarbon dates (9150 B.P. and 9700 B.P.) for sites within the *lomas* vegetation of the Peru coast in the tenth millennium B.P. No specific description of the dated sites is available, but Engel's general description of the *lomas* camps (the great majority of which are dated after 9000 B.P.) indicates that they contain quantities of marine shellfish in addition to abundant projectile points, suggesting a mixed hunting and shellfishing economy.

In Bolivia, the only putatively early industry has been isolated at the Viscachani site, a surface site scattered over an area of approximately 10 hectares on an ancient beach terrace in the west central highlands of the department of La Paz. Crude, bifacial, percussion-flaked implements have been found here and have been attributed an early date by a number of scholars (Menghin 1955; Muller-Beck cited in Krieger 1964; Ibarra Grasso 1955, 1957); but finer work, known to be of later date, also occurs and questions arise as to whether the crude stone work can be segregated or whether the assemblage as a whole must be considered of relatively late date. Various authorities (Lanning and Hammel 1961; Lynch 1967; Patterson and Heizer 1965) argue that there is no basis for segregation. In any case there is no independent evidence of high antiquity for any portion of the collection. Aside from the presence of projectile points there are apparently no indications of food gathering activities associated with this assemblage.

For most of north and central Chile there is only one archaeological site for which great antiquity can be established on the basis of radiocarbon determinations. At Laguna de Tagua Tagua (Casamiquela et al. 1967; Montané 1968; Mostny 1968) excavations have yielded a stone industry of flake tools, scrapers, and choppers associated with the bones of mastodon, horse, and deer and with the abundant remains of smaller fauna including birds, frogs, fish, and small mammals. The assemblage has been dated at 11,400 B.P. No projectile points have been found, but it has been noted (cf. Forbis 1974) that stone tools recovered resemble artifacts associated with fishtail points farther south and it is possible that the industry represents an incomplete sample of a culture of the type which elsewhere contains these points. The position of the site is also interesting since it occurs near a pass in a ring of low hills surrounding a lake, a location presumably chosen as an ideal position from which to hunt animals approaching the lake to drink (Montané 1968).

Surface collections in Chile have produced a number of industries which may be of early date, although direct confirmation of their antiquity is lacking. Around the Salar de Atacama in Chile, Le Paige (1959, 1960, 1964) and Barfield (1961) have isolated an industry of crudely chipped basalt tools of the biface or handaxe type, roughly comparable to the Chivateros materials from Peru. The industry referred to as Ghatchi I has been criticized by a number of authorities (cf. Lynch 1967, 1974; Montané 1972), who argue that there is no proof of antiquity and that the Ghatchi I industry represents selective collecting of crude tools from an assemblage that also contains finer work. Lanning (1970; see also Patterson 1968) has also isolated a series of industries from surface collections surrounding the Salar de Talabre in northern Chile, which he compares to the Red Zone-Chivateros sequence of the Peruvian coast. The Chuqui complex, which Lanning compares to the Red Zone assemblage, consists of a series of burins or scrapers mostly produced by unifacial chipping along the edges of thin tabular stone blocks. I assisted Lanning in the collection of the Chuqui materials and found them unconvincing as evidence of human occupation, though no other explanation could be found for what was clearly percussion flaking in localized concentrations where natural sources of such flaking could not be identified. Subsequently Grove (MS) has mapped the Chuqui sites, showing them to follow a linear pattern corresponding to early historic wagon trails; and his suggestion that the chipping results from this source seems to me the most probable explanation of this "industry." The prehistoric nature of this assemblage has also been challenged by Montané (1972) and by Lynch (1974). In the same region Lanning has isolated a crude bifacial industry with handaxe-like materials comparable to those from Ghatchi I and Chivateros. In this case the artifactual nature of the material is not in dispute and the collections represent a complete, not a selected, sample of the material. However, the assumption of great antiquity is based only on the typological comparisons with Chivateros.

Similar materials have been found at Tarapacá in northern Chile by True et al. (1970), although they assign a date of no more than 8000 B.P. to the crude bifacial work and note also that the material, which is occasionally found isolated, also occurs in sites that are clearly of later date. Finally, at Taltal, along the Chilean coast, crude stonework has been found (Capdeville 1922; Krieger 1964) and attributed great antiquity. But at this site excavations by Bird (1965) have established that the crude stonework is associated throughout the deposits with fine pressure-flaked projectile points and knives of later date. Bird also notes that similar crude stonework occurs in the industries of all subsequent periods along the Peru coast, up to and including the Inca period.

Near the southern tip of Chile along the Strait of Magellan, excavations by Junius Bird (1938, 1946, 1970) have yielded early radiocarbon dates on human activity and at the same time have provided two of the few sites in South America where fishtail points have been excavated *in situ* in reasonably full cultural context. At Fell's Cave, a rockshelter overlooking a river in a manner reminiscent of North American hunting camps, Bird excavated fishtail points in association with extinct fauna, including sloth and native horse as well as guanaco and a variety of small fauna. The lowest level here has two radiocarbon dates, 11,000 B.P. and 10,700 B.P. A second level in the cave produced only modern fauna accompanied by bone projectile points, although Emperaire et al. (1963) report stone projectile points from this level as well. At nearby Palli Aike Cave similar fishtail points were also associated with bones of horse and sloth, but here the radiocarbon date is only 8700 B.P. At both sites ground or pecked discoidal objects form part of the assemblage. These have been characterized by Krieger (1964) as manos, leading him to suggest that the cave occupations might represent an early Archaic rather than a Paleo-Indian habitation. Bird's (1970) description of the objects makes it clear, however, that they could not have served a grinding function. One is of lava so porous that it could not have served as a mano. Moreover, none of the discs shows any signs of abrasion which would indicate its use as a grinding tool.

Excavation in the Cañadon de las Cuevas on the Estancia de Los Toldos in southern Argentina provide a third series of early carbon dates from Patagonia (Menghin 1952a,b; Cardich et al. 1973). The lowest levels at the site, which have been dated back to 12,600 B.P., contain only an industry of flakes and scrapers in association with guanaco bones. Overlying this industry are artifacts of the Toldense complex associated with the tooth of an extinct horse and the bones of varied modern fauna including guanaco, rhea, and a number of smaller animals. This industry, dated roughly between 9000 and 11,000 B.P., contains a variety of projectile point types, including large triangular points and other points that are considered related to the fishtail forms at Fell's Cave (Bird 1970; Schobinger 1972, 1974).

For all of eastern and southern South America outside Patagonia, there is only one site which can reliably be dated by carbon-14 to the period prior to 9000 B.P. At Lagoa Santa and Confins caves in Minas Gerais, southern Brazil, extinct Pleistocene fauna including horse, giant cave bear, glyptodon, sloth, and mastodon have long been assumed to be associated with human artifacts and human skeletal remains known as Confins man (Lund 1950; Walter 1948). Recent excavations here by Hurt (1960, 1962, 1964, 1966) have failed to establish direct association between human artifacts and the extinct animals. However, his excavations at Lagoa Santa Rock Shelter 6 have produced a series of radiocarbon dates between 9000 and 10,000 B.P. associated with a stone tool industry that includes projectile points and pitted anvil stones (perhaps indicative of the cracking of nuts or the preparation of other vegetable foods) along with a variety of scrapers, choppers, and axelike forms, some of which are shaped by grinding. It is noteworthy that the sites are caves overlooking a lake, and, as Hurt concludes, it seems probable on the basis of the dates that the occupants of the caves hunted extinct animals even though no direct associations have yet been established.

A number of other archaeological sites and assemblages in southeastern South America which lack carbon-14 dates have been assigned ages of 9,000 years or more based on their association with extinct fauna, their relationship to dated geological features, their stylistic resemblance to well-dated assemblages, or their apparently primitive nature. In Brazil the Alice Boer site (Conceição cited in MacNeish et al. 1975) has produced a flake industry which MacNeish compares to the Ayacucho materials (from the region of the same name in highland Peru), dated to the period 12,000 to 15,000 B.P. Overlying this industry is a second level which MacNeish compares to the Lagoa Santa materials and suggests may date to the period between 12,000 and 9000 B.P. The original reference, however, is unpublished, and so no further detail is available at present.

In Argentina there are two sites which apparently display associations between human artifacts and extinct Pleistocene megafauna. In Buenos Aires province, Austral (1972) reports the association between lanceolate projectile points and other chipped stone tools and a variety of fauna including glyptodon and llama. The site is estimated to be terminal Pleistocene in date on the basis of geological features. Bones of glyptodon have also been identified in association with human artifacts at a second location in Buenos Aires province in deposits estimated to be of about the same age (Schobinger 1974).

Schobinger (1972) has mapped the distribution of fishtail points from a number of locations in Argentina, Uruguay, and southern Brazil. No dates are available for these finds other than those mentioned above, but the widely accepted stylistic unity of the artifacts suggests that they represent a true cultural horizon, implying the distribution of human populations through much of southeastern South America between 12,000 and 9000 B.P. No organic remains are associated with most of these finds, but the points themselves, as well as their distribution on the game-rich pampas of the southeast and the association between similar points and extinct fauna in other regions, may indicate that these populations had a basically hunting oriented economy.

In the same regions of southeastern South America there are a number of other sites and complexes, most occurring in poorly defined contexts and lacking definitive stylistic elements, which are assigned by Krieger (1964) to a pre-projectile point stage or by Willey (1971) to his broadly defined "biface" and "flake" traditions. The best known and most widely debated of these industries are probably the Caracaraña complex and the Ampajango complex, both of Argentina (González and Lorandí 1959; Cigliano 1962). If Krieger and Willey are correct in their classifications, the series of sites and industries which they enumerate provides a considerable corpus of data in support of the assumption that South America was colonized at a relatively early date. Both Krieger's pre-projectile point stage and Willey's flake and biface traditions, however, are subject to question. Moreover, the antiquity of most of the individual assemblages which they cite has been the object of considerable controversy (cf. Lynch 1974 and references cited therein). In the absence of more direct evidence, the antiquity of these assemblages must be considered questionable. More important for present purposes is the fact that almost all of these assemblages lack associated organic remains from which their economy could be reconstructed. Moreover, the assemblages generally lack diagnostic tool types such as projectile points or grinding stones which could provide reliable indirect information about their economies. It is noteworthy, however, that like the fishtail points these industries are confined primarily to open regions of southeastern South America, a game-rich area which supported hunting oriented economies until the very recent past.

As far as I have been able to determine, the sites and complexes described above represent an essentially complete documentation of archaeological materials dated, or considered to date, prior to 9000 B.P. in South America. It is clear that much of this material is controversial. Despite the controversies, however, certain patterns emerge from the data and a number of conclusions can be drawn about the economy of the Paleo-Indian groups.

There are, for example, a number of indications that hunting played a major part in the economy of the early inhabitants of South America, regardless of which of the questionable early assemblages one chooses to accept. Lynch (1976) has noted that the early sites are confined primarily to relatively open (and game-rich) environments. His conclusion, of course, is intended to apply only to the relatively small group of sites which he considers to be reliably dated, but it is noteworthy that inclusion of the sites he questions does not markedly affect the known distribution of early man in South America. It is also important that a number of the sites discussed above, such as the shelters overlooking bodies of water or camps placed along natural game trails, seem to have been chosen for their strategic position in the hunting of large game.

The significance of hunting is also suggested by the nature of the tool kits described from these early sites. Projectile points are prominent in many of the early assemblages; these hunting tools are much more common than tool forms such as grindstones or fishing equipment which would indicate other economic pursuits. (The relative frequency of these functionally diagnostic tool types is markedly altered and even reversed in later sites in many regions.) Projectile points are admittedly absent from many of the early sites; but it should be pointed out that many of the industries which have been characterized as "pre–projectile point," "flake," or "biface" industries because they lack well-made, finely flaked projectile points nonetheless do contain unifacial forms or large and crudely flaked forms which their excavators have characterized as projectile points of one type or another.

The organic refuse also indicates the importance of hunting in the

Paleo-Indian economy. Mammal bone is the major item in the refuse at almost all sites where organic remains are preserved, including sites where no projectile points are found. Moreover, large mammals, including guanaco, horse, deer, sloth, mastodon, and glyptodon, appear typically to provide the bulk of the total meat weight.

There are a number of sites at which associations between early human activities and now extinct Pleistocene fauna can be established. These associations support the contention that human hunting was at least one ingredient in the extinction of many of these animals. The fact that the dated associations between man and extinct fauna all precede 9000 B.P., with the exception of one marginally later date at Palli Aike Cave, suggests that the wave of extinctions was essentially complete by this date. The role of human hunting in the extinction of Pleistocene fauna is also supported by finds from two additional sites in southern South America: Mylodon Cave in southern Chile and the Gruta del Indio in Argentina, where no associated human artifacts occur, but where radiocarbon dates on sloth dung suggest that these animals made their final appearance in the eleventh millennium B.P., contemporaneous with the dated human associations known in other locations. (See Long and Martin 1974.)

It is also clear, however, that the Paleo-Indian economy was by no means limited to the hunting of large game. The organic refuse found at many of the sites listed above indicates that a variety of small fauna, including some nonmammalian forms, were also hunted, trapped, or gathered. Small mammals, including rodents, rabbits, and small carnivores, seem to have formed a significant part of the Paleo-Indian diet. In addition, small birds, amphibians, and reptiles occur occasionally in the refuse along with some invertebrate fauna. The collection of snails is documented in at least two well-dated early sites.

On the other hand, aquatic resources do not seem to have been important in the Paleo-Indian economy. Fishbone occurs in some early sites but it is never plentiful. Identifiable fishing equipment is virtually unknown in sites dating before 9000 B.P. The bones of freshwater reptiles are found occasionally in these early sites but never in the quantities in which they occur at some later sites. The bones of marine mammals are not reported from any of the early sites, although marine hunting became a significant activity along many portions of the coast in later periods. Finally, it should be noted that marine molluscs are relatively rare (or absent entirely) in the Paleo-Indian economy. Marine molluscs are reported in a small number of early sites, but it may be significant that all of these are sites whose antiquity has been challenged. Even if these questionable early sites are accepted, however, there is no indication that the collection of shellfish even remotely approached the dimensions it later assumed. Prior to 9000 B.P. there is no evidence whatsoever of the large shell middens which dot almost the entire coast of South America just prior to the florescence of agricultural economies.

The absence of early coastal shellmounds may, of course, be a function of the destruction of early coastal sites by rising sea level (Fairbridge 1976), but, as I argued for other continents, there are several lines of evidence to suggest that this is not a sufficient explanation. Along the Pacific Coast, at least, tectonic uplift since the Late Pleistocene has raised and preserved old beaches and marine terraces where Paleo-Indian activity would presumably be preserved (Bryan 1973; Lynch 1976; Sarma 1974; Bird 1965). In addition, as has been noted for other continents, the coastal economies which later emerge are accompanied by the rise of intensive riverine economies whose late appearance cannot be explained by related factors of preservation. This co-occurrence of marine with riverine or lacustrine economies, which we have encountered repeatedly, is unlikely to be coincidental; therefore it seems probable that the late development of the marine economies is historic fact and not simply a reflection of differential preservation.

The role of plant foods in the Paleo-Indian economy is difficult to assess because of factors of preservation. It seems probable that some use was made of plant foods, at least foods favored for dietary variety, and there is even some direct evidence for the use of vegetable foods, such as the preserved remains at Guitarrero Cave. On the other hand, there is strong evidence that plant foods did not have the economic importance they later assumed. The most important point is that grinding tools, which become quite prominent in many regions of the continent in later pre-agricultural periods, are almost unknown before 9000 B.P., although specimens from Guitarrero and from Tres Ventanas and Quiqche caves may date slightly before this time.

One other pattern concerning the economy and distribution of the Paleo-Indian sites is worthy of note. Despite speculations to the contrary by various scholars (cf. Ranere 1972; Evans 1964; Carneiro 1968), there is no archaeological evidence of Paleo-Indian activity in the entire area of north central South America which comprises the basin of the Amazon River and its tributaries and encompasses the lowland tropical rain forest biome (PRONAPA 1970; Willey 1971; Lathrap various; Lynch 1976). The absence of such early sites may reflect the paucity of research in this region or the problems of archeological survey and recovery there. The tropical rain forest is notoriously difficult to survey and is subject to a very rapid evolution of landforms, which tends to destroy or bury early archaeological evidence (cf. Lathrap 1968a, 1970). In addition, it has been pointed out (PRONAPA 1970) that workable stone is rare in the rain forest; tools tend to be made of perishable materials and preceramic occupations would therefore leave few traces.

On the other hand, a number of geographers (Hester 1966; Sauer 1944; Lothrop 1961) have noted that the climate and topology of South America would have favored a north to south movement of early populations. Such a pattern of movement along the Andes, with lateral expansion into semiarid portions of the western coast and into the pampas (grasslands) of southeastern South America, would explain the known distribution of early sites. As Lynch (1976) has pointed out it is unlikely that the forest zone would have been penetrated until the "carrying capacity" of the open environments had been reached. Even Lathrap (1968b) has argued that the tropical forest is unlikely to have been exploited by early hunting and gathering populations, and his suggestion that contemporary hunting and gathering groups in the Amazon Basin are displaced riverine farmers of recent origin rather than indigenous hunter-gatherers of long standing may be of some significance. The key argument concerning the tropical forest, however, is that it is known to be an area exceedingly poor in huntable land mammals. Both modern populations in this area and those known archaeologically relied very heavily on riverine resources (fish and riverine mammals and reptiles) as sources of protein (Lathrap various; Steward 1948; Sauer 1952; Carneiro 1961). As Lynch has pointed out (1976), the rain forest would not have been penetrated by populations lacking a pattern of substantial dependence upon fish as a source of protein, and there is no evidence of the existence of such a pattern anywhere in South America in the early period now under discussion. Anticipating evidence discussed below, I would point out that the known archaeological occupation of the Amazon does in fact coincide with the appearance of riverine (and coastal) sites in other, better documented portions of the continent. Again, despite the patterns of poor preservation in the rain forest, this pattern is unlikely to be entirely coincidental.

Economic Trends after 9000 B.P.

In the millennia following 9000 B.P., a fairly clear trend occurs through much of South America toward a more complete use of diverse environmental niches and the food sources they offer. Various scholars (see Willey 1971; Lynch 1967; Lanning and Hammel 1961; Lanning 1963, 1967a) have called attention to the increasing importance of the gathering portion of the economy after this date. Hunting unquestionably continues to be significant; many sites such as high altitude camps in the Peruvian Andes continue to exist as specialized hunting camps and in some regions, such as the Argentine pampas, specialized hunting economies continue largely unaltered except for a shift in the choice of exploited species following the extinction of the Pleistocene megafauna. In most regions and at most sites, however, hunting is increasingly supplemented or replaced by the exploitation of other resources. The remaining specialized hunting camps appear now to be nothing more than one phase in an increasingly complex economic cycle. Trends toward greater use of vegetable foods are evidenced not only by the increasing frequency of their remains in the organic refuse but also by the appearance of grinding tools (both mortars and pestles and manos and metates). At the same time coastal and riverine resources come into increasing use around almost the entire circumference of the continent. Parallels to the Archaic cultures of North America are obvious and have been drawn by a number of scholars (cf. Willey 1971; Lynch 1967).

Although social trends are more difficult to interpret from archaeological data than are economic patterns, the distribution of artifact styles and the nature of habitation sites may provide some clues to the evolution of social organization in adjustment to the use of the new resources. At least along the western portions of the continent with their relatively complete archaeological record, it seems possible to identify two roughly sequential phases of socioeconomic development among populations still largely dependent on wild resources. In the earlier of these phases, dating very roughly between about 9000 and 5000 B.P., the stylistic uniformity of sites in various ecological zones, along with their temporary nature and the systematic variations in the frequency of certain tool classes from site to site, suggests patterns of seasonal movement or transhumance combined with increasingly careful scheduling in the exploitation of seasonally abundant resources (Lynch 1967; Willey 1971; Lanning 1963, 1967a; MacNeish et al. 1975). Transhumant behavior itself may not be new; Lynch (1971) has suggested that such movements are implied by the marked altitudinal range of even the earliest dated camps in the Callejón de Huaylas in Peru. However, it is clear that both the geographical range of such activities and the variety of ecological zones included increase after 9000 B.P.

In the second of these socioeconomic phases, dating primarily after about 5000 B.P., sedentary communities emerge, some using agriculture, but most apparently still primarily dependent on wild food resources. If the interpretations put forward in chapter 3 are correct, we may be witnessing two stages in the adjustment of growing populations. These populations appear first to expand the range of their environmental exploitation and to schedule their activities and movements with increasing care in order to maximize the availablility of preferred but scarce resources, using secondary resources only to the extent necessitated by increased demand. These same groups are later forced to settle in the vicinity of the less desirable resources as further increases in demand require their use on a regular basis and the efficient, intensive exploitation which full-time specialized populations can bring to bear. What is striking in the South American pattern is that while the earliest evidence of domestication occurs in highland Peru fairly early in this sequence of Archaic development (in the earlier transhumant phase), it is only much later that domestication spreads to much of the rest of South America; agriculture becomes economically significant in most regions only after sedentary or semisedentary exploitation of low-priority wild resources is well established.

The Western Cordillera and Desert Coast

Unfortunately, we know very little about the circumstances surrounding the earliest emergence of domesticated plants in the Peruvian

Andes. At Guitarrero Cave, interpreted by Lynch as a seasonal base camp forming part of a transhumant cycle (Lynch 1971; Lynch and Kennedy 1970), domestic common and lima beans occur at about 7600 B.P. (Kaplan et al. 1973). This is at the very end of the Guitarrero II phase discussed above, an undifferentiated 3,000-year period with a mixed hunting and gathering economy, in which the use of plant resources is attested both by the remains of the plants themselves and by the presence of a single grindstone. Since only the preliminary reports are as yet available, we know very little about the nature of the plants used or about quantitative or qualitative trends in the Guitarrero economy within the preceding 3,000-year period. We also lack knowledge of the evolution of subsistence and settlement patterns during the same period in the immediate vicinity. For the present, economic trends leading to the appearance of domesticates at Guitarrero will have to be inferred from those of surrounding areas.

In the Ayacucho region of the Peruvian highlands, the economic sequence has been described in somewhat greater detail. MacNeish's descriptions of the late pre-agricultural phases, however, are as yet only preliminary and rather impressionistic. From his descriptions (MacNeish et al. 1975) a picture emerges in which the predominantly hunting oriented economies of earlier periods (described above) give way to seasonal scheduling of varied activities; this transition is accompanied by a decline in hunting and the emergence of trapping and plant gathering as the primary activities prior to the appearance of the first cultigens. Grindstones and wild seeds (unidentified) occur in Jaywa phase refuse (9100–7800 B.P.), going along with an increasing proportion of small mammal bones and a decline in the frequency (except in certain seasons) of large hunted fauna. These latter changes continue a trend already observed in the earlier, Puente phase. The distribution of sites suggests use of a wider range of microenvironments; and, for the first time, larger concentrations of population are indicated by MacNeish's definition of macroband camps in addition to the microband camps identified from previous periods. In the Piki phase (7800–6550 B.P.) a large number of camps are identified covering a still broader range of microenvironments. In addition, MacNeish notes the tendency for wet season camps to become both larger and of longer duration than those of preceding periods. Although seasonal hunting camps are still identifiable, the economy appears to be one geared primarily to trapping and plant gathering, as evidenced by the frequency of the bones of small mammals as well as of grindstones and preserved seeds. MacNeish has identified domestic quinoa and squash from Piki phase refuse, and he suggests that incipient cultivation as well as incipient control of the guinea pig population supplemented the gathering economy. As discussed above, however, the provenience and/or identification of the domestic plants has been questioned, and it may be only in the subsequent Chihua and Cachi phases, perhaps at about 5000 B.P., that agriculture is firmly established in Ayacucho.

The circumstances surrounding the emergence of agriculture on portions of the coast of Peru are somewhat better defined, both because the prevailing desert conditions result in organic preservation of consistently high quality and because the coast is one of the most intensely studied regions on the continent. Even here, however, we are forced to extrapolate from a limited sample of archaeological data in an attempt to reconstruct the general pattern of events. On the basis of the known evidence, the economic prehistory of the coast between 9000 B.P. and the emergence of full-scale farming can be divided roughly into two phases, the first of which saw the expansion of (predominantly mobile) hunting and gathering groups into the zones of *lomas* vegetation which dot the coast south of Trujillo, and the second of which witnessed the emergence of (predominantly sedentary) coastal fishing and shellfishing populations. While the *lomas* camps have so far been studied only along selected portions of the central and southern coasts of Peru, the shoreline villages are well documented, and have been reasonably well described along the entire length of the coast. It is only after the establishment of these villages that farming seems to have emerged as a significant economic factor. Its growing importance is evidenced by the gradually increasing quantity and variety of vegetable refuse in the coastal middens and by the slow expansion of sedentary populations inland along the major river valleys. (See Lanning 1967a,b; Patterson 1971a,b; Patterson and Moseley 1968; Moseley 1972, 1975; Cohen 1975b; Moseley and Willey 1973; Engel, various.)

The *lomas* vegetation (Weberbauer 1936; Ferreyra 1953) is a loosely knit community consisting predominantly of herbaceous annual plants which include a number of species bearing edible seeds, tubers, and rhizomes. There is some evidence from archaeological sites and modern remnant vegetation that in the past the *lomas* community may have included a larger proportion of woody plants than it does today. This vegetation is highly seasonal, depending on the moisture from winter fog in otherwise almost totally arid regions of the coastal desert. The markedly seasonal nature of the *lomas* resources, which works in opposition to the seasonal flow of coastal rivers, has led various workers (cf. Cohen 1975b; Patterson 1971a,b; Moseley 1972) to assume that the *lomas* camps were predominantly winter camps of populations who spent the rest of their time in the coastal river valleys. One or two sites occur in adjacent coastal river valleys contemporary with the *lomas* occupations and these are presumed to represent the postulated summer camps; however, most such camps, if they ever existed, are now lost under subsequent cultivation. An alternative possibility is that some or all of the *lomas* camps are seasonal occupations of transhumant populations who spent their summers in the highlands at sites such as Lauricocha (Lynch 1967; Lanning 1963; Willey 1971). In at least two cases, however, one in the Lurín Valley and one in the Paloma Quebrada of the Chilca Valley, where *lomas* vegetation occurs in close proximity with other wild food sources, permanent occupation seems to have been established solely on the basis of the use of these varied wild foods (Patterson 1971a, b; Engel 1971).

Minor occupation of the *lomas* may be traceable back as far as 10,000 B.P., if Patterson's estimated dates for the Conchitas complex from the Lurín Valley (cited above) are correct. Otherwise, as indicated above, Engel (1973) has provided two radiocarbon dates suggesting penetration of the *lomas* environment shortly before 9000 B.P. However, the great bulk of the dated sites fall between 8500 B.P. and about 4500 B.P. The best defined of the *lomas* complexes are those of the central coast, including the Chilca Valley, the Lurín Valley, the Chillón Valley, and the adjoining region of Ancón. The archaeological assemblages here include projectile points and a variety of chipped stone tool forms, as well as manos and metates, and, more rarely, mortars and pestles. The refuse is almost entirely that of wild plant and animal species, gourds being the only possible domesticate until late in the sequence. Protein appears to have been supplied by a mixture of game, presumably hunted as it grazed on the *lomas* vegetation; large snails indigenous to the *lomas* vegetation itself; and a variety of fish and shellfish, which were apparently imported to the *lomas* camps from the shoreline. The few contemporary campsites known from the river valleys display comparable refuse although the proportions of items in the refuse may vary.

At Ancón the dated sequence of *lomas* occupations suggests a number of interesting trends (Cohen 1975b). Early sites dated between 8000 and 7000 B.P. display a relatively large quantity of mammal bone in the refuse and woody plant material is also quite abundant. Conversely, both shellfish and grinding tools are relatively rare. In succeeding phases the quantity of mammal bone declines progressively and that of shellfish increases. Simultaneously the frequency of grinding equipment increases significantly as does the frequency of small seeds, particularly the seeds of wild grasses. Wood also becomes progressively scarcer in the *lomas* sites.

The whole sequence of change is suggestive of the progressive degradation of the *lomas* environment and the increasing exhaustion of the favored wild game. Apparently greater reliance was gradually placed on secondary resources and the *lomas* populations turned more and more to the sea to supply their protein and perhaps their caloric needs as well. By 4500 B.P. the *lomas* zones were abandoned altogether in favor of habitats along the shore, suggesting that the *lomas* zones were no longer adequate to feed the existing population or at least that they could no longer provide a quality diet at a level of labor cost competitive with that offered by the littoral zone.

It is interesting that the first evidence of the (small-scale) use of domestic food plants occurs at the very end of the sequence of *lomas* occupations. Domestic squash first occurs in the *lomas* camps at about 5000 B.P., and lima beans are reported from a coastal village of about the same age. I find it striking that although domestication technology was practiced in adjacent regions of the highlands approximately 3,000 years before, the first domesticates on the Peruvian coast occur just as the *lomas* camps were being abandoned; and I have argued that these domesticates were in essence an emergency food introduced to bolster a badly degraded hunting and gathering economy at a time of economic crisis (Cohen 1975b).

The reasons for the apparent degradation and abandonment of the *lomas* environment are not entirely clear. Both Lanning (1967a,b) and Engel (1970, 1973) have tended to see the rise and decline of the *lomas* sites as linked to climate change, arguing that the *lomas* area was used during the relatively moist Altithermal period and abandoned when drier conditions destroyed much of its resource potential. Various geographers (cf. Craig and Psuty 1968; Craig in Lynch 1974) have argued that no significant climate change can be established however. In addition, I have suggested (Cohen 1975b) that the particular pattern of the use and decline of the *lomas* is more in keeping with the assumption of human degradation of the environment. Both Patterson (1971a,b) and Moseley (1972) have also recognized population growth and human overexploitation as ingredients in the decline of the *lomas*. One additional piece of evidence that may support demographic interpretation is the fact that known river valley sites display economic trends paralleling those of the *lomas*. In the Lurín Valley a site related to the early *lomas* camps has refuse which indicates that hunting still predominated over shellfish collecting in the economy (Patterson 1971b), whereas in the Chilca Valley, at a site contemporary with the later *lomas* camps, shellfish collecting and the exploitation of other littoral resources had a clear primacy among economic activities (Patterson 1971b; Donnan 1964). Since the plant and animal communities of the river valleys are dependent on a different source of moisture and subject to a different climate regime than that which governs the *lomas* vegetation, it seems unlikely that the pattern of economic change can be attributed to changes in the coastal climate.

The extent to which the pattern of events on the central coast can be extrapolated to explain events elsewhere on the coast of Peru is not yet clear. It is evident, however, that the pattern of increasing use of marine resources in the *lomas* camps led ultimately to the establishment of permanent shoreline villages on the central coast and that similar permanent fishing villages emerged all along the Peruvian coast at about the same time. Prior to 4500 B.P., the only permanent villages in Peru were those associated with *lomas* vegetation under the favored circumstances discussed above and possibly also Chilca Village 1, which may have been a permanent settlement by about 5700 B.P. However, between 4500 B.P. and 4000 B.P. shell middens representing permanent fishing settlements became very widespread on the north and central coasts of Peru. Along the southern Peruvian coast similar, but smaller, shell middens occurred about 500 years later.

The first of these coastal villages to be adequately described was the Huaca Prieta midden in the Chicama Valley in northern Peru. Bird (1948, 1951) found a village made up of semi-subterranean houses, whose total permanent population could be estimated at a few hundred individuals. The rich refuse consisted primarily of shell, fishbone, and the bones of sea mammals including sea lion, porpoise, and whale; there were also a variety of plant remains including squash, peppers (*Capsicum* sp.), lima beans, jack beans, achira, gourds, and cotton; a variety of (cultivated?) fruits including lucumas and *ciruelas* (*Bunchosia* sp.); and a variety of wild roots and tubers (Towle 1961). The artifact assemblage lacked projectile points but did contain crude stonework as well as a variety of fishing equipment including nets, sinkers, and gourd floats.

Since Bird's pioneering excavations, more than 30 similar coastal middens have been identified along the Peruvian coast from Chicama in the north to Nazca in the south (see Lanning 1967; Moseley 1975). These include Huaca Negra in the Virú Valley (Willey 1953; Bird 1948; Strong and Evans 1952); Las Haldas south of the Casma Valley (Ishida et al. 1960; Fung Pineda 1969); Playa Culebras at the mouth of the Río Culebras (Engel 1957a,b); Huarmey in the Huarmey Valley (Kelley and Bonavía 1963); Aspero near the port of Supe (Moseley and Willey 1973; Willey and Corbett 1954); Río Seco north of the Chancay Valley (Wendt 1964); various sites at Ancón and Ventanilla north of the Chillón Valley (Lanning 1967a,b; Patterson and Moseley 1968; Moseley 1975; Cohen 1975b); Chuquitanta (El Paraísou) in the Chillón Valley (Patterson and Lanning 1964; Engel 1967); Chira-Villa in the Rimac Valley (Lanning 1967a); Asia in the Asia or Omas Valley (Engel 1963); the mounds at the Laguna de Otuma west of Ica (Engel 1957a; Craig and Psuty 1971); Casavilca in the Ica Valley (Engel 1957a,b); and the mounds on San Nicholas Bay south of the Nazca River (Strong 1957).

The sites in question all lack pottery, at least in their earliest phases, but most contain twined cotton textiles, a horizon marker for the late preceramic period along the Peruvian coast. Otherwise, they show a rather pronounced degree of variation in size, organization, and style. Lanning (1967) in particular has called attention to the degree of cultural diversity among what are otherwise economically and ecologically very similar sites.

Some of the sites, particularly those along the south coast such as Otuma and Casavilca, are small shallow middens considered to be seasonal campsites of small hunting and gathering populations numbering no more than 50 to 100 persons. Others such as Aspero, Culebras, Las Haldas, and Chuquitanta are large, substantial sites which are presumed to be permanent occupations for populations of 1,000 or more individuals. At Aspero, for example, the refuse covers an area of more than 13 hectares with an average depth of about 2 meters. At the Chuquitanta site, where most of the refuse has been destroyed, the original midden is estimated to have covered approximately 50 hectares. Many of the larger sites including Aspero, Río Seco, Culebras, and Chuquitanta also have large and well-planned architectural features indicative of corporate planning and labor management. At some of these sites the major architecture is clearly for ceremonial or other public purposes, while at others the major investment was made in domestic architecture.

The economy of all of these sites, however, is substantially the same. The hunting of land mammals is almost unknown. The bone of such mammals is exceedingly rare in all of the coastal middens, and projectile points are generally lacking except in the small (seasonal?) sites of the southern coast such as those at Casavilca, Otuma, and San Nicholas. There is no evidence of domestic camelids from the sites of the period on the coast, but domestic guinea pigs do appear to have been kept at some of the sites such as Culebras, where house construction included the type of double wall enclosures which were used to keep guinea pigs in historic times. All of the sites tested also display plant domestication at least to the extent that domesticated cotton is present, and in general a variety of cultivated food crops is found as well. The domesticated plants include those mentioned above at Huaca Prieta, plus a variety of other minor fruits including guavas (*Psidium guajava*) and *pacae* (*Inga* sp.). In addition, sweet potatoes, peanuts, and maize are present at some of the sites just prior to the arrival of pottery. There are several indications, however, that agricultural products are not the major ingredients of the economy of the period and that it is not agriculture but seafood that determines the sedentary patterns and supports the large, well-organized populations (cf. Lanning 1967a; Moseley and Willey 1973; Moseley 1975). In the first place, most of the known sites display no evidence of maize or any other crop (grain or tuber) which is likely to have served as an economic staple. Moreover, even where present in the refuse (maize occurs at Aspero, Culebras, Huarmey, and Las Haldas), these potential staples do not generally occur in sufficient quantity to indicate that they accounted for the major part of the diet. The overwhelming bulk of the refuse at almost all of the sites consists of marine products including shellfish; fish, sharks and rays; crabs and tunicates; shore birds including cormorants, gulls and pelicans; and marine mammals including whales, dolphins, seals and sea lions. Various edible

forms of seaweed are also common. The refuse content may be misleading. As I have suggested elsewhere (Cohen 1972-4), plant remains may
be quantitatively underrepresented in these coastal middens even when
a wide variety of plant foods is preserved, since the preservation of
plant material is sporadic while that of shell and bone is more constant.
It is thus possible that vegetable foods were more important than the
refuse indicates. However, the importance of marine resources is also
attested indirectly in two other ways. First, fishing equipment including nets, lines, sinkers, hooks, and floats is prominent in the industrial
refuse at the sites. More important, the site locations seem very commonly to have been chosen with reference to the availability of coastal
resources rather than with an eye to maximum agricultural productivity.
Río Seco, Las Haldas, and the Ancón sites, for example, are all in desert
areas outside of the coastal river valleys, where no cultivable land
exists. The cultivated plants in these sites must have been tended in
adjacent river valleys on a part-time basis or looked after by the populations of subsidiary sites. At Aspero, cultivation would have been possible on an adjacent river floodplain; but the area available for farming
in the vicinity is very limited and would have been insufficient to provide a staple crop for the estimated population of the site (Moseley and
Willey 1973). An exception may occur at Chuquitanta which is located
near a relatively broad area of the floodplain of the lower Chillón River.
Here the tiny remnants of refuse still preserved contain a smaller proportion of shellfish than at other sites; it is questionable, however,
whether any potential agricultural staple was under cultivation. Patterson (1971a,b) indicates that maize was present in preceramic levels at
the site, but neither I (Cohen 1975b) nor Moseley (Moseley and Willey
1973) has been able to substantiate this and maize is unknown at other
contemporary sites on the central coast. In any case, there seems to be
general agreement (Moseley 1975; Lanning 1967a; Patterson 1971a,b)
that it is only after about 3900 B.P., well after the establishment of
most of these coastal sites, that agriculture becomes important enough
to supplant coastal harvesting as the major focus of the economy of the
Peruvian coast.

In the southern Andes (western Bolivia, northwestern Argentina, and
Chile) a pattern of economic trends is found that is roughly parallel to
events already described in highland and coastal Peru. After 9000 B.P.
inland and highland sites in this area display an increase in the range of
foods, particularly vegetable foods, which are exploited. At a slightly

later time, probably between 7000 and 6000 B.P., coastal sites based on the intensive exploitation of marine resources seem to have emerged along virtually the entire extent of the Chilean coast. The major difference between this area and Peru is that in the south agriculture spread at a relatively late date, the first signs of domestication occurring long after coastal exploitation patterns were well established. Agriculture seems to have spread in the southern Andes beginning only at about 3000 B.P. or perhaps slightly earlier. Unfortunately the data on early agricultural sites are as yet too sparse for the spread of agriculture in this region to be documented fully. However, by the historic period, agriculture had spread to all of the southern Andean area except some highland regions and the southern tip of South America, where historic tribes such as the Alacaluf, Chono, and Yahgan of western Tierra del Fuego continued the pattern of marine exploitation into the historic period.

Inland, the archaeological sequence is probably best described at a series of related sites in the provinces of Cordoba and San Luis in northwestern Argentina. One of these, the Ayampitín site (González 1952), is an open campsite where large projectile points along with other chipped stone artifacts occur with manos and milling stones. The economy of the site is interpreted as one of guanaco hunting combined with the processing of wild seeds. Surface finds of similar projectile points suggest that the culture complex was widespread in the highland valleys of northwestern Argentina (Cigliano 1964, 1965; González 1952). At Intihuasi Cave (González 1960) a very similar assemblage of points, milling stones, and manos occurs in an excavated layer (the lowest occupation layer in the cave, Intihuasi IV), accompanied by two carbon-14 dates suggesting an age of about 8,000 years. Bones of deer and guanaco occur, and González interprets the site as indicating a long-term pattern of seasonal reoccupation by hunting and gathering groups.

Overlying the Ayampitín layer at Intihuasi Cave are three later layers which span the late preceramic and early ceramic period in Argentina. Manos and milling stones continue to be common throughout the sequence but the large projectile points are gradually replaced by smaller points. Blunt-ended bone projectile points also occur, which may be indicative of emphasis on small game. Despite the presence of pottery in the uppermost level in the cave, there is no direct evidence of agriculture at any point in the sequence.

At Ongamira Cave (González 1941, 1952, 1960; Menghin and González

1954) a sequence of four superimposed layers, totalling about two meters of refuse, appears to parallel the later phases of the Intihuasi occupation. The two lower levels at Ongamira, variously estimated to have begun at about 5000 B.P. (Menghin and González 1954) or at about 3000 B.P. (González 1960), contain a combination of small chipped stone projectile points, blunt bone projectile points, and milling stones and manos. The assemblage is considered roughly comparable to the last preceramic occupation at Intihuasi. The two upper levels contain pottery in addition to manos and milling stones, mortars and pestles, and a variety of small projectile points. These levels are considered to equate roughly with the ceramic occupation at Intihuasi Cave.

The date for the arrival of pottery in northwestern Argentina is not yet firmly established. The earliest carbon dates for ceramic sites in this region come from El Piquete and Las Cuevas, where dates of 2750 B.P. and 2500 B.P. respectively suggest the diffusion of ceramic technology early in the third millennium B.P. (Schobinger 1974). The date for the arrival of agriculture is equally uncertain. Remains including beans, potatoes, maize, and peppers have been found in at least one preceramic context, the Cave of Huachichocana, suggesting that agriculture may have arrived in portions of northwestern Argentina 3,000 years ago or more (Alicia Fernandez Distel in Schobinger 1974). However, as far as I can determine, the dates on this assemblage are not firm and it is unclear that the preceramic levels at Huachichocana predate the early dated ceramic assemblages discussed above. At present, too little is known about early ceramic and/or agricultural sites to permit us to determine how widely or how evenly pottery or agriculture had spread in northwestern Argentina in the third millennium B.P.

In Bolivia and highland Chile a series of industries have been reported which, although most are undated and unstratified, appear to be roughly contemporary with the preceramic portions of the Intihuasi and Ongamira sequences. The Tulan, Puripica, Cebollar, and Tambillo assemblages (Le Paige 1959, 1960, 1964; Barfield 1961; Nuñez 1965) display a variety of projectile point types which, roughly in the order listed, appear to span the transition from the large willow leaf forms to the later, smaller point types. Rough grinding stones are associated with the Puripica and Tambillo assemblages, and masonry house foundations suggestive of sedentary settlements are associated with Puripica.

In the Tarapacá region of Chile, True et al. (1970) have identified at least two patterns of presumed late preceramic occupation, although the temporal relationship between them is not clear. One pattern consists of campsites which clearly represent temporary occupation by small groups in contact with the coast, since fishbone, dried fish, and shellfish occur in the refuse along with the bones of guanaco. The second complex, consisting of slightly larger sites, displays a combination of small projectile points, chipped stone tools of various forms, and milling stones. A single radiocarbon determination indicates a date of 4700 B.P. for the second pattern.

Pollard (1971; Pollard and Drew 1975) reports that mortars and pestles occur in late preceramic occupations along the Río Loa in the Atacama region of highland Chile which also contain preserved edible rhizomes of wild plants. Extrapolating from dated carbon samples, he suggests that early ceramic phases along the Río Loa begin by about 2800 B.P. Milling stones and manos are found along with mortars and pestles in the earliest sites that have pottery. No cultigens occur in these locations, however, and they are interpreted as campsites of hunting and gathering populations. In the second ceramic phase, beginning about 2500 B.P., small, semi-permanent villages occur which are still without agriculture although possibly based on the herding of llamas. The refuse indicates the processing of a variety of wild plant foods, including *algorroba* (*Prosopis* sp.), the rhizomes of various wild plants, and the fruits of cactus. Maize agriculture and full sedentism occur only after about 1850 B.P.

Along the coast of Chile, the northernmost of the major marine-oriented sites is the Quiani site, near Arica, which has been excavated by Junius Bird (1943, 1946; see also Bird 1965; Nuñez 1965). The refuse at the site consists primarily of the remains of shellfish, fish, crabs, shore birds, porpoises, and sea lions. The only bone from land mammals is what occurs in the form of tools. Stone projectile points occur throughout the refuse, as do various types of fishhooks. In addition, lava bowls, manos, and mortars are found, suggesting that some processing of vegetable foods occurred. There is no evidence of domestication, however, until the very end of the archaeological sequence, when maize and cotton occur. Radiocarbon dates from the lower levels of the site indicate that the occupation began a little more than 6,000 years ago, but the known distribution of maize and cotton in adjacent regions indicates that the late, agricultural level can be no more than

4,000 years old. The site therefore either represents a long-term pattern of seasonal reoccupation of the coast by a transhumant population or, as Lanning (1974a) has suggested, it represents one of the earliest permanent occupation sites in South America.

Similar shellmounds which appear to be approximately contemporary with the Quiani site have also been excavated at Punta Pichalo and at Taltal, south of Arica (Bird 1943, 1946, 1965). At Punta Pichalo, maize and cotton again occur at the top of the sequence, but at this site they occur with pottery. Comparisons between the pottery and the well-dated ceramic sequence in Peru suggest that the agricultural layers at Punta Pichalo are no more than about 2,000 years old (Willey 1971).

Farther south along the Chilean coast, at Guanaqueros, north of the Río Choapa (Schiappacasse and Niemeyer 1964; Iribarren 1956), and at other sites at the mouth of the river itself (Gajardo Tobar 1962-63), large shell middens occur which contain numerous small projectile points as well as harpoon points and both shell and composite fishhooks. The sites also contain doughnut stones and manos as well as *tacitas* or mortars ground into large slabs of stone partially buried in the ground. *Tacitas* are also reported at Sitio Alacranes I, a shell midden south of the Río Choapa (Silva 1964), and in the large shell midden at Las Cenizas, slightly to the south near Valparaiso (Gajardo Tobar 1958-59). Whether these unusual tools were used to process seafood or vegetable foods is not clear, although Iribarren (1962) notes that they were used by later agricultural populations to process plant foods and Schiappacasse and Niemeyer (1964) note that the distribution of the *tacitas* conforms to the distribution of edible wild plant resources along the Chilean coast, suggesting their use in plant processing. Willey (1971), using stylistic comparisons suggests that the Río Choapa sites are roughly contemporary with the shell middens farther north.

South of Valparaiso a number of additional coastal shell middens as well as scattered finds of fishhooks and other artifacts related to these coastal occupations have been reported by a number of scholars (Bird 1943; Berdichewsky 1964; Ortiz Troncoso 1964; Menghin 1959-60). In the extreme south, what may be the earliest of the coastal occupations is reported from Englefield Island in the Sea of Otway (Emperaire and Laming 1961), where a shell midden with the remains of numerous marine mammals has been radiocarbon dated to the period between 8500 and 9000 B.P., a date which Willey (1971), however, considers to be too early for the cultural assemblage represented. Even further south, at the Beagle Channel sites, Bird (1938) has identified a Shell Knife culture with a marine pattern of exploitation which precedes the arrival of the historic Yahgan in the area. Bird estimates the entire time depth of this sequence as no more than 2,000 years, while Willey (1971) prefers an estimate of about 4000 B.P. for the beginning of the Beagle Channel sequence.

Northwestern South America

In northwestern South America (Ecuador, Colombia, Venezuela, and Guyana) the archaeological record suggests a sequence of events roughly paralleling those observed in Peru and Chile. Agriculture appears to have emerged in most of the northwest only after the development of intensive systems of marine exploitation, and it seems—although this is subject to somewhat more controversy—that agriculture became economically significant only after sedentary or at least semisedentary coastal populations were well established.

The parallelism between the Peruvian sequence and that of the northwest is particularly striking because the two areas are subject to different climate regimes, suggesting that changes in climate cannot be used to explain the observed events. Unlike the coasts of Peru and Chile, which are dominated by deserts, most of the northwestern coast of South America is a relatively moist environment in which mangrove trees are common. The coastal and marine resources are typically species adapted to warm ocean temperatures and the mangrove environment as opposed to the species suited to the colder water and rocky or sandy shorelines which predominate farther south. The boundary between these two climate regimes occurs today in Ecuador, but there is a good deal of evidence indicating that this boundary has fluctuated. In particular, portions of the coast of Ecuador which are today included in the coastal desert show signs of having been dominated by mangrove at various periods in the prehistoric past (Sarma 1974).

On the coast of Ecuador, the earliest sites to display evidence of significant exploitation of coastal resources are those of the Vegas complex of Guayas province. Lanning (1967a, 1970, 1974a) has suggested that the Vegas sites represent populations that moved seasonally between the coast and the banks of inland creeks and rivers on the coastal plain. The coastal sites, which Lanning presumes represent the dry season, consist largely of densely packed shell along with small quantities of the bones of land mammals. The wet season riverine sites are interpreted as plant gathering and fishing sites, although as far as I can determine there is no direct evidence to support this interpretation of their economy. There is no evidence of domestication in either coastal or riverine sites.

Sarma (1974) provides three radiocarbon dates for Vegas sites, ranging between 8800 B.P. and 7200 B.P., and he argues that there was a gap in the archaeological sequence along this portion of the Ecuadoran coast thereafter. However, at least one radiocarbon date on a late or terminal Vegas site places the complex at about 4700 B.P. (Bischof 1972). This date, along with the recent discovery of preceramic layers underlying the ceramic phases in the Valdivia middens (discussed below; see Bischof and Viteri Gamboa 1972), suggests that occupation may have been continuous.

A large series of radiocarbon dates suggests that at or shortly before 4700 B.P. the temporary preceramic campsites along the Guayas coast were superceded by permanent settlements with pottery. These settlements were originally characterized by Meggers, Evans, and Estrada in terms of two overlapping ceramic sequences, the Valdivia and Machalilla sequences, dated between approximately 5000 B.P. and 3500 B.P. (Meggers 1966; Meggers and Evans 1962; Meggers et al. 1965; Evans and Meggers 1957; Evans et al. 1959; Estrada and Evans 1963). According to their reconstruction, both the Valdivia and the Machalilla middens represent a predominantly littoral economy based on the harvesting of shellfish, fish (including some offshore species, which would have necessitated the use of boats), crustacea, and small reptiles. The middens also contain deer bones indicating hunting, as well as manos and milling stones indicating the processing of some sort of vegetable foods. Plant remains are not preserved. These authors suggest that some cultivation might have occurred as a relatively minor supplement to the diet; however, by their reconstruction it was only after 3500 B.P. that maize was introduced and that agriculture became the basis of subsistence along the Ecuadoran coast. Willey (1971) largely supports the same hypothesis, arguing that maize arrived from Mesoamerica, probably by sea, at about this date.

The Meggars, Evans, and Estrada model has, however, been challenged in several respects. The chronology of the Valdivia-Machalilla sequence has been questioned by a variety of authorities (Bischof 1972; Bischof and Viteri Gamboa 1972; Hill 1972-74), but the changes proposed are not major and not of great importance for present purposes. They may have a significant bearing on the interpretation of the development and spread of pottery in South America.) Of more immediate importance is the assertion by Lathrap (1973; see also Zevallos 1971) that the Valdivia sequence represents people already possessing well-established maize agriculture. Lathrap would argue that the use of coastal resources was a secondary or even marginal pursuit in the Valdivia economy. His reassessment is based primarily on what he (and Lanning) perceive to be the relative scarcity of shell in Valdivia middens compared to those of preceding (and some following) periods and on the recent discovery of a number of inland Valdivia sites where coastal resources could not have been the determining factor in the choice of locations and could not have provided the economic basis for the apparently sedentary Valdivia populations. The early dated arrival of maize in Peru (discussed above) certainly supports the assumption that maize is likely to have occurred before 3500 B.P. in coastal Ecuador, but the question of the relative importance of agriculture and marine resources as determinants of Valdivia settlement patterns is as yet unresolved. In any case, in later periods a mixed farming and marine economy seems to have persisted, with farming gradually assuming greater importance.

In Colombia, late preceramic complexes are rare and only poorly defined. Most of the evidence of preceramic populations comes from surface finds of projectile points and other chipped stone tools to which no dates can be attributed, and the sites lack organic refuse. Except for the projectile points none of the tools are distinctive enough in design to permit reliable reconstruction of economic patterns.

Beginning about 5000 B.P., however, a series of ceramic complexes has been defined which appears to represent a period of pre-agricultural sedentary or semisedentary occupation focusing on riverine and coastal habitats along the Caribbean coast (Reichel-Dolmatoff 1965a,b, 1971; Bischof 1966, 1972). The environment as described by Reichel-Dolmatoff (1971) is one rich in food-gathering potential for populations willing to make use of the broad range of aquatic resources provided by the coastal rivers, oxbow lakes, and the sea itself. The earliest of these ceramic phases, to judge by its radiocarbon date, of between 5000 and

4000 B.P., is the Puerto Hormiga phase. This phase has been defined primarily on the basis of excavations at the site of the same name located on the bank of a former outlet of the Río Magdalena, adjoining an area of mangrove swamp a short distance from the ocean. The large Puerto Hormiga midden consists primarily of clamshells in addition to fishbone, crabshell, and the bones of reptiles (a mixture of marine and riverine species). The bones of a few mammals, mostly rodents, also occur. Vegetable refuse is not preserved, but the use of vegetable foods (which Reichel-Dolmatoff presumes to have been wild) is indicated by the frequency of nutting stones and small grinding slabs. A similar coastal economy is reported from other shellmounds in the same region representing later ceramic phases dated between 4000 and 3000 B.P.

Much less is known about the interior at this time although there is some evidence of related sites on riverine locations well removed from the coast. The Bucarelia site, 150 kilometers inland along the Magdalena River, has pottery which recalls that of the coastal sites, and its economy appears to be one of aquatic hunting and fishing in the river and in adjoining oxbow lakes (Reichel-Dolmatoff 1971). Another similar site has been identified at the Isla de los Indios, a small island at the confluence of the Magdalena and César rivers (G. and A. Reichel-Dolmatoff 1946–50).

Agriculture, involving first manioc and later maize, is assumed by Reichel-Dolmatoff to have developed in Colombia only after 3000 B.P. Budares or pottery manioc griddles, implying the processing of bitter manioc, occur at the Malambo site on the lower Río Magdalena, which has been radiocarbon dated between 3100 B.P. and 2000 B.P. (Angulo Valdez 1962; Reichel-Dolmatoff 1965). At Momíl, on the Sinú River (G. and A. Reichel-Dolmatoff 1956; Reichel-Dolmatoff 1957), manioc griddles occur in the lower levels of the site, while in upper levels manos and metates become abundant, suggesting that an early manioc farming complex was replaced by one focused on maize cultivation. No carbon-14 dates are available here and there is some controversy about the dating of the site. Reichel-Dolmatoff would place the whole Momíl sequence within the third millennium B.P., while Lathrap (1970) would begin the sequence at least 1,000 years earlier.

The emergence of the agricultural economy in Colombia occurs in correspondence with a change in the system of aquatic exploitation. This is admittedly puzzling given the model developed in this book. The coastal region seems largely to have been abandoned and shellfish to have disappeared from the diet even in locations where these resources had previously been used. Neither Malambo nor Momíl contains significant shell refuse. At the same time, however, the riverine and lacustrine orientation of the new agricultural villages is so marked and the refuse of other aquatic resources, particularly freshwater fish and reptiles, is so abundant, that Reichel-Dolmatoff (1959, 1965) has argued that it was primarily freshwater fishing and aquatic hunting rather than agriculture that provided the character and stability of these sedentary populations.

It is worth noting that Reichel-Dolmatoff (1965) has also argued that the later switch from manioc to maize (which is richer in protein) may have been motivated by the growing inadequacy of the protein supply available from these animals.

Along the Caribbean coasts of Venezuela and Guyana it is also possible to trace a series of marine-oriented habitations preceding and accompanying the spread of early agriculture (Cruxent and Rouse 1958; Rouse and Cruxent 1963; Evans and Meggers 1960). In contrast to the pattern reported in Colombia, however, the coastal economies seem to have continued in much of this region after the inception of agriculture and to have persisted alongside the agricultural economy until the historic period (cf. Sanoja 1966; Rouse 1948; Kirchoff 1948). The earliest shellmounds from the region are preceramic and are generally attributed to a range of 3000 to 7000 years B.P. (Rouse and Cruxent 1963; Willey 1971). Only a few of the sites have been radiocarbon dated, however, others being attributed to this period on the basis of stylistic similarities to dated sites on this or other portions of the coast. On the central coast of Venezuela, Rouse and Cruxent have defined a complex of sites known as El Heneal, based on their excavations at two shellmounds, El Heneal and Cerro Iguanas. Both lack pottery but contain a stone industry consisting primarily of "edge grinders" as well as hammerstones and anvil stones. El Heneal has been dated by carbon-14 to approximately 3500 B.P., whereas Cerro Iguanas has provided three dates suggesting an age of approximately 5,800-5,400 years. Farther east on the Venezuelan mainland and on adjoining Cubagua Island the same archaeologists have excavated a number of additional shellmounds which they refer to as the Manicuaroid series. Carbon-14 dating suggests that the earliest of these sites dates to about

4300 B.P., but a series of later carbon dates and occasional pieces of pottery both suggest that the last phases of the sequence overlapped the later ceramic and agricultural occupations of the region.

The first evidence of incipient agriculture in Venezuela occurs shortly after the appearance of the coastal shellmounds. At Rancho Peludo in western Venezuela (Rouse and Cruxent 1963) two fragments of pottery griddles were found, possibly suggesting the processing of bitter manioc in what the authors otherwise describe as a hunting and gathering economy. A single date for this site suggests that the occupation may extend back to 4800 B.P.; but the bulk of the radiocarbon determinations attributed to the portions of the site sequence in question range between 3800 and 2400 B.P., and this time range may be a more accurate estimate of the age of early pottery and agriculture in western Venezuela.

There is no evidence that agriculture played a significant part in the Venezuelan economy until about 3000 B.P. Ceramic traditions with abundant *budares* or manioc griddles become common in eastern Venezuela after this date. The earliest of these agricultural, ceramic complexes is the Saladoid complex (3000 to 2700 B.P.) known from the lower levels of the Saladero site on the lower Orinoco River. In western Venezuela, maize rather than manioc provided the main agricultural staple (Sanoja 1966), but here too well-established maize cultivation appears just after 3000 B.P.: maize farming is in evidence among the Caño del Oso people of the Hato de la Calzada site, dated to about 2900 B.P. (Zucchi 1973). Significantly, both the Saladero site and the Hato de la Calzada site demonstrate that agriculture emerged in a context in which freshwater resources, particularly fish, formed a significant portion of the protein supply.

In Guyana, preceramic shellmounds have been designated the Alaka phase by Evans and Meggers (1960). Potsherds confined to the uppermost levels of the middens indicate contact with dated ceramic cultures in Venezuela and provide a terminal date for these cultures of about 1500 B.P. Evans and Meggers place the start of the sequence at no more than 2000 B.P., but Willey (1971) estimates that it must have begun at least a thousand years earlier. Meggers and Evans suggest that domestic plants gradually superseded shellfish as the focus of the Alaka economy even before the Alaka culture itself was superseded by pottery-making tropical forest cultivators.

Eastern South America

In eastern and southern Brazil, outside the zone of tropical rain forest, economic trends roughly parallel those of the other margins of the continent. Inland the picture of economic change is not particularly clear. Preceramic campsites have been defined in a number of locations (cf. T. O. Miller 1969; Laming and Emperaire 1959a,b; Chmyz 1963, 1967; Blasi 1965; Bórmida 1964a; Hurt 1964, 1968; Silva 1967; E. T. Miller 1967; PRONAPA 1970; Willey 1971; Lanning 1974b). Unfortunately most of these are poorly dated and the chronological relationships among them are only beginning to be worked out. More important, the sites provide few clues as to their economy. Organic preservation is poor and most of the tools recovered are not distinctive enough to permit functional interpretation. The persistence of projectile points suggests that hunting continued to be of some importance; and the occurrence of shell fishhooks at some inland sites suggests fishing. In addition, flaked and polished axes occur, but there is little clue to their use. Grinding tools are scarce or absent.

The significant trend in eastern Brazil occurs along the coast, where intensive exploitation of marine resources is evident beginning at about 7000 B.P. Large shellmounds (known locally as *sambaquís*) ranging between 10 and 25 meters in depth are common along the coast south of Rio de Janeiro, and similar mounds occur in smaller numbers north of the city as well (Fairbridge 1976; PRONAPA 1970; Menghin 1962; Serrano 1946; Bigarella 1965; Emperaire and Laming 1956; Laming 1960; Hurt and Blasi 1960; Stuckenrath 1963; Hurt 1966, 1968, 1974; Rauth and Hurt 1960). These mounds consist primarily of the remains of various mollusc species, along with fishbones, the bones of marine mammals, and some bones of land mammals. The tool kit includes the same range of chipped stone tools and polished or semipolished axes which occurs inland. It seems probable that the middens represent regular seasonal transhumance of inland populations, although they have also been interpreted as representing sedentary coastal populations. The oldest of the *sambaquís* so far dated is the Sambaquí do Maratua on the Bay of Santos, São Paulo, which has produced radiocarbon dates of 7800 and 7300 B.P. Another *sambaquí* from the Bay of Antonina in Paraná has yielded a date a little in excess of 6,000 years. Most of the dated *sambaquís,* however, are somewhat younger, as shown by their radiocarbon dates: the Sambaquí do Gomes, also near the Bay of Antonina in Paraná, 4500 B.P.; the Sambaquí de Saquarema, between 4000 and 4500 B.P.; the Sambaquí do Macedo, between 3200 and 3500 B.P.; the Sambaquí de Forte Marechal Luz on San Francisco Island, Santa Catarina, from 3700 B.P. to about 800 B.P.; and the Sambaquí da Ilha Ratos, approximately 1500 B.P. It is clear from some of the more recent dates as well as from the presence of recent and even postcolonial potsherds on the upper layers of some of the shellmounds that the occupation and exploitation of the coast persisted into the recent past, well after the arrival of ceramic and agricultural populations.

Exactly when agricultural economies emerged in (or spread to) eastern Brazil, however, is an open question. Late preceramic sites provide no evidence of domestication, nor do early ceramic sites, which appear in the region beginning about 2800 B.P. (PRONAPA 1970). The first reasonably firm evidence of agriculture in this region occurs only about 1500 B.P., when pottery of the type associated with the historic Tupi-Guaraní first appears in Brazil. The sites of this ceramic complex contain *budares,* suggesting that manioc cultivation was a major part of their economy. (As indicated above, the presence of *budares,* which can be used to process other vegetable foods, does not necessarily prove that manioc was under cultivation. But, in this case, their presence in combination with a type of pottery made by people known historically to have been farmers does seem to provide a reasonable estimate of the minimum age of manioc cultivation in eastern Brazil.)

In contrast to the long but spotty archaeological record in eastern Brazil, Paraguay, which lies in the Gran Chaco depression, has produced very little in the way of archaeological evidence. Despite a number of archaeological surveys (Nordenskiold 1902–03; Biro de Stern 1944; Márquez Miranda 1942; Rydén 1948; Fock 1961, 1962), there is no evidence whatever of preceramic occupation in most of this area (Willey 1971). The known ceramic archaeology has yielded little economic evidence, although the riverine distribution of pottery and the presence of shell refuse mounds associated with pottery in some locations (Boggiani 1900) indicate a riverine fishing and shellfishing economy. The Chaco tribes of the ethnographic present seem to have subsisted primarily on the basis of the processing of wild plant foods, particularly *algorroba* (*Prosopis* sp.), palm nuts and shoots, and wild fruits, supplemented with seasonal fishing, some hunting, and the gathering of small animals and insects (Métraux 1946). By 1800 A.D. most of the Chaco tribes seem to have also practiced some agriculture.

The lack of early history in the Chaco area may reflect poor preservation (although this is an arid environment) or the paucity of modern archaeological work. However, given the natural limits of the region and the impoverished economy of its historic groups, it is possible that this area simply was not occupied until the pressure of population forced people into what was clearly a marginal environment.

Unlike the Gran Chaco, the Argentine pampas and Patagonia seem to have provided rich hunting for human populations throughout the history of human occupation in South America. As discussed above, the earliest known occupations of the region at Fell's Cave and related sites indicate the hunting of extinct Pleistocene fauna. Once the fauna were extinct the inhabitants of the region seem to have turned to modern fauna, particularly guanaco and rhea, retaining their plains-oriented hunting emphasis until the very recent past. The Magellan I-V sequence from these caves (Bird 1938) documents a history of minor stylistic change within what was predominantly a hunting economy from 11,000 B.P. to the historic period. Stylistically related finds occur through much of Argentine Patagonia. After the extinction of the Pleistocene fauna, the only indications of population pressure on the hunting economy are the appearance of possible late preceramic and early ceramic coastal shellmounds in northern Patagonia and the Argentine pampas (Cigliano 1966a,b; Lanning 1974b) and the recorded hunting of sea mammals and riverine fishing by historic tribes in the area (Lothrop 1928, 1946; Serrano 1946). Agriculture never penetrated this area in the prehistoric period, nor is there archaeological or ethnographic evidence of significant dependence on plant processing among its tribes.

In northern Argentina and Uruguay, however, along the lower Paraná and lower Uruguay rivers, similar hunting cultures do seem to have been replaced by riverine groups in the recent prehistoric past. Ethnographic descriptions of the northern riverine tribes depict groups relying primarily on fishing and secondarily on hunting and plant gathering, with farming playing only a minor part in the economy. Such groups seem to have coexisted with more fully agricultural populations (Lothrop 1932, 1946). The archaeological evidence, though sparse and poorly dated, does indicate some time depth to this riverine pattern. Ceramic sites of the Malabrigo phase on the Paraná River in Santa Fé province, Argentina, include refuse of shell and fishbone, as well as the bones of birds and reptiles and the charred remains of palm nuts (Frenguelli and Aparicio 1923; Aparicio 1948; Serrano 1954). Similar riverbank sites are reported from a number of other locations in the region (Willey 1971).

The Amazon Basin

The tropical forest zone in the basin of the Amazon River and its tributaries is potentially one of the most important areas in South America for the study of early agriculture because of the number of cultigens presumed to have been developed east of the Andes; but it is an area for which the archaeological record is extremely poor. As a result, the reconstruction of economic prehistory in this region is somewhat more speculative than for the other regions discussed. It was pointed out earlier that there is no evidence of Paleo-Indian occupation in the forest zone. I suggested that this pattern was probably not simply a matter of poor preservation, since the tropical forest would have been an inhospitable environment for early hunters and gatherers and since there is no evidence, even in zones of better preservation, that the Paleo-Indians possessed the type of fishing oriented economy which would have been necessary for survival in the rainforest.

The earliest archaeological sites in the Amazon Basin date to only about 4000 B.P. Lathrap (1970) has identified a number of settlements dating to the fourth millennium B.P. on the eastern fringes of the Amazon Basin along the upper Amazon tributaries of eastern Peru. The Tutishcainyo ceramic assemblage from the banks of Yarinacocha, an oxbow lake along the upper Ucayalí, is dated to approximately 4000 B.P. by a combination of detailed ceramic comparisons to dated sequences in the Peruvian highlands and extrapolation backwards from later radiocarbon dated assemblages in the Yarinacocha area itself. By Lathrap's interpretation, Tutishcainyo pottery already represents a fairly well developed tropical forest culture, involving sedentary occupation by a group of a few hundred fishermen and manioc farmers oriented to a riverine environment. The economy can be reconstructed only from indirect sources, however. Shells of freshwater molluscs are used as pottery temper, and fishbone and fish scales occur as occasional inclusions in pottery, indicating the use of riverine resources. The cultivation of manioc is inferred from the presence of pottery vessels characteristically associated with manioc beer, though griddles for processing bitter manioc are absent. Lathrap also refers to the Cobichaniqui complex along the Río Nazaratequi in eastern Peru, which has provided several radiocarbon dates ranging from about 3400 to about 3700 B.P. Here again, however, the only reasonably direct economic evidence is in the form of shell temper in ceramic wares and in the riverine orientation of the site. Lathrap has also called attention to early ceramics at the Cave of Owls in the Huallaga Valley, which he places in the fourth millennium B.P. on the basis of stylistic comparisons with other dated sites.

Outside eastern Peru, none of the dated ceramic assemblages has proved to be quite so old. In eastern Ecuador, Evans and Meggers (1968) have defined an early Yasuní complex which contains pottery that some authorities (cf. Willey 1971) compare to the early Tutishcainyo wares. The radiocarbon date here, however, is only about 2000 B.P. At the opposite end of the continent, on Marajó Island at the mouth of the Amazon, Meggers and Evans (1957; see also Evans and Meggers 1968) have identified a similar ceramic complex, named the Ananatuba phase, which has been carbon dated to 2900 B.P. Between the two extremes, at Ponta do Jauarí on the lower middle Amazon near Alemquer, Hilbert (1968) has identified pottery resembling that of Ananatuba in middens consisting primarily of the remains of freshwater shellfish, along with fishbones, and the bones of tapir, pig, alligator, and turtle. No dates are available from Ponta do Jauarí, but it is presumed to date to the third or fourth millennium B.P. on the basis of its stylistic similarities to the other early ceramic complexes.

It should be noted, incidentally, that preceramic occupations of the Amazon Basin are all but unknown; none has yet been firmly established. Hilbert (1968) does refer, however, to the existence of possible preceramic shellmounds on the Atlantic coast just south of the mouth of the Amazon as well as to others in the Amazon Valley at Santarém and Alemquer.

The strikingly late dates for the earliest known occupations of the Amazon may, of course, be a function of poor preservation and poor recognition, as discussed above. It is noteworthy that most Amazonian sites are identifiable primarily through their pottery, and it is therefore likely that preceramic sites would go largely if not totally undetected. It is also noteworthy, however, that the fourth millennium dates from the Amazon Basin are roughly contemporary with the appearance of the majority of the coastal and riverine sites in other areas that have more fully preserved archaeological sequences. I suspect that major penetration of the Amazon would only have taken place as a result of the same pressures that forced populations into riverine habitats in other areas. On this basis I would argue that significant occupation of the heart of the Amazon Basin may not have preceded the known, dated sites by any significant period of time.

Lathrap (1970), however, has made a strong case from indirect evidence (linguistics, the distribution of ceramic styles, and the distribution of known tropical cultigens) that the occupation of the Amazon Basin, although not extending to the Paleo-Indian period, is considerably older than the present archaeological record indicates. In a recent paper (Lathrap in press) he has argued that the major hearth of domestication in South America centered on the alluvial floodplains of Amazonia and northern South America by 6000 to 7000 B.P. and that significant population densities would have been achieved there by 5000 B.P.

Lathrap's argument is both convincing and significant up to a point. As he notes, the appearance of cultigens such as manioc, peanuts, and sweet potatoes on the coast of Peru between 3000 and 4000 B.P. and the appearance of griddles for processing manioc at the same time or slightly earlier in Venezuela and Colombia argues strongly for the existence of an earlier period during which these cultigens were under domestication east of the Andes. I would differ with Lathrap, however, if he intends to imply that this evidence entails the early or dense populating of Amazonia as a whole, and I disagree with his contention that cultivation was fully established in the Amazon Basin earlier than its known occurrence on the margins of the continent. The key point, as noted above, is that most of the lowland tropical crops appear to have developed on the margins of, rather than within, the rain forest proper. Moreover, to the extent that their origins are well defined, the various lowland tropical cultigens seem to have been derived from different areas, suggesting synchronous parallel pressures for domestication in a number of these marginal environments. These crops may not have been brought into the heart of the Amazon or combined to form a single, well-defined agricultural complex until long after each was developed as a cultigen. Their diffusion through the Amazon Basin need not have occurred any earlier than their diffusion to coastal regions.

I would tentatively propose the following model for cultural development and early domestication in lowland tropical America, noting first that the model is speculative, derived *from* the assumptions used in this book and not intended as independent corroboration of those assumptions.

The earliest colonizers of South America, arriving at least 11,000–12,000 years ago and possibly much earlier, seem to have confined themselves to relatively open, game-rich environments, migrating south along the Andean chain and then expanding on to the plains of southeastern South America. Subsequently these groups were forced, presumably by their own population growth in combination with the extinction of some of their favored prey, to expand the range of resources exploited and their own geographical range. This expansion is quite clearly documented, beginning about 9000 B.P., in the exploitation of coastal *lomas* environments in Peru and the gradually increasing use of marine resources. I would suggest that movement eastward from the Andes occurred in a roughly symmetrical manner and that we will eventually find evidence of penetration of the *margins* of Amazonia dating back 8,000 years or more. As this movement progressed, the low density of game in the forest environment would have necessitated a shift to the use of riverine resources, paralleling economic changes on the coasts. At the same time, as Lathrap (in press) has suggested, the low density of individual plant species in the tropical forest would have led to human efforts to increase artificially the numbers of selected plants. Such efforts would have led to incipient domestication of a variety of lowland tropical plants in a number of regions. This process must have begun at least 8,000 years ago, if we are to account for the arrival of lima beans at Guitarrero Cave. The coalescence of such patterns of incipient cultivation into an evolved tropical forest culture with intensive agriculture and fishing is likely to have proceeded slowly, however. Both the evolution of this economic pattern and the concomitant geographical expansion of human populations into the heart of the rain forest are likely to have occurred only as fast as was made necessary by the buildup of population pressure. Hence, fully developed tropical forest cultivators may in fact only have become widespread in Amazonia 4,000 to 5,000 years ago. If Lathrap (1968b, discussed above) is correct, this initial colonization of most of Amazonia would have occurred through the movements of farming and fishing populations along the rivers. Colonization of the inter-riverine forest environment may have been even more recent, resulting from a buildup of population pressure along the rivers which gradually forced some tropical forest communities to abandon the rivers and revert to a more generalized type of hunting, gathering, and small-scale farming that was suited to the hinterlands.

7 SUMMARY AND CONCLUSIONS

In the first chapter, two questions were posed about the origins of agriculture. I asked, first, why human populations successfully adapted to a hunting and gathering life-style would abandon it in favor of farming; and, second, I asked why so many of the world's people undertook this economic transformation at approximately the same time. It was pointed out that these questions were particularly difficult to answer since the evidence suggested that primitive agriculture had no economic advantage over hunting and gathering except that it provided a greater total number of calories for each unit of space. The data indicated, in fact, that the adoption of agriculture probably resulted in an increased per-capita work load and a decline in the quality of the diet. It seemed likely, therefore, that agriculture would only have been adopted under conditions in which the demand for calories was increasing or at least where the demand was out of balance with the productive potential of the existing economy. It was suggested that any number of factors might be cited to account for increasing demand or ecological imbalance in particular locations, but I argued that only one possible explanation—actual population growth—could account for increasing demand or ecological imbalance which appeared to span a wide range of ecological zones and to embrace a large number of different cultures. I argued, in essence, that the nearly simultaneous adoption of agricultural economies throughout the world could only be accounted for by assuming that hunting and gathering populations had saturated the world approximately 10,000 years ago and had exhausted all possible (or palatable) strategies for increasing their food supply within the constraints of the hunting-gathering life-style. The only possible reaction to further growth in population, worldwide, was to begin artificial augmentation of the food supply.

This conclusion in turn necessitated a reevaluation of existing assumptions about the equilibrium presumed to exist between human populations and their resources. I argued that, rather than stabilizing at some optimum level or "carrying capacity," the human population as an aggregate has grown continuously, requiring more or less continuous redefinition of the ecology of the species as a whole. On a somewhat more controversial plane, I argued that the growth of population and the buildup of population pressure could have occurred very gradually and very evenly throughout the world despite the acknowledged existence of a host of variables tending to create irregular patterns of population growth and decline on the local level. Briefly, I argued, using ethnographic analogies, that mechanisms for the redistribution of population exist among hunter-gatherers and that these mechanisms are highly effective in equalizing population pressure among local groups both within culturally homogeneous units and across cultural barriers. Using archaeological evidence, I argued further that such mechanisms were considerably more widespread in the past than they are today. It was thus possible that populations throughout the world approached critical densities and adopted broad-spectrum economies and then agriculture at approximately the same time.

Finally, the means of identifying population pressure in early pre-history were discussed and the prehistory of the world's major regions was reviewed, providing evidence of the continuous evolution of human economies, not in the direction dictated by technological advances and improvements in the quality of life, but rather in the direction dictated by ever-increasing demand. The record appears to show that human populations on each continent first concentrated fairly heavily on the exploitation of large mammalian fauna—a prized but scarce resource—and then shifted gradually toward broader spectrum economies geared to more plentiful but less palatable resources. In each case, domestication techniques were then focused on plant species chosen not for their palatability but for their ability to provide large quantities of storable calories or storable protein in close proximity to human settlements.

The Problem of Local Variation

My primary aim has been to account for the origins of agriculture. A case has been made for a *general* explanation which accounts, satisfactorily I believe, for the *general* distribution of agricultural origins in time and in space. It has been my contention that the broad similarity of events in various regions of the world calls for such an explanation despite the historical particularism preferred by many of my colleagues. It is clear, however, that the explanation which has been offered is not yet adequate to account for the *particular* sequences of local events in which the more general pattern is played out nor for the regional irregularities in agricultural origins. I cannot yet explain satisfactorily why, given the overall buildup of population pressure, agriculture began slightly earlier in some regions than in others nor why it was developed independently in some regions and diffused to others. I have suggested a number of possible types of explanations in the course of this book. It has been suggested, for example, that slight irregularities in the development of population pressure may account for the patchwork pattern of agricultural origins; it has also been suggested that agriculture may first have begun where developing population pressure intersected the distribution of particular environmental phenomena or the distribution of relatively palatable, easily manageable and responsive potential domesticates. Conversely, agriculture may have been delayed or even postponed indefinitely in some regions by the relative intractability of potential local domesticates or by the inhospitability of the local environment to cultigens available by diffusion.

David Harris (1972, in press) has provided a somewhat more refined list of variables which may help to explain irregularities in the pattern of agricultural development in greater detail. He notes, for example, that the spatial distribution of resources over the landscape may influence the mobility structure of the local human population, affecting both its rate of population growth and its probable response to increasing demand. He contrasts simple and complex ecosystems, as defined in terms of the number of different species and the number of individuals of each species found within a unit area. He argues that complex systems, with their variety of species, are likely to generate sedentism and domestication, whereas simple systems have tended historically to encourage mobile reliance on individual migratory species, a pattern not conducive to emergent agriculture. Even more explicitly, he has suggested that the most propitious zones for emergent agriculture may have been the marginal or transitional zones between major ecosystems, where maximum local and seasonal variation was available.

Harris also calls attention to the possible role of environmental changes in triggering sedentism and domestication on the local level, but he makes the interesting observation that the *rate* as well as the type of environmental change may be significant. The critical variable may thus not be the environmental change itself so much as the rate of change relative to the potential effective rates of various cultural responses. Very gradual changes may have no significant effect on the equilibrium of a population; too rapid change may simply necessitate outmigration. In contrast environmental changes of intermediate rate and duration may effectively trigger changes in economic activities. Presumably, the rate of environmental change might also help select the type of economic activity undertaken and thus help determine whether domestication occurs.

A number of technological variables may also help to influence the direction of response to population pressure. As Harris points out, the techniques employed by a population as it intensifies its use of certain resources may help to determine whether actual domestication results and whether efforts at intensification are self-reinforcing (positive feedback) or ultimately self-limiting. He notes, for example, following Wilke et al. (1972), that certain patterns of grass harvesting such as the hand and sickle harvesting methods employed in the Middle East tended automatically to select for useful mutants, thus reinforcing the expenditure of human labor devoted to these grasses. Equally intense exploitation of grasses among the Shoshone of the New World failed to result in domestication because the harvest technique employed produced no such selection.

Harris also calls attention to a number of variables among potential domesticates themselves which may help to account for differential patterns in the emergence of economies based on domestication. He suggests, for example, that various tree species, although used as intensively as some cereals, resisted domestication not only because of a long life span that inhibited human selection but also because of their tendency to cross-pollinate, which would discourage human efforts to promote or control genetic variation. Domestication was also inhibited among species whose life cycles resisted human manipulation in other ways. Species of migratory fish or marine mammals, although exploited as staples in some regions, tended to resist domestication since critical portions of their life cycles were outside human control. In a slightly different vein, Harris suggests that domestication may be determined by the degree to which certain species are structurally preadapted for particular cultural practices. Thus he suggests that among tuberous plants differences in toxicity and in the capacity of the tuber to persist underground prior to harvest may help to explain why some taxa were domesticated and others were not.

Finally, amplifying on the work of Flannery (1968), Harris suggests that the seasonality of potential domesticates may have influenced patterns of domestication. Where the seasonal exploitation of potential domesticates tended to conflict with the harvesting or processing of other significant or valued food sources, domestication may have been inhibited. In contrast, complementary seasonal patterns among various resources or the lack of marked seasonality among competing resources may have helped to promote domestication.

Despite such refinements in our awareness of the variables controlling domestication, however, it is clear that a good deal of work still needs to be done before patterns of historical development can be analyzed on a local level. The thrust of this book and of other recent studies docs, however, suggest the direction which this work might most profitably take. The model presented here suggests that we should focus our attention not on the "invention" or "diffusion" of agriculture nor on the conditions under which invention and diffusion are likely to take place. Rather we need to be concerned with the conditions under which domestication will come to be viewed as profitable by hunting and gathering populations. The primary variable here appears to be the ratio between the size of the human population inhabiting a particular space and the productivity or potential productivity of different economic strategies in that space. Since population density appears to be the central variable, continuing refinement in our knowledge of the factors governing hunting and gathering populations would appear to be essential to the building of more accurate reconstructions of prehistoric events on a local scale. A good deal of work has, of course, already been done along these lines. Unfortunately, the work most often cited by students of cultural evolution has tended to focus heavily on theoretical or idealized mechanisms of population limitation rather than on the analysis of the quantitative operation of actual mechanisms. The systems models which are prevalent in anthropology need to be supplemented with actual statistical models of choices made by potential parents among hunting and gathering groups under specified conditions. In particular, it seems necessary that we locate and quantify the foci of "slippage" or imperfect operation in population control systems. In order to accomplish this end, it will be necessary for evolutionary anthropologists to focus their attention more acutely on the individual case studies and on the techniques of microanalysis which are available from our colleagues in the demographic sciences. Unfortunately, of course, potential informants representing hunting and gathering populations are becoming exceedingly scarce.

Similar quantifications must also be applied to the other variables in the equation. As Harris (in press) has pointed out, if environmental changes are to be viewed as potentially significant variables on the local level it will be necessary to redefine generalized notions of environmental change into specific, quantifiable statements about the precise nature and rate of change in the natural productivity of specific resources whose impact on the human economy can then be measured.

In addition, it is clear that much more detailed data must be made available concerning the quantitative parameters involved in economic decisions. Further work on labor costs and productivity for particular wild and domestic plants needs to be done before even rational decisions on the part of prehistoric man can be understood or reconstructed with any scientific rigor. Recent work cited by Flannery (1973) on the relative productivity of domestic maize in various stages of development in comparison with the natural productivity of certain wild legume plants in Mexico is of considerable importance in this regard. Flannery has already shown that the relative productivity figures can be directly related to actual economic developments in the Tehuacán sequence, and it seems probable that similar figures for a range of plants in a variety of environments would contribute significantly to a precise understanding of the distribution of early cultivation in time and space.

Unfortunately, simple rational computation of labor costs and caloric outputs will probably also prove insufficient as a device to predict prehistoric behavior and it will be necessary to supplement this data with rank-order analyses of the cognitive variables involved in decision-making among hunting and gathering groups. Such cognitive studies in conjunction with accurate data on labor costs and productivity should enable us to predict with considerable accuracy the conditions under which the transition to agricultural economies would occur. Here again, however, the lack of potential informants is critical.

In conclusion, two other points should be raised. The data presented here have implications which go far beyond an explanation of the origins of agriculture and reflect on the very nature of man as an animal and on the nature of his adaptive patterns. First, the archaeological record indicates fairly clearly that the process of ecological adjustment on the part of human populations has been a continuous one. The apparent stability of ethnographic populations is misleading, either because modern "primitive" groups are anomalous historically or because they are being observed over too brief a time span to reveal their inherent growth. The archaeological data suggest that we must revise our model building or, more particularly, revise our systems models and break away from our assumption that human cultures are inherently stable systems until jostled by some "outside event" such as climate change or technological discovery. Clearly, successful human systems are designed for more or less continuous change, and, as I have suggested above, such change is apparently in part responsible for their competitive success. Natural selection seems to have selected for unstable systems.

At the same time, but on a somewhat less sophisticated level, it should be clear that culture growth cannot be equated with progress as it is in the public mind and in much of the older literature. The data suggest that for much of human history labor costs for food went up; that man moved into more and more inhospitable environments; and that he forced himself to concentrate on less and less palatable foods. Our technology, which we are inclined to view as a great liberating force, appears in historical perspective to be more of a holding action. Rather than progressing, we have developed our technology as a means of approximating as closely as possible the old status quo in the face of our ever-increasing numbers. With this realization cultural evolution is tied just a little more securely to biological evolution as an adaptive process and the gap between man and animal is narrowed proportionately.

This altered conception of our own evolution, should help us to respond more constuctively to the present "food crisis." Our prevailing historical optimism about our own progress has been badly shaken by the realization that we are "suddenly" being asked to make do with less—in particular with less meat—and to substitute foods which are available or can be produced in quantity but which we now define as unpalatable. The prospect of altering our food supply might be more constructively faced if we realized that it is the prevailing notion of "progress" rather than the contemporary "crisis" which is the historical anomaly. Perhaps it will aid us in our economic transition to realize that human populations once faced the notion of eating oysters and later the prospect of eating wheat with much the same enthusiasm that we now face the prospect of eating seaweed, soy protein, and artificial organic molecules.

BIBLIOGRAPHY

Acheson, R. M. 1959. Effects of starvation, septicaemia and chronic illness on the growth cartilage plate and metaphysis of the immature rat. *J. Anat.* 93:123–30.

Ackerknecht, E. H. 1948. Medicine and disease among Eskimoes. *The Eskimo, CIBA Symp.* 10:916–20.

Adams, R. M. 1966. *The evolution of urban society*. Chicago: Aldine.

Agogino, G. A. and Frankfortor, W. D. 1960. A Paleo-Indian bison kill in northwestern Iowa. *Amer. Antiquity* 25:414–15.

Agogino, G. A. and Rovner, I. 1964. Paleo-Indian traditions. *Archaeo.* 17:237–43.

Aikens, C. M. 1970. *Hogup Cave*. Univ. Utah Anthropol. Papers 93.

Alexander, H. L. 1963. The Levi site: a Paleo-Indian campsite in central Texas. *Amer. Antiquity* 28:510–28.

Alexander, J. and Coursey, D. C. 1969. The origins of yam cultivation. In Ucko and Dimbleby, eds., pp. 405–25.

Allan, W. 1965. *The African husbandman*. Edinburgh: Oliver and Boyd.

Alland, A. and McKay, B. 1973. The concept of adaptation in biological and cultural evolution. In J. Honigman, ed. *Handbook of social and cultural anthropology,* pp. 143–78. Chicago: Rand McNally.

Allchin, B. 1973a. Blade and burin industries of west Pakistan and western India. In Hammond, ed. pp. 39–50.

———. 1973b. Problems and perspectives in South Asian archaeology. In Hammond, ed. pp. 1–11.

Allchin, F. R. and Allchin, B. 1968. *The birth of Indian civilization*. Baltimore: Penguin.

Ammerman, A. J. 1975. Late Pleistocene population dynamics: an alternative view. *Human Ecol.* 3:219–34.

Ammerman, A. J. and Cavalli-Sforza, L. L. 1973. A population model for the diffusion of early farming in Europe. In Renfrew, ed., pp. 343–57.

Anderson, E. 1952. *Plants, man and life*. Boston: Little, Brown.

Angel, J. L. 1972. Ecology and population in the eastern Mediterranean. *World Archaeol.* 4:88–105.

———. 1975. Paleoecology, paleodemography and health. In Polgar, ed. pp. 167–90.

Angulo Valdez, C. 1962. Evidence of the Barrancoid Series in northern Colombia. In Wilgus, ed. *The Caribbean: Contemporary Colombia*, pp. 35–46. Gainesville: Univ. Florida.

Aparicio, F. de 1928. Notas para el estudio de la arqueología del sur de Entre Ríos. *Anal., Fac. Cienc. y Educ.* 3:1–63. Paraná.

——. 1948. The archaeology of the Paraná River. In Steward, ed., vol. 3, pp. 57–68.

Arkell, A. J. 1949. *Early Khartoum*. London: Oxford Univ. Press.

——. 1953. *Shaheinab*. London: Oxford Univ. Press.

Arkell, A. J. and Ucko, P. J. 1965. Review of predynastic developments in the Nile Valley. *Current Anthropol.* 6:145–166.

Arnold, B. A. 1957. Late Pleistocene and recent changes in land forms, climate, and archaeology in central Baja California. *Univ. Cal. Pub. Geog.* 10:201–318.

Asch, N. et al. 1972. *Paleoethnobotany of the Koster site: the archaic horizons.* Ill. State Mus. Res. Papers 6. Rep. of Invest. 24.

Ascher, R., and Ascher, M. 1965. Recognizing the emergence of man. *Science* 147:243–50.

Austral, A. G. 1972. El yacimiento de Los Flamencos II; la coexistencia del hombre con fauna extinguida en la región pampeana. Soc. Argentina Antropol. *Relaciones* 6:203–09.

Aveleyra Arroyo de Anda, L. 1950. *Prehistoria de México*. México D. F.: Ediciones Mexicanas.

Baerreis, D. A. 1951. *The preceramic horizons of northeastern Oklahoma*. Univ. Mich. anthropol. papers 6.

——. 1959. The Archaic as seen from the Ozark region. *Amer. Antiquity* 24: 270–75.

Baggerly, C. 1954. Waterworn and glaciated stone tools from the thumb district of Michigan. *Amer. Antiquity* 27:93–100.

Baker, H. G. 1970. *Plants and Civilization* (2nd ed.). Belmont:Wadsworth.

Bakker, E. M. van Zinderen. In press. Paleoecological background in connection with the origins of agriculture in Africa. In Harlan et al., eds., in press.

Bakker, E. M. van Zinderen and Clark, J. D. 1962. Pleistocene climates and cultures in northeastern Angola. *Nature* 196:639–42.

Balikci, A. 1968. The Netsilik Eskimoes: Adaptive processes. In Lee and DeVore, eds., pp. 78–82.

Barbour, E. H. and Schultz, C. B. 1932. *The Scottsbluff bison quarry and its artifacts*. Nebr. State Mus. Bull. 34:1.

Barfield, L. 1961. Recent discoveries in the Atacama Desert and the Bolivian Altiplano. *Amer. Antiquity* 27:93–100.

——. 1971. *Northern Italy before Rome*. New York: Praeger.

Barker, G. W. W. 1973. Cultural and economic change in the prehistory of central Italy. In Renfrew, ed., pp. 359–90.

——. 1975. Prehistoric territories and economies in central Italy. In Higgs, ed., pp. 111–76.

Barnicot, N. A. 1969. Human nutrition: evolutionary perspectives. In Ucko and Dimbleby, eds., pp. 525–30.

Bartholomew, G. A. and Birdsell, J. B. 1953. Ecology and the protohominids. *Amer. Anthropol.* 55:481–98.

Bartlett, K. 1943. The primitive stone industry of the Little Colorado Valley, Arizona. *Amer. Antiquity* 8:266–68.

Bar-Yosef, O. 1975. Archaeological occurrences in the Middle Pleistocene of Israel. In Butzer and Isaac, eds., pp. 571–604.

Bates, M. 1955. *The prevalence of people*. New York: Scribner's.

Baumhoff, M. A. and Heizer, R. F. 1965. Postglacial climate and archaeology in the desert west. In Wright and Frey, eds., pp. 697–707.

Beadle, G. 1972. The mystery of maize. *Field Mus. Natur. Hist. Bull.* 43:2–11.

———. In press. The origins of *Zea mays*. In Reed, ed., in press.

Beardsley, R. K. 1948. Culture sequences in central California archaeology. *Amer. Antiquity* 14:1–28.

Bell, R. E. 1960. Evidence of a fluted point tradition in Ecuador. *Amer. Antiquity* 26:103–06.

———. 1965. *Archaeological investigations at the site of El Inga, Ecuador*. Quito: Casa de la Cult.

———. 1974. *Investigation of the El Inga complex and preceramic occupations of highland Ecuador*. Norman: Univ. Okla.

Bell, R. E. and Baerreis, D. A. 1951. A survey of Oklahoma archaeology. *Bull. Texas Archaeol. and Paleontol. Soc.* 22:7–100.

Benedict, B. 1972. Social regulation of fertility. In Harrison and Boyce, eds., 1972.

Bennyhoff, J. A. 1950. California fish spears and harpoons. *Univ. Cal. Rec.* 9: 295–337.

———. 1958. The Desert West. A trial correlation of culture and chronology. In *Current views on Great Basin archaeology*. Reports of the Univ. of Cal. Archaeol. Surv. 42.

Benson, E. P., ed. 1968. *Dumbarton Oaks Conference on the Olmec*. Washington: Dumbarton Oaks.

Bentzen, R. 1962. The Powers-Yonkee Bison trap. *Plains Anthropol.* 7:113–18.

Berciu, D. 1967. *Romania before Burebista*. London: Thames and Hudson.

Berdichewsky, B. 1964. Arqueología de la desembocadura del Aconcagua y zonas vecinas de la costa central de Chili. In *Arqueología de Chile Central y Areas Vecinas*. Pub. de los Trabajos Presentados al Tercer Congr. Int. de Arqueología Chilena. Santiago: Imprenta "Los Andes."

Berger, R. and Libby, W. F. 1967. UCLA Radiocarbon dates VI. *Radiocarbon* 9:477–504.

Biberson, P. 1961. *Le paléolithique inférieure du Maroc atlantique*. Rabat: Service des antiquités du Maroc Mém. 17.

Bibikov, S. N. 1950. Pozdneisii paleolitkryma. Materialy po cerverticnomu Periodu. *SSSR* 2:122–123.

Binford, L. R. 1968. Post Pleistocene adaptations. In Binford and Binford, eds., pp. 313–41.

———. 1972. Contemporary model building: paradigms and the current state of Paleolithic research. In Clarke, ed., pp. 109–166.

Binford, L. R. and Binford, S. R. 1966. A preliminary analysis of functional variability in the Mousterian or Levallois facies. *Amer. Anthropol.* 68(2 pt. 2): 238–295.

———. eds. 1968. *New perspectives in archaeology*. Chicago: Aldine.

Binford, S. R. 1968. A structural comparison of disposal of the dead in the Mousterian and Upper Paleolithic. *Southwestern J. Anthropol.* 24:139–54.

Bird, J. B. 1938. Antiquity and migrations of the early inhabitants of Patagonia. *Geogr. Rev.* 28:250–75.

———. 1943. Excavations in Northern Chile. *Amer. Mus. Natur. Hist. Anthropol. Papers* 38:171–318.

———. 1946. The archaeology of Patagonia. In Steward, ed., Vol. 1, pp. 17–24.

———. 1948. Preceramic culture in Chicama and Viru. In W. Bennett, ed., *A reappraisal of Peruvian archaeology*. Soc. Amer. Archeol. Mem. 4, pp. 21–28.

———. 1965. The concept of a 'Pre-Projectile Point' cultural stage in Chile and Peru. *Amer. Antiquity* 31:262–70.

———. 1970. Paleo-Indian discoidal stones from southern South America. *Amer. Antiquity* 35:205–08.

Birdsell, J. B. 1953. Some environmental and cultural factors influencing the structure of Australian aboriginal populations. *Amer. Natur.* 87:171–207.

———. 1958. On population structure in generalized hunting and collecting populations. *Evolut.* 12:189–205.

———. 1968. Some predictions for the Pleistocene based upon equilibrium systems among recent hunters. In Lee and DeVore, eds., pp. 229–40.

———. 1972. *Human evolution*. New York: Rand McNally.

Biro de Stern, A. 1944. Hallazgos de alfarería decorada en el territorio del Chaco. Soc. Argentina Antropología, *Relaciones* 4:157–161.

Bischof, H. 1966. Canapote—an early ceramic site in northern Colombia. *Actas y memorias, 36th Int. Congr. of Americanists* 1:484–91.

———. 1972. The origins of pottery in South America: recent radiocarbon dates from southwest Ecuador. *Atti del XL Congr. Int. degli Americanisti* 1:269–81.

Bischof, H. and Viteri Gamboa, J. 1972. Pre-Valdivia occupations on the southwest coast of Ecuador. *Amer. Antiquity* 37:548–51.

Black, D. 1933. *Fossil man in China*. Memoirs of the Geological Survey of China. Series A:3.

Blasi, O. 1965. Os indícos arqueológicos do Barrãcao e Dionísio Cerquerira, Paraná Santa Catarina. *Arq. Mus. Paranaense, Arqueología* 2.

Boggiani, G. 1900. Compendio de etnografía Paraguaya moderna. *Rev. Inst. Paraguayo* 3:23–25, 27, 28.

Bökönyi, S. 1970. Animal remains from Lepenski Vir. *Science* 167:1702–04.

Bonwick, J. 1870. *Daily life and origin of the Tasmanians*. London: Low and Merston.

Borden, C. E. 1966. Radiocarbon and geological dating of the lower Fraser Canyon archaeological sequence. In *Proceedings of the 6th. International Conference on Radiocarbon and Tritium Dating, Pullman, 1965*.

Bordes, F. 1953. Essai de classification des industries "moustériennes." *Bull. Soc. Préhist. Française* 50:457–66.

———. 1961a. Mousterian cultures in France. *Science* 134:803–10.

———. 1961b. *Typologie du paléolithique ancien et moyen*. Publ. Inst. Préhist. Univ. Bordeaux. Mèm. 1.

———. 1968. *The Old Stone Age*. London: Wiedenfeld and Nicholson.

Bordes, F. and Prat, F. 1965. Observations sur les faunes du Riss et du Wurm I en Dordogne. *L'Anthropol.* 69:31–45.

Bórmida, M. 1964a. Las industrias líticas precerámicas del Arroyo Catalán Chico y del Río Cuareim. *Riv. Sci. Prehist.* 19:Fasc. 1–4.

——. 1964b. El cuareimense. In *Homenaje a Fernando Márquez-Miranda.* Madrid: Ediciones Castilla, S.A., pp. 105–28.

Borns, H. W. 1971. Possible Paleo Indian migration routes in northeastern North America–a geological approach. *Maine Archaeol. Soc. Bull.* 2:33–39.

Bose, S. 1964. Economy of the Onge of Little Andaman. *Man in India* 44:298–310.

Boserup, E. 1965. *The conditions of agricultural growth.* Chicago: Aldine.

Bouchud, J. 1954. Le renne et le problème des migrations. *L'Anthropol.* 58:79–85.

Bourlière, F. 1963. Observations on the ecology of some large African mammals. In Howell and Bourlière, eds., pp. 43–54.

Braidwood, R. J. 1957. Jericho and its setting in Near Eastern prehistory. *Antiquity* 31:73–80.

——. 1958. Near Eastern prehistory. *Science* 127:1419–30.

——. 1960. The agricultural revolution. *Sci. Amer.* 203:130–48.

——. 1967. *Prehistoric Men.* 7th edition. Glenview: Scott, Foresman.

Braidwood, R. J. and Braidwood, L. 1950. Jarmo: a village of early farmers in Iraq. *Antiquity* 24:189–95.

Braidwood, R. J. et al. 1971. Beginnings of village farming communities in southeastern Turkey. *Proc. Nat. Acad. Sci.* 68:1236–40.

Braidwood, R. J. and Howe, B. 1960. *Prehistoric investigations in Iraqi Kurdistan.* Univ. of Chicago, Oriental Inst. stud. in ancient Oriental civilization 31.

——. 1962. Southwestern Asia beyond the lands of the Mediterranean littoral. In Braidwood and Willey, eds., pp. 132–46.

Braidwood, R. J. and Reed, C. 1957. The achievement and early consequences of food production: a consideration of the archaeological and natural historical evidence. *Cold Spring Harbor Symp. Quant. Biol.* 22:19–31.

Braidwood, R. J. and Willey, G. R., eds. 1962. *Courses toward urban life.* Chicago: Aldine.

Bray, R. T. 1956. The culture complexes and sequence at the Rice site(23sn200) Stone County, Mo. *Missouri Archaeol.* 18:46–134.

Bronson, B. 1972. Farm labor and the evolution of food production. In Spooner, ed. pp. 190–218.

——. 1975. The earliest farming: demography as cause and consequence. In Polgar, ed. pp. 53–78.

Brothwell, D. R. 1969. Human nutrition:evolutionary perspectives. In Ucko and Dimbleby, eds., pp. 531–46.

——. 1972. Paleodemography of earlier British populations. *World Archaeol.* 4:75–87.

Browman, D. L. and Munsell, D. A. 1969. Columbia Plateau prehistory: cultural development and impinging influences. *Amer. Antiquity* 34:249–64.

Broyles, B. J. 1966. Preliminary Report: The St. Albans site(46Ka27) Kanawha County West Virginia. *West Virgina Archaeol.* 19:1–42.

——. 1971. *The St.Albans site, Kanawha County, West Virginia.* West Virginia Geol. and Econ. Surv. Rep. Archaeol. Invest. 3.

Brush, C. F. 1965. Pox pottery:earliest identified Mexican ceramic. *Science* 149: 194–95.

Bryan, A. L. 1965. *Paleo-American prehistory.* Pocatello:Idaho State Univ. Occasional Papers 16.

——. 1969. Early man in America and the late Pleistocene chronology of western Canada and Alaska. *Current Anthropol.* 10:339–67.

——. 1973. Paleoenvironments and cultural diversity in late Pleistocene South America. *Quaternary Res.* 3:237–56.

——. 1975. Paleoenvironments and cultural diversity in late Pleistocene South America: a rejoinder to Vance Haynes and a reply to Thomas Lynch. *Quaternary Res.* 5:151–59.

Bryan, K. 1939. Stone cultures near Cerro Pedernal and their geological antiquity. *Texas Archaeol. and Paleontol. Soc. Bull.* 11:9–42.

Burch, E. S. 1972. The caribou/wild reindeer as a human resource. *Amer. Antiquity* 37:339–68.

Bürgl, H. 1957. Artefactos paleolíticos de una tumba en Garzón(Huila). *Rev. Colombiana de Antropología* 6:7–28.

Bushnell, G. H. S. 1971. Text of lecture delivered November 5, 1970 at the University of Illinois, Urbana. *J. Steward Anthropol. Soc.* 3:56–65.

Butler, B. R. 1959. Lower Columbia Valley archaeology: a survey and appraisal of some major archaeological resources. *Tebiwa* 2:6–24.

——. 1961. *The Old Cordilleran culture in the Pacific Northwest.* Occasional Papers no. 5, Idaho State Coll. Mus.

——. 1966. Comments on prehistory. In d'Azevedo et al. eds., *The current status of anthropological research in the Great Basin.* Reno:Desert Research Institute, Soc. Sci. and Humanities Pub. 1.

——. 1969. Some large thin bifaces from southeastern Idaho. *Tebiwa* 12:62–68.

Butzer, K. 1971. *Environment and archaeology.* (2nd edition). Chicago: Aldine.

Butzer, K. and Isaac, G. eds., 1975. *After the Australopithecines.* The Hague: Mouton.

Byers, D. S. 1954. Bull Brook—a fluted point site in Ipswich, Massachusetts. *Amer. Antiquity* 19:343–51.

——. 1959. The eastern archaic:some problems and hypotheses. *Amer. Antiquity* 24:233–56.

——. ed. 1967. *The prehistory of the Tehuacan Valley,* vol. 1. Austin: Univ. Texas Press.

Calderón, V. 1964. *O sambaquí da Pedra Oca.* Instituto de Ciencias Sociais 2. Salvador: Univ. Bahía.

Caldwell, J. 1958. *Trend and tradition in the prehistory of the eastern United States.* Amer. Anthropol. Ass. Mem. 88.

——. 1962. Eastern North America. In Braidwood and Willey, eds., pp. 288–307.

Callen, E. O. 1967. The first New World cereal. *Amer. Antiquity* 32:535–38.

Cambron, J. W. and Hulse, D. C. 1960. An excavation on the Quad site. *Tenn. Archaeol.* 16:14–26.

Cambron, J. W. and Waters, S. A. 1959. Flint Creek Rock Shelter pt. 1. *Tenn. Archaeol.* 15:73-87.

Campbell, E. W. C. 1949. Two ancient archaeological sites in the Great Basin. *Science* 109:340.

Campbell, E. W. C. and Campbell, W. H. 1935. The Lake Mohave site. In *The Archaeology of Pleistocene Lake Mohave: a symposium.* Southwest Mus. Papers 11, pp. 9-43.

Campbell, J. M. 1961. The Kogruk complex of Anaktuvuk Pass, Alaska. *Anthropol.* 3:1-18.

———. 1963. Ancient Alaska and Paleolithic Europe. In F. H. West, ed. *Early man in the western American Arctic: a symposium.* Univ. Alaska Anthropol. Papers 10:2.

Campo, M. van, and Bouchud, J. 1962. Flore accompagnant le squellete d'enfant mousterien découvert au Roc de Marsal, commune du Bugue (Dordogne) et premier étude de la faune du gisement. *Compt. Rend. Hebdomadaires Acad. Sci.* 254:897-99.

Camps, G. 1969. *Amekni: néolithique ancien du Hoggar.* Paris: CRAPE mèm. 10.

———. 1971. A propos du néolithique ancien de la Mediterranée occidentale. *Bull. Soc. Prehist. Francaise* 68: CRSM Fasc. 2:48-50.

Camps-Fabrer, H. 1966. *Matière et art mobilier dans la préhistoire nordafricaine et saharienne.* Paris: CRAPE mèm. 5.

Capdeville, A. 1922. Apuntes para la arqueología de Taltal. *Bol. Acad. Nac. Hist.* 4:115-18.

Capes, K. H. 1964. *Contributions to the prehistory of Vancouver Island.* Idaho State Univ. Mus. Occasional Papers 15.

Cardich, A. 1958. *Los yacimientos de Lauricocha, nuevas interpretaciones de las prehistoria peruana.* Studia Praehistorica 1. Centro Argentino de Estudios Prehistóricos.

Cardich, A. et al. 1973. Sequencia arqueológica y chronología radiocarbónica de la cueva 3 de Los Toldos (Santa Cruz, Argentina). *Relaciones de la Soc. Argentina de Antropol.* 7:85-123.

Carneiro, R. 1961. Slash and burn cultivation among the Kuikuru and its implications for cultural development in the Amazon Basin. In W. Johannes, ed. The evolution of horticultural systems in South America. *Anthropol.* supp. pub. 2, pp. 47-67.

———. 1967. On the relationship between size of population and complexity of social organization. *Southwestern J. Anthropol.* 23:234-43.

———. 1968. The transition from hunting to horticulture in the Amazon Basin. In *Proc. VIII Congr. Anthropol. Ethnol. Sci.,* pp. 224-48.

———. 1970. A theory of the origin of the state. *Sci.* 169:733-38.

———. 1972. From autonomous villages to the state: a numerical estimation. In Spooner, ed., 1972, pp. 64-77.

Carneiro, R. F., and Hilse, D. F. 1966. On determining the probable rate of population growth during the neolithic. *Amer. Anthropol.* 68:177-81.

Carr-Saunders, A. M. 1922. *The population problem: a study in human evolution.* Oxford: Clarendon Press.

Carter, G. F. 1951. Man in America: a criticism of scientific thought. *Sci. Monthly* 73:297-307.

———. 1957. *Pleistocene man at San Diego*. Baltimore: Johns Hopkins Univ. Press.

Casamiquela, R. 1961. Dos nuevos yacimientos patagónicos de la cultura Jacobaccense. *Univ. Nac. La Plata, Fac. cienc. natur. y mus., seccion anthropol.* 5:171-78.

Casamiquela, R. et al. 1967. *Convivencia del hombre con el mastodonte en Chile central*. Mus. Nac. Hist. Natur., Noticiario Mensual 132.

Cassidy, C. M. 1974. Determination of the nutritional and health status in skeletal populations. Paper presented to the annual meeting of the American Association of Physical Anthropologists.

Caton-Thompson, G., and Gardner, E. W. 1934. *The desert Fayum*. London: Royal Anthropoligical Society.

Chang, K. C. 1962. China. In Braidwood and Willey, eds., 1962, pp. 177-202.

———. 1968. *The archaeology of ancient China*. New Haven: Yale Univ. Press.

———. 1970. The beginnings of agriculture in the Far East. *Antiquity* 64:175-85.

———. 1973. Radiocarbon dates from China: some initial interpretations. *Current Anthropol.* 14:525-28.

Chard, C. S. 1959. New World origins, a reappraisal. *Antiquity* 33:44-49.

———. 1969. *Man in prehistory*. New York: McGraw-Hill.

———. 1974. *Northeast Asia in prehistory*. Madison: Univ. Wisconsin Press.

Chernysh, O. P. 1961. *Palaeolitvěna Stojanka Molodove V*. Kiev: Akad. Nauk.

Childe, V. G. 1944. Archaeological ages as technological stages. *J. Roy. Anthropol. Inst.* 74:7-24.

———. 1951. *Man makes himself*. New York: Mentor.

Chmyz, I. 1963. Nota prévia sobre a jazida PR-UV-1(63): Kavales. *Rev. Mus. Paulista* n.s. 14:493-512.

———. 1967. Dados parciais sobre a arqueologia do valedo Río Paranapanema. In *Programa Nacional de Pesquisas Arqueológicas: Resultados preliminares do Primeiro Ano 1965-1966*, pp. 59-78. Mus. Paraense Emílio Goeldi Pub. Avulsas 6.

Cigliano, E. M. 1962. *El Ampajanguense*. Pub. 5, Inst. Antropol. Univ. Nacional del Litoral, Rosario Argentina.

———. 1964. El precerámico en el N.W. Argentino. In *Arqueología de Chile central y areas vecinas*. Publicación de los trabajos presentados al Tercer Congreso Internacional de Arqueología Chilena, pp. 191-98. Santiago: Imprenta "Los Andes."

———. 1965. Dos nuevos sitios precerámicos en la Puna de Argentina: Turilar, Departmento de Susques, Provincia de Jujuy. *Etnía* 2:6-8.

———. 1966a. La cerámica temprana en America del Sur. El yacimiento de Palo Blanco(Partido de Berisso, Province de Buenos Aires, Argentina.) *Ampurias* 28:163-70.

———. 1966b. Contribución a los fechados radiocarbónicas Argentinos(I). *Rev. Mus. la Plata*(n.s.) 6:1-16.

Clark, C. and Haswell, M. 1966. *The economics of subsistence agriculture*. New York: St. Martin's Press.

Clark, J. D. 1950. The newly discovered Nachikufu culture of Northern Rhodesia. *South African Archaeol. Bull.* 5:2.
———. 1951. Bushmen hunters of the Barotse Forests. *Northern Rhodesia J.* 1: 56–65.
———. 1954. An early upper Pleistocene site at the Kalambo Falls on the Northern Rhodesia/Tanganyika border. *South African Archaeol. Bull.* 9:51–56.
———. 1959. *The Prehistory of Southern Africa.* London: Penguin.
———. 1960. Human ecology during the Pleistocene and later times in Africa south of the Sahara. *Current Anthropol.* 1:307–24.
———. 1965. Prehistoric cultures in Angola. In *Actas del V. Congreso Panafricano de prehistoria y de estudio del cuaternario 1963*, pp. 225–309.
———. 1966. *The distribution of prehistoric cultures in Angola.* Diamang: Companhia de Diamantes de Angola. Publicações culturais 73.
———. 1967. *An atlas of African prehistory.* Chicago: Univ. Chicago Press.
———. 1969. *The Kalambo Falls prehistoric site.* London: Cambridge Univ. Press.
———. 1970. *The Prehistory of Africa.* New York: Praeger.
———. 1971. A re-examination of the evidence for agricultural origins in the Nile Valley. *Proc. Prehist. Soc.* 37:34–79.
———. 1972. Mobility and settlement patterns in sub-Saharan Africa: a comparison of late prehistoric hunter/gatherers and early agricultural occupation units. In Ucko et al., eds.
———. 1975. A comparison of the late Acheulian industries of Africa and the Middle East. In Butzer and Isaac, eds., pp. 605–60.
———. In press. Prehistoric populations and pressures favoring plant domestication. In Harlan et al., eds., in press.
Clark, J. G. D. 1948. The development of fishing in prehistoric Europe. *Antiquaries J.* 28:440–85.
———. 1952. *Prehistoric Europe: the economic basis.* London: Cambridge Univ. Press.
———. 1953. The economic approach to prehistory. *Proc. Br. Acad.* 39:215–38.
———. 1954. *Excavations at Starr Carr.* London: Cambridge Univ. Press.
———. 1962. *World prehistory:an outline.* London: Cambridge Univ. Press.
———. 1965. Radiocarbon dating and the expansion of farming culture from the Near East over Europe. *Proc. Prehist. Soc.* 31:58–73.
———. 1967. *The stone age hunters.* New York: McGraw-Hill.
———. 1968. The economic impact of the change from late glacial to post glacial conditions in northern Europe. In *8th Congress of Anthropological and Ethnological Sciences*, pp. 241–44.
———. 1969. *World prehistory: a new outline.* London: Cambridge Univ. Press.
———. 1972. Starr Carr: a case study in Bioarchaeology. Reading: Addison-Wesley Module 10.
Clark, J. G. D. and Piggott, S. 1965. *Prehistoric societies.* New York: Knopf.
Clarke, D. L. 1968. *Analytical archaeology.* London:Methuen.
———. ed. 1972. *Models in archaeology.* London:Methuen.
Cleland, C. K. 1966. *The prehistoric animal ecology and ethnozoology of the upper Great Lakes region.* Anthropol. Papers 29, Univ. Mich. Mus. Anthropol.

Coale, A. 1974. The history of human population. *Sci. Amer.* 231:41–51.

Coe, J. 1964. The formative cultures of the Carolina piedmont. *Trans. Amer. Phil. Soc.* n.s. 54:5.

Coe, M. D. 1962. *Mexico.* New York: Praeger.

———. 1966. *The Maya.* New York: Praeger.

Coe, M. D. and Flannery, K. V. 1964. Microenvironments and Mesoamerican prehistory. *Science* 143:650–54.

———. 1967. *Early culture and human ecology in south coastal Guatemala.* Smithsonian Contrib. Anthropol. 3.

Cohen, M. N. 1972–74. Some problems in the quantitative analysis of vegetable refuse illustrated by a Late Horizon site on the Peruvian coast. *Ñawpa Pacha* 10–12: 49–60.

———. 1975a. Archaeological evidence of population pressure in pre-agricultural societies. *Amer. Antiquity* 40:471–74.

———. 1975b. Population pressure and the origins of agriculture: an archaeological example from the coast of Peru. In Polgar, ed., pp. 79–122.

Coles, J. M. and Higgs, E. S. 1969. *The archaeology of ancient man.* New York: Praeger.

Collins, D. 1969. Culture traditions and environment of early man. *Current Anthropol.* 10:267–96.

Colvinaux, P. A. 1967. Quaternary vegetational history of arctic Asia. In Hopkins, ed., pp. 207–31.

Conklin, H. C. 1957. *Hanunoo agriculture.* Rome: United Nations FAO.

Cook, D. 1975. Changes in subsistence base: the human skeletal evidence. Paper presented to the annual meetings. Amer. Anthropol. Assoc.

Cook, S. F. 1972. *Prehistoric demography.* Addison-Wesley Modular Publication 16.

Cooke, C. K. 1963. Report on excavations at Pomongwe and Tshangula caves, Matopo Hills, Southern Rhodesia. *South African Archaeol. Bull.* 18:73–151.

Cooke, H. B. S. 1958. Observations relating to Quaternary environments in east and southern Africa. Geological Society of South Africa: Alex du Toit memorial lecture 5. *Trans. Geolog. Soc. South Africa* 60.

Cornwall, I. W. 1964. *The world of ancient man.* New York: John Day.

Correal, G. 1974. *Tequendema Cave, Bogota, Colombia.* Ph.D. dissertation. Bogota.

Coursey, D. G. In press. The origins and domestication of yams in Africa. In Harlan et al., eds., in press.

Cowgill, G. L. 1975a. On the causes and consequences of ancient and modern population changes. *Amer. Anthropol.* 77:505–25.

———. 1975b. Population pressure as a non-explanation. In Swedlund, ed., pp. 127–31.

Craig, A. K. and Psuty, N. P. 1968. *The Paracas papers: studies in marine and desert ecology 1: reconaissance report.* Occasional Publications No. 1 of the Dept. of Geogr., Florida Atlantic Univ.

———. 1971. Paleoecology of shellmounds at Otuma, Peru. *Geogr. Rev.* 61: 125–32.

Cressman, L. S. 1942. *Archaeological researches in the northern Great Basin.* Washington: Carnegie Institution Publication 538.

Cressman, L. S. and Bedwell, S. F. 1968. Report to the secretary of the Dept. of the Interior on archaeological research on public lands in northern Lake County, Oregon. MS.

Cressman, L. S. et al. 1960. Cultural sequences at the Dalles, Oregon. *Trans. Amer. Phil. Soc.* 50:10.

Crook, W. W. and Harris, R. K. 1952. Trinity aspect of the archaic horizon: the Carrollton and Elam foci. *Bull. Texas Archaeol. and Paleontol. Soc.* 23:7–38.

——. 1957. Hearths and artifacts of early man near Lewisville, Texas and associated faunal material. *Bull. Texas Archaeol. and Paleontol. Soc.* 28:7–97.

Cruxent, J. M. 1961. Huesos quemados en el yacimiento prehistórico Muaco, Edo. Falcón. *Inst. Venezolano Invest. Cient. Dep. Anthropol. Bol. Informativo* 2: 20–21.

——. 1962. Artifacts of Paleo-Indian type, Maracaibo, Zulia, Venezuela. *Amer. Antiquity* 27:576–79.

——. 1967. El Paleo-Indio en Taima-Taima, Estado Falcón, Venezuela. *Acta Cient. Venezolana* suppl. 3:3–17.

——. 1970. Projectile points with Pleistocene mammals in Venezuela. *Antiquity* 44:223–25.

Cruxent, J. M. and Rouse, I. 1958. *An archaeological chronology of Venezuela.* Washington: Pan American Union Soc. Sci. Monogr. 6.

Curr, E. M. 1965. *Recollections of squatting in Victoria, then called the Port Phillip District from 1841-1851.* Melbourne: Melbourne Univ. Press.

Cutler, H. C. and Whitaker, T. W. 1961. History and distribution of the cultivated cucurbits in the Americas. *Amer. Antiquity* 26:4:469–85.

——. 1967. Cucurbits from the Tehuacan caves. In Byers, ed., pp. 212–19.

——. 1969. A new species of cucurbita from Ecuador. *Ann. Missouri Botanical Garden* 55:392–96.

Dahlberg, A. A. 1960. The dentition of the first agriculturalists (Jarmo, Iraq). *Amer. J. of Phys. Anthropol.* 18:243–56.

Daniels, S. 1969. The Middle and Late Stone Age. In T. Shaw, ed. *Lectures on Nigerian Prehistory and Archaeology.* Ibadan: Ibadan Univ. Press. pp. 23–29.

Daugherty, R. D. 1956. Archaeology of the Lind Coulee site, Washington. *Proc. Amer. Phil. Soc.* 100:3.

——. 1962. The intermontane western tradition. *Amer. Antiquity* 28:144–50.

David, N. 1973. On upper paleolithic society, ecology and technological change: the Noaillian case. In Renfrew, ed., pp. 277–303.

Davies, O. 1968. The origins of agriculture in West Africa. *Current Anthropol.* 9:479–82.

Day, G. M. 1953. The Indians as an ecological factor in the northeastern forest. *Ecology* 34:329–46.

Day, K. C. 1964. Thorne Cave, northeastern Utah: Archaeology. *Amer. Antiquity* 30:50–59.

Deacon, H. J. 1969. Plant remains from Melkhoutboom Cave, South Africa.

Actes du VIe Congrès Panafrican de préhistoire et de l'étude du Quaternaire, Dakar 1967.

———. 1970. The Acheulian occupation of Amanzi Springs, Uithenge District, Cape Province. *Annals of the Cape Provincial Mus.* 8:89–189.

———. 1975. Demography, subsistence and culture during the Acheulian in southern Africa. In Butzer and Isaac, eds., pp. 543–70.

Deevy, E. 1960. The human population. *Sci. Amer.* 204:194–204.

———. 1965. Pleistocene nonmarine environments. In Wright and Frey, eds. pp. 643–57.

———. 1968. Measuring resources and subsistence strategy. In Lee and DeVore, eds., pp. 94–95.

deHeinzelin de Braucourt, J. 1957. *Les Fouilles d'Ishango.* Exploration du Parc National Albert, 1950. Vol. 2. Brussels, Institut des parcs nationaux du Congo.

deLumley, H. 1969. A paleolithic camp at Nice. *Sci. Amer.* 220:42–50.

———. 1975. Cultural evolution in France in its paleoecological setting during the middle Pleistocene. In Butzer and Isaac, eds. pp. 745–808.

Denham, W. W. 1974. Population structure, infant transport and infanticide among Pleistocene and modern hunter-gatherers. *J. Anthropol. Res.* 30:101–98.

Dennell, R. W. and Webley, D. 1975. Prehistoric settlement and land use in southern Bulgaria. In Higgs, ed., pp. 97–110.

deTerra, H. et al. 1949. *Tepexpan Man.* Viking Fund Pub. in Anthropol. 11.

Devereux, G. 1955. *A study of abortion in primitive societies.* New York: Julian.

———. 1967. A typological study of abortion in 350 primitive, ancient and preindustrial societies. In Rosen, ed., *Abortion in America.* Boston: Beacon.

deWet, J. M. J. et al. In press. Evolutionary dynamics of sorghum cultivation. In Harlan et al., eds., in press.

Dick, H. W. 1965. *Bat Cave.* School of Amer. Res. Monogr. 27.

Dimbleby, G. W. 1967. *Plants and Archaeology.* London: John Baker.

Divale, W. T. 1972. Systematic population control in the middle and upper paleolithic: inferences based on contemporary hunter-gatherers. *World Archaeol.* 4:222–43.

Doggett, H. 1970. *Sorghum.* London: Longmans.

Dolukhanov, P. M. 1973. The neolithization of Europe: a chronological and ecological approach. In Renfrew, ed., pp. 329–42.

Donnan, C. B. 1964. An early house from Chilca, Peru. *Amer. Antiquity* 30: 137–44.

Drucker, P. 1955a. *Indians of the Northwest Coast.* Anthropol. Handbook 10. New York: Amer. Mus.

———. 1955b. Sources of northwest coast culture. In *New Interpretations of aboriginal American culture history*, pp. 59–81. Washington: Anthropol. Soc. Washington.

Dumond, D. E. 1965. Population Growth and Cultural Change. *Southwestern J. Anthropol.* 21:302–25.

———. 1975. The limitation of human population: a natural history. *Science* 187: 713–21.

Dunn, F. L. 1968. Epidemiological factors: health and disease in hunter-gatherers. In Lee and Devore, eds., pp. 221–28.

Edwards, W. E. 1967. The late Pleistocene extinction and diminution in size of many mammalian species. In Martin and Wright, eds., pp. 141–54.

Ellis, D. V. In press. The advent of food production in West Africa. In Harlan et al., eds., in press.

Emperaire, J. and Laming, A. 1956. Les sambaquis de la côte méridionale du Bresil, Campagnes de Fouilles(1954–1956). *J. Soc. Amér.* 45:5–163.

———. 1961. Les gisements des Iles Englefield et Vivian dans la Mer d'Otway, Patagonie Australie. *J. Soc. Amér.* 50:5–75.

Emperaire, J. et al. 1963. La Grotte Fell et autres sites de la région volcanique de la Patagonie Chilienne. *J. Soc. Amér.* 52:167–252.

Engel, F. 1957a. Early sites on the Peruvian coast. *Southwestern J. Anthropol.* 13:54–68.

———. 1957b. Sites et établissments sans céramique de la côte Péruvienne. *J. Soc. Amér.* 49:7–35.

———. 1963. A preceramic settlement on the central coast of Peru: Asia. *Trans. Amer. Phil. Soc.* 53:3.

———. 1967. Le complex précéramique d'el Paraiso(Perou). *J. Soc. Amér.* n.s. 1:43–95.

———. 1970. Exploration of the Chilca Canyon, Peru. *Current Anthropol.* 11:55–58.

———. 1971. D'Antival à Huarangal. *L'Homme* 11:39–57.

———. 1973. New Facts about pre-Columbian life in the Andean lomas. *Current Anthropol.* 14:271–80.

Estrada, E. and Evans, C. 1963. Cultural development in Ecuador. In Meggers and Evans, eds., pp. 77–88.

Evans, C. 1964. Lowland South America. In Jennings and Norbeck, eds., pp. 419–50.

Evans, C. and Meggers, B. J. 1957. Formative period cultures in the Guayas Basin, coastal Ecuador. *Amer. Antiquity* 22:235–46.

———. 1960. *Archaeological investigations in British Guiana.* Washington: Bur. Amer. Ethnol. Bull. 177.

———. 1968. *Archaeological investigations on the Rio Napo, eastern Ecuador.* Washington: Smithsonian Contributions to Anthropol. 6.

Evans, C., Meggers, B. J., and Cruxent, J. M. 1959. Preliminary results of archaeological investigations along the Orinoco and Ventuari Rivers, Venezuela. *Actas, 33rd Int. Congr. of Americanists* 2:359–69.

Evans, C., Meggers, B. J., et al. 1959. *Cultura Valdivia.* Guayaquil: Publ. Mus. Victor Emilio Estrada 6.

Evans, G. L. 1961. *The Friesenhahn Cave.* Austin: Texas Mus. Mem. Bull. 2.

Evans, L. G. 1969. The exploitation of molluscs. In Ucko and Dimbleby, eds., pp. 479–84.

Ewing, J. F. 1949. The treasures of Ksar'Akil. *Thought* 24:255–88.

Eyre, E. J. 1845. *Journals of expeditions of discovery in central Australia and*

overland from Adelaide to King George's Sound in the years 1840-1841.
London: Boone.

Faegri, K. and Iverson, J. 1964. *Textbook of Pollen Analysis.* New York: Hafner.

Fagan, B. M. 1960. The Glentyre Shelter and Oakhurst re-examined. *South African Archaeol. Bull.* 15:80-94.

——. 1974. *Men of the Earth.* Boston: Little, Brown.

Fagan, J. L. 1975. A supposed fluted point from Fort Rock Cave, and error of identification and its consequences. *Amer. Antiquity* 40:356-57.

Fairbridge, R. W. 1976. Shellfish eating preceramic Indians in coastal Brazil. *Science* 191:353-59.

Fairservis, W. 1971. *The roots of ancient India.* New York: Macmillan.

Faris, J. C. 1975. Social evolution, population, and production. In Polgar, ed., pp. 235-72.

Fauré, H. et al. 1963. Formations lacustres du Quaternaire supérieur du Niger oriental, diatomites et âges absolus. *Bull. Bur. Rech. Geol. Min.* 3:41-63.

Ferguson, G. J. and Libby, W. F. 1964. UCLA radiocarbon dates III. *Radiocarbon* 6:318-39.

Ferreyra, R. 1953. *Communidades vegetales de algunas lomas costaneras del Péru.* Lima: Estación Experimental Agricola de La Molina, Bol. 53.

Fitting, J. E. 1968. Environmental potential and post-glacial readaptation in eastern North America. *Amer. Antiquity* 33:441-45.

——. 1970. *The archaeology of Michigan.* Garden City: Natural History Press.

Fitzhugh, W. 1972. The eastern archaic: commentary and northern perspective. *Pennsylvania Archaeol.* 42:1-19.

Flannery, K. V. 1965. The ecology of early food production in Mesopotamia. *Science* 147:1247-56.

——. 1966. The postglacial "readaptation" as viewed from Mesoamerica. *Amer. Antiquity* 31:800-06.

——. 1967. Vertebrate fauna and hunting patterns. In Byers, ed., pp. 132-77.

——. 1968. Archaeological systems theory and early Mesoamerica. In Meggers, ed., *Anthropological Archaeology in the Americas*, pp. 67-86. Washington: Anthropol. Soc. Washington.

——. 1969. Origins and ecological effects of early domestication in Iran and the Near East. In Ucko and Dimbleby, eds., pp. 73-100.

——. 1973. The origins of agriculture. In Siegel, Beals, and Tyler, eds., *Annual Review of Anthropology* 2:271-310.

Flannery, K. V. et al. 1967. Farming systems and political growth in ancient Oaxaca. *Science* 158:445-54.

Flight, C. 1970. Excavations at Kintampo. *West Afr. Arch. News.* 12:71-72.

Flint, R. F. 1957. *Glacial and Pleistocene Geology.* New York: Wiley.

——. 1959. Pleistocene climates in eastern and southern Africa. *Bull. Geol. Soc. Amer.* 70:343-74.

Fock, N. 1961. Inca imperialism in northwest Argentina and Chaco burial forms. *Folk* 3:67-90.

——. 1962. Chaco pottery and Chaco history, past and present. In *34th International Congress of Americanists*, pp. 477-84.

Forbis, R. G. 1974. The paleoamericans. In Gorenstein et al., eds., pp. 11-28.

Fowler, M. L. 1959. Modoc Rock Shelter: an early archaic site in southern Illinois. *Amer. Antiquity* 25:476-522.

Freeman, L. 1964. *Mousterian Developments in Cantabrian Spain.* Ph.D. dissertation. Univ. Chicago.

———. 1966. The nature of the Mousterian facies in Cantabrian Spain. *Amer. Anthropol.* 68:2 (part 2):230-37.

———. 1973. The significance of mammalian faunas from the paleolithic occupations of Cantabria, Spain. *Amer. Antiquity* 38:3-44.

———. 1975. Acheulian sites and stratigraphy in Iberia and the Maghreb. In Butzer and Isaac, eds., pp. 661-744.

Freeman, L. G. and Butzer, K. 1966. The Acheulian station at Torralba (Spain): a progress report. *Quaternaria* 8:9-21.

Freitas, C. A. 1942. Alfarería del delta del Río Negro. *Rev. Hist.*, secunda epoca, Año 36, 13:363-418.

Frenguelli, J. and de Aparicio, F. 1923. Los paraderos de la margen derecha del Río Malabrigo. Paraná. *Anal. Fac. Cienc. y Educ.* 1:7-112.

Fryxell, R. and Daugherty, R. D. 1963. *Late glacial and post glacial geological and archaeological chronology of the Columbia Plateau.* Pullman, Wash.: Washington State Univ., Lab. Anthropol., Report of Investigations 23.

Fryxell, R. et al. 1965. Scabland tracts, loess soils and human prehistory. In *Guidebook for Field Conference E: Northern and Middle Rocky Mountains,* pp. 79-89. Int. Ass. Quaternary Res., 7th Congress.

Fung Pineda, R. 1969. Las Aldas: su ubicación dentro del proceso histórico del Peru antiguo. *Dedalo* 5:9-10.

Fung Pineda, R. et al. 1972. El taller lítico de Chivateros, valle de Chillón. *Rev. Mus. Nacional (Lima)* 38:61-72.

Funk, R. et al. 1970. Caribou and Paleo Indians in New York State: a presumed association. *Amer. J. Sci.* 268:181-86.

Gabel, C. 1963. Lochinovar Mound: a Late Stone Age camp-site in the Kafue Basin. *South African Archaeol. Bull.* 18:40-48.

———. 1965. *Stone age hunters of the Kafue—the Gwisho A site.* Boston: Boston Univ. Press.

Gade, D. W. 1970. Ethnobotany of cañihua (*Chenopodium pallidicaule*), rustic seed crop of the altiplano. *Econ. Bot.* 24:55-61.

Gajardo Tobar, R. 1958-59. Investigaciones acerca de las piedras con tacitas en la zona central de Chile. *Anal. Arqueol. y Etnol.* 14, 15. Argentina: Universidad Nacional de Cuyo, Mendoza.

———. 1962-63. Investigaciones arqueológicas en la desembocadura del Río Choapa (Prov. de Coquimbo, Chile): La Cultura de Huentelauquén. *Anal. Arqueol. y Etnol.* 17, 18:7-57. Argentina: Univ. Nac. Cuyo, Mendoza.

Galinat, W. C. 1971. The origin of maize. *Annu. Rev. Genet.* 5:447-78.

Garn, S. et al. 1968. Lines and bands of increased density. *Med. Radiography and Photogr.* 44:58-89.

Garn, S. et al. 1969. Subperiosteal gain and endosteal loss in protein-calorie malnutrition. *Amer. J. Phys. Anthropol.* 30:153-55.

Garrod, D. 1958. The Natufian culture: the life and economy of a mesolithic people in the Near East. *Proc. Br. Acad.* 43:211-27.

Garrod, D. and Bate, D. M. A. 1937. *The stone age of Mount Carmel.* vol. 1. Oxford: Clarendon Press.

Glover, I. C. 1973. Late stone age traditions in South East Asia. In Hammond, ed., pp. 51-65.

González, A. R. 1939. Excavaciones en el túmulo del Paraná Pavón. *Rev. Geogr. Amer.* 12:151-53.

———. 1941. Restos arqueológicos del abrigo de Ongamira. *Congr. Hist. Argentina del Norte y Centro (1941)* 1:143-58.

———. 1947. *Investigaciones arqueológicas en las nacientes del Paraná Pavón.* Publicaciones del Instituto Arqueología, Lingüística y Folklore 17. Univ. Nac. Córdoba, Argentina.

———. 1952. Antiguo horizonte precerámico en las sierras centrales de la Argentina. *Runa* 5:110-33.

———. 1960. La estratigrafía de la gruta de Intihuasi (Prov. de San Luis, R.A.) y sus relaciones con otros sitios precerámicos de Sudamerica. *Rev. Inst. Anthropol.* 1. Argentina: Univ. Nac. Córdoba.

———. 1967. Una excepcional pieza de mosaico del N.O. Argentina. *Etnia* 6:1-28.

González, A. R. and Lorandí, A. M. 1959. Hallazgos arqueológicos a las orillas del Río Caracaraña. *Revista Inst. Antrop.* 1:161-222.

Goodwin, A. J. H. et al. 1938. Archaeology of the Oakhurst Shelter. *Trans. Roy. Soc. South Africa* 25:3.

Gorenstein, S. et al., eds. 1974. *Prehispanic America.* New York: St. Martin's Press.

Gorman, C. 1969. Hoabinhian: a pebble tool complex with early plant associations in Southeast Asia. *Science* 163:671-73.

———. In press. A priori models and Thai prehistory: a reconsideration of the beginnings of agriculture in southeastern Asia. In Reed, ed., in press.

Gould, R. A. 1971. Uses and effects of fire among the western desert Aborigines. *Mankind* 8:14-24.

Graham, J. A. and Heizer, R. F. 1968. Man's antiquity in North America: views and facts. *Quaternaria* 9:225-37.

Gramsch, B. 1971. Zum problem des Uebergangs vom Mesolithikum zum Neolithikum im Flachland zwischen Elbe und Oder. In Schlette, F., ed. *Evolution und Revolution im alten Orient und in Europa.* Berlin: Akademie-Verlag.

Grey, Sir G. 1841. *Journals of two expeditions of discovery in northwest and western Australia, during the years 1837, 38 and 39.* London: Boone.

Griffin, J. B. 1964. The northeast woodlands area. In Jennings and Norbeck, eds., pp. 223-58.

———. 1967. Eastern North American archaeology: a summary. *Science* 156: 175-91.

Groot, G. 1951. *The prehistory of Japan.* New York: Columbia Univ. Press.

Grobman, A. et al. 1961. *Races of maize in Peru.* Washington NAS-NRC Pub. 915.

Grove, R. B. 1970. Investigations of the Chuqui complex. MS.

Gruhn, R. 1961. *The archaeology of Wilson Butte Cave, south-central Idaho.*

Idaho State Coll. Mus. Occasional Papers 6.

———. 1965. Two early radiocarbon dates from the lower levels of Wilson Butte Cave, south-central Idaho. *Tebiwa* 8:57.

Guilday, J. E. 1967. Differential extinction during late Pleistocene and recent times. In Martin and Wright, eds., pp. 121–40.

Guthe, A. 1967. *The Paleo Indians of Tennessee.* Eastern States Archaeol. Federation Bull. 26.

Hagedorn, R. and Jakel, D. 1969. Bemerkungen zur Quartaren Entwicklung des Reliefs im Tibesti-Gebirge (Tchad). Bull. ASEQUA 23:25–42. Dakar.

Hainline, J. 1965. Cultural and biological adaptation. *Amer. Anthropol.* 67:1174–97.

Haldane, J. B. S. 1957. Natural selection in man. *Acta Genet. et Statist. Med.* 6:321–32.

Hallum, S. J. 1973. Comment. *Current Anthropol.* 14:57.

Hammond, N. 1974. Paleolithic mammalian faunas and parietal art in Cantabria: a comment on Freeman. *Amer. Antiquity* 39:618–19.

Hammond, N., ed. 1973. *South Asian archaeology.* Park Ridge: Noyes Press.

Harlan, J. R. 1967. A wild wheat harvest in Turkey. *Archaeol.* 20:197–201.

———. 1971. Agricultural origins: centers and non-centers. *Science* 174:468–74.

———. In press. The origins of cereal cultivation in the Old World. In Reed, ed., in press.

Harlan, J. R. and Zohary, D. 1966. Distribution of wild wheats and barley. *Science* 153:1074–80.

Harlan, J. R. et al. In press. Plant domestication and indigenous African agriculture. In Harlan et al., eds., in press.

Harlan, J. R. et al., eds. In press. *The origins of African plant domestication.* The Hague: Mouton.

Harner, M. 1970. Population pressure and the social evolution of agriculturalists. *Southwestern J. Anthropol.* 26:67–86.

Harper, K. T. and Adler, G. M. 1970. The macroscopic plant remains of the deposits of Hogup Cave, Utah and their paleoclimatic implications. In Aikens, ed., pp. 215–40.

Harrington, M. R. 1933. *Gypsum Cave, Nevada.* Los Angeles: Southwest Mus. Papers 8.

———. 1957. *A Pinto site at Little Lake, California.* Los Angeles: Southwest Mus. Papers 17.

Harrington, M. R. and Simpson, R. D. 1961. *Tule Springs, Nevada with other evidence of Pleistocene man in North America.* Los Angeles: Southwest Mus. Papers 18.

Harris, D. 1969. Agricultural systems, ecosystems and the origins of agriculture. In Ucko and Dimbleby, eds., pp. 3–16.

———. 1972. The origins of agriculture in the tropics. *Amer. Sci.* 60:180–93.

———. 1973. The prehistory of tropical agriculture: an ethnoecological model. In Renfrew, ed., pp. 391–417.

———. In press. Alternative pathways toward agriculture. In Reed, ed., in press.

Harris, M. 1971. *Culture, man and nature*. New York: Crowell.

Harrison, G. A. and Boyce, A. J. 1972. *The structure of human populations*. Oxford: Clarendon Press.

Hassan, F. 1973. On methods of population growth during the neolithic. *Current Anthropol.* 14:535–42.

———. 1974. Population growth and cultural evolution. *Rev. Anthropol.* 1:205–12.

———. 1975a. Determination of the size, density and growth rate of hunting-gathering populations. In Polgar, ed., pp. 27–52.

———. 1975b. Nutrition and agricultural origins in the Near East. Paper read at the 74th annual meeting of the American Anthropological Association. San Francisco.

———. 1975c. Reply to Stephens. *Current Anthropol.* 16:289–90.

Haury, E. W. 1950. *The stratigraphy and archaeology of Ventana Cave, Arizona*. Tucson: Univ. Arizona Press.

———. 1960. Association of fossil fauna and artifacts of the Sulphur Springs stage, Cochise culture. *Amer. Antiquity* 25:609–10.

Haury, E. W. et al. 1953. Artifacts with mammoth remains, Naco, Arizona. *Amer. Antiquity* 19:1–24.

Haury, E. W. et al. 1959. The Lehner mammoth site, southeastern Arizona. *Amer. Antiquity* 25:2–30.

Hawkes, J. G. 1967. The history of the potato. *J. Roy. Hort. Soc.* 92:207–24, 249–62, 288–302, 364–65.

———. 1969. The ecological background of plant domestication. In Ucko and Dimbleby, eds., pp. 17–34.

Hawley, A. H. 1973. Ecology and population. *Science* 179:1196–1201.

Hayden, B. 1972. Population control among hunter-gatherers. *World Archaeol.* 4:205–21.

———. 1975. The carrying capacity dilemma. In Swedlund, ed., pp. 11–21.

Haynes, C. V. 1964. Fluted projectile points: their age and dispersion. *Science* 145:1408–1413.

———. 1966. Elephant Hunting in North America. *Sci. Amer.* 214:104–12.

———. 1967. Carbon 14 dates and early man in the New World. In Martin and Wright, eds., pp. 267–86.

———. 1969. The earliest Americans. *Science* 166:709–15.

———. 1973. The Calico site: artifacts or geo-facts. *Science* 181:305–10.

———. 1974. Paleoenvironments and cultural diversity in late Pleistocene South America: a reply to A. L. Bryan. *Quaternary Res.* 4:378–82.

Haynes, C. V. and Hemmings, E. T. 1968. Mammoth-bone shaft wrench from Murray Springs, Arizona. *Science* 159:186–87.

Heiser, C. B. 1965. Cultivated plants and cultural diffusion in nuclear America. *Amer. Anthropol.* 67:930–49.

———. 1973. *Seeds to Civilization*. San Francisco: Freeman.

Heizer, R. F. 1951. An assessment of certain radiocarbon dates from Oregon, California and Nevada. In *Radiocarbon Dating*, pp. 23–25. Soc. for Amer. Archaeol. Mem. 8.

——. 1952. A review of problems in the antiquity of man in California. In *Symposium on the antiquity of man in California*, pp. 13–17. Univ. California Archaeol. Surv. Rep. 16.

——. 1955. Primitive man as an ecological factor. *Kroeber Anthropol. Soc. Papers* 13:1–31.

——. 1956. Recent cave explorations in the lower Humboldt Valley, Nevada. In *Papers on California archaeology* 42:50–57. University of California Archaeological Survey Report 33.

——. 1958. Prehistoric central California: a problem in historical-devolopmental classification. In *California Archaeological Papers* 66:19–26. Univ. California Archaeol. Sur. Rep. 41.

——. 1964. The western coast of North America. pp. 117–148. In Jennings and Norbeck, eds.

Heizer, R. F. and Krieger, A. D. 1956. The archaeology of Humboldt Cave, ·Churchill County, Nevada. *Univ. California Pub. in Amer. Archaeol. and Ethnol.* 47:1–190.

Heizer, R. F. and Napton, L. K. 1970. Archaeological investigations in Lovelock Cave, Nevada. In Heizer and Napton, eds.

Heizer, R. F. and Napton, L. K., eds. 1970. *Archaeology and the prehistoric Great Basin lacustrine subsistence regime as seen from Lovelock Cave, Nevada.* Contributions of the Univ. California Archaeol. Res. Facility 10.

Helbaek, H. 1960. Cereals and weed grasses in phase A. Appendix 2 in Braidwood and Braidwood *Excavations in the Plain of Antioch 1*. Chicago: Univ. Chicago Press.

——. 1964. First impressions of the Çatal Hüyük plant husbandry. *Anatolian Stud.* 14:122.

——. 1966. Pre-pottery neolithic farming at Beidha. *Palestine Exploration Quart.* 98:62.

——. 1969. Plant Collecting, dry-farming and irrigation agriculture in prehistoric Deh Luran. In Hole, et al., pp. 383–428.

——. 1970. Plant husbandry of Hacilar: a study of cultivation and domestication. In Mellaart, ed., pp. 189–244.

Hemmings, E. T. and Haynes, C. V. 1969. The Escapule mammoth and associated projectile points, San Pedro Valley, Arizona. *J. Arizona Acad. Sci.* 5:184–88.

Hester, J. J. 1960. Late Pleistocene extinctions and radiocarbon dating. *Amer. Antiquity* 26:58–77.

——. 1966. Late Pleistocene environments and early man in South America. *Amer. Natur.* 100:377–88.

——. 1967. The agency of man in animal extinctions. In Martin and Wright, eds., pp. 169–92.

——. 1972. *Blackwater locality no. 1*. Pub. Fort Burgwin Res. Center 8.

Hester, T. R. 1973. *Chronological ordering of Great Basin prehistory*. Univ. California, Archaeol. Res. Facility Contrib. 17.

Hibben, F. C. 1941. Evidences of early occupation in Sandia Cave, New Mexico and other sites in the Sandia-Manzano region. *Smithsonian Miscellaneous Collections* 99:23.

Hiernaux, J. 1963. Some ecological factors affecting human populations in sub-Saharan Africa. In Howell and Bourlière, eds., pp. 534–46.

Higgs, E. 1961. Some Pleistocene faunas of the Mediterranean coastal area. *Proc. Prehist. Soc.* 27:144–54.

Higgs, E., ed. 1972. *Papers in Economic Prehistory.* London: Cambridge Univ. Press.

———. (ed.) 1975. *Paleoeconomy.* London: Cambridge Univ. Press.

Higgs, E. and Jarman, M. R. 1969. The origins of agriculture; a reconsideration. *Antiquity* 43:31–43.

———. 1972. The origins of animal and plant husbandry. In Higgs, ed., pp. 3–14.

Higgs, E. et al. 1967. The climate, environment and industry of stone age Greece. pt. III. *Proc. Prehist. Soc.* 33:1–29.

Hilbert, P. P. 1968. *Archäologischen Untersuchungen am Mittleren Amazonas.* Marburger Stud. zur Vokerkunde, vol. 1, Berlin: Dietrich Reimer Verlag.

Hill, B. 1972–74. A new chronology of the Valdivia ceramic complex from the coastal zone of the Guayas Province, Ecuador, *Ñawpa Pacha* 10–12:1–32.

Ho, P. T. In press. The indigenous origins of Chinese agriculture. In Reed, ed., in press.

Hobler, P. M. and Hester, J. J. 1969. Prehistory and environment in the Libyan desert. *South African Archaeol. Bull.* 23:120–30.

Hodgkinson, C. 1845. *Australia, from Port MacQuarie to Moreton Bay with description of the natives.* London: Boone.

Hole, F. and Flannery, K. V. 1967. The prehistory of southwestern Iran: a preliminary report. *Proc. Prehist. Soc.* 33:147–206.

Hole, F. et al. 1969. *Prehistory and human ecology of the Deh Luran Plain.* Mem. Mus. Anthropol., Univ. Michigan, 1.

Hopf, M. 1969. Plant remains and early farming in Jericho. In Ucko and Dimbleby, eds. pp. 355–60.

Hopkins, D. M. 1959. Cenozoic history of the Bering land bridge. *Science* 129: 1519–28.

Hopkins, D. M., ed. 1967. *The Bering land bridge.* Stanford: Stanford Univ. Press.

Hopkins, M. L. and Butler, B. 1961. Remarks on a notched fossil bison ischium. *Tebiwa* 4:10–18.

Houart, G. L. 1971. *Koster: a stratified archaic site in the Illinois Valley.* Illinois State Mus. Rep. Invest. 22.

Hough, W. 1926. *Fire as an agent in human culture.* U.S. Nat. Mus. Bull. 139.

Howard, E. V. 1935. Occurrence of flints and extinct animals in pluvial deposits near Clovis, New Mexico. *Proc., Philadelphia Acad. Nat. Sci.* 87:299–303.

Howard, G. D. and Willey, G. R. 1948. *Lowland Argentine archaeology.* Yale Univ. Publ. in Anthropol. 39.

Howell, F. C. 1959. Upper Pleistocene stratigraphy and early man in the Levant. *Proc. Amer. Phil. Soc.* 103:1–65.

———. 1966. Observations on the earlier phases of the European lower paleolithic. *Amer. Anthropol.* 68(2 pt. 2) 88–201.

Howell, F. C. and Bourlière, F. eds. 1963. *African ecology and human evolution.* Chicago: Aldine.

Howell, F. C. and Clark, J. D. 1963. Acheulian hunter gatherers of sub-Saharan Africa. In Howell and Bourlière, eds., pp. 458-533.

Hugot, H. J. 1968. The origins of agriculture: Sahara. *Current Anthropol.* 9: 483-88.

Huntingford, G. 1955. The economic life of the Dorobo. *Anthropos* 50:605-84.

Hunziker, A. T. and Planchuelo, A. M. 1971. Sobre un nuevo hallazgo de *Amaranthus caudatus* en tumbas indígenas de Argentina. *Kurtziana* 6:63-67.

Hurt, W. R. 1960. The cultural complexes from the Lagoa Santa region, Brazil. *Amer. Anthropol.* 62:569-85.

———. 1962. New and revised radiocarbon dates from Brazil. *Mus. News* 23:1-4. W. H. Over Museum, Univ. South Dakota.

———. 1964. Recent radiocarbon dates for central and southern Brazil. *Amer. Antiquity* 30:25-33.

———. 1966. Additional radiocarbon dates from the sambaquís of Brazil. *Amer. Antiquity* 31:440-41.

———. 1968. The preceramic occupations of central and southern Brazil. In *37th international congress of Americanists*, vol 3, pp. 275-97.

———. 1971. Late Pleistocene and Holocene evidence for climatic changes in the Sabana de Bogotá and their human ecological implications. Paper presented to the annual meetings of the Society for American Archaeology.

———. 1974. *The interrelationships between the natural environment of four sambaquís, coast of Santa Catarina, Brazil.* Indiana Univ. Mus. Occasional Papers and Monographs 1.

Hurt, W. R. and Blasi, O. 1960. *A sambaquí do Macedo A.52.13-Paraná, Brasil.* Univ. Paraná, Dep. Antropol. Arqueol. 2.

Hurt, W. R. et al. 1972. Preceramic sequences in the El Abra rock shelters, Colombia. *Science* 175:1106-08.

Ibarra Grasso, D. E. 1955. Hallazgos de puntas paleolíticas en Bolivia. *Anal.: 31st. Int. Congr. of Americanists* 2:561-68.

———. 1957. El paleolitício inferior en América. *Cuadernos Amer.* 16:135-75.

Iribarren, C. J. 1956. Investigaciones arqueológicas en Guanaqueros. *Pub. Mus. y Soc. Arqueol. Serena.* Bol. 8:10-22.

———. 1962. Correlaciones entre las piedras tácitas y la cultura de El Molle. La Tortorita, sitio arqueológico en el Valle de Elquí. *Pub. Mus. y Soc. Arqueol. Serena.* Bol. 12:39 ff.

Irving, W. N. 1971. Recent early man research in the north. In Shutler, ed., pp. 68-82.

Irwin, H. T. 1971. Developments in early man studies in western North America, 1960-1970. In Shutler, ed., pp. 42-67.

Irwin, H. T. et al. 1968. *University of Colorado investigations of paleolithic and epipaleolithic sites in the Sudan, Africa.* Univ. Utah, Anthropol. Papers 10.

Irwin, H. T. and Wormington, M. 1970. Paleo Indian tool types in the Great Plains. *Amer. Antiquity* 35:124-34.

Irwin, C. et al. 1962. Wyoming muck tells of battle: ice age man vs. mammoth. *Nat. Geogr.* 121:829-37.

Irwin-Williams, C. 1963. Explorations and excavations near Valsequillo, Puebla, Mexico. *Amer. Phil. Soc. Yearbook* (1963):550-53.

———. 1967. Associations of early man with horse, camel and mastodon at Hueyatlaco, Valsequillo (Puebla), Mexico. In Martin and Wright, eds., pp. 337–50.

———. 1969. Comments on the associations of archaeological materials and extinct fauna in the Valsequillo region, Puebla, Mexico. *Amer. Antiquity* 34: 82–83.

———. MS. Paleo-Indian and archaic cultural systems in the southwestern United States. Manuscript prepared for the forthcoming *Handbook of North American Indians.*

Irwin-Williams, C. and Haynes, C. V. 1970. Climatic change and early population dynamics in the southwestern United States. *Quaternary Res.* 1:59–71.

Irwin-Williams, C. et al. 1973. Hell Gap: Paleo Indian occupation on the high plains. *Plains Anthropol.* 18:40–53.

Isaac, E. 1970. *Geography of domestication.* Englewood Cliffs: Prentice Hall.

Isaac, G. 1968. Divisions within the Acheulian of eastern Africa: some interpretative suggestions. Paper presented to the annual convention of the American Anthropological Association.

———. 1969. Studies of early cultures in East Africa. *World Archaeol.* 1:1–28.

———. 1971. The diet of early man: aspects of archaeological evidence from lower and middle paleolithic sites in Africa. *World Archaeol.* 2:378–99.

———. 1972. Early phases of human behavior: models in lower paleolithic archaeology. In Clarke, ed., pp. 167–99.

———. 1975a. Stratigraphy and cultural patterns in East Africa during the middle range of Pleistocene time. In Butzer and Isaac, eds., pp. 495–542.

———. 1975b. Sorting out the muddle in the middle. In Butzer and Isaac, eds., pp. 875–88.

Ishida, E. et al. 1960. *Andes 1: University of Tokyo scientific expedition to the Andes.* Tokyo: Kadokawa.

Iversen, J. 1967. Naturens ud vikling siden sidste istid. Forskere og Nethoder. In *Danmarks Natur* 1, pp. 343–445. Copenhagen.

Jacobi, R. M. 1973. Aspects of the "Mesolithic Age" in Great Britain. In Koslowski, ed., pp. 237–66.

Jarman, M. R. and Webley, D. 1975. Settlement and land use in Capitania, Italy. In Higgs, ed., pp. 177–222.

Jelinek, A. J. 1965. The Upper Paleolithic revolution and the peopling of the New World. *Michigan Archaeol.* 11:85–88.

———. 1967. Man's role in the extinction of Pleistocene faunas. In Martin and Wright, eds., pp. 193–200.

Jennings, J. D., ed. 1956. The American Southwest: a problem in cultural isolation. In Wauchope, ed. *Seminars in Archaeology. Amer. Antiquity* 22:2 pt. 2, pp. 59–127.

———. 1957. *Danger Cave.* Univ. Utah Anthropol. Papers 27.

———. 1964. The desert west. In Jennings and Norbeck, eds., pp. 149–74.

———. 1974. *Prehistory of North America.* (2nd ed.) New York: McGraw-Hill.

Jennings, J. D. and Norbeck, E. 1955. Great Basin prehistory: a review. *Amer. Antiquity* 21:1–11.

———. (eds.) 1964. *Prehistoric man in the New World.* Chicago:Univ. Chicago Press.

Johnston, F. et al. 1942. *The Boylston Street fishweir.* Andover: Papers of the Peabody Foundation for Archaeol. 2.

Jolly, A. J. 1972. *The evolution of primate behavior.* New York:Macmillan.

Jones, P. R. and Dean, R. F. 1956. The effects of kwashiorkor on the development of the bones of the hand. *J. Trop. Pediat.* 2:51–68.

Judge, W. J. and Dawson, J. 1972. Paleo Indian settlement technology in New Mexico. *Science* 176:1210–16.

Kaplan, L. 1965. Archaeology and domestication in American *Phaseolus* (beans). *Econ. Bot.* 19:358–68.

———. 1967. Archaeological *Phaseolus* from Tehuacan. In Byers, ed., pp. 201–11.

Kaplan, L. et al. 1973. Early cultivated beans(*Phaseolus vulgaris*) from an intermontane Peruvian valley. *Science* 179:76–77.

Kelley, D. H. and Duccio Bonavia B. 1963. New evidence for preceramic maize on the coast of Peru. *Ñawpa Pacha* 1:39–42.

Kelley, J. C. 1959. The desert cultures and the Balcones phase:archaic manifestations in the Southwest and Texas. *Amer. Antiquity* 25:276–88.

Kennerly, T. E. 1956. Comparisons between fossil and recent species of the genus *Perognathus. Texas J. Sci.* 8:74–86.

Keyfitz, N. 1966. How many people have lived on earth. *Demogr.* 3:581–82.

Kirchoff, P. 1948. The food gathering tribes of the Venezuelan Llanos. In Steward, ed., 1946–59, vol. 4, pp. 445–68.

Kirkby, A. 1973. *The use of land and water resources in the past and present valley of Oaxaca, Mexico.* Univ. Michigan Mus. Anthropol., Mem. 5.

Klein, R. G. 1969a. Mousterian Cultures in European Russia. *Science* 165:257–265.

———. 1969b. *Man and culture in the late Pleistocene.* San Francisco: Chandler.

———. 1969c. The Mousterian of European Russia. *Proc. Prehist. Soc.* 35:77–111.

———. 1974. Environment and subsistence of prehistoric man in the southern Cape Province, South Africa. *World Archaeol.* 5:249–84.

Klima, B. 1954. Paleolithic huts of Dolni Věstonice, Czechoslovakia. *Antiquity* 109:4–14.

———. 1962. The first ground plan of an upper paleolithic loess settlement in middle Europe and its meaning. In Braidwood and Willey, eds., pp. 193–210.

Kornfield, W. J. 1972. Significado de la industria lítica de Paiján. *Bol. Seminario Arqueol.* 13:52–141.

———. 1974. Horizonte paleo-indio en los Andes centrales. Paper read at Segundo Congresso Peruano del Hombre y la Cultura Andina, Trujillo.

Koslowski, S. K. 1973. Introduction to the history of Europe in the early Holocene. In Koslowski, ed., pp. 331–66.

———. (ed.) 1973. *The Mesolithic in Europe.* Warsaw:Warsaw Univ. Press.

Kowalski, K. 1967. The Pleistocene extinction of mammals in Europe. In Martin and Wright, eds., pp. 349–64.

Krapovickas, A. 1969. The origin, variability and spread of the groundnut (*Arachis hypogaea*). In Ucko and Dimbleby, eds., pp. 427–42.

Kraybill, N. In press. Pre-agricultural tools for the preparation of foods in the Old World. In Reed, ed., in press.

Kretzoi, M. and Vértes, L. 1965. Upper Biharian(Intermindel) pebble industry occupation site in western Hungary. *Current Anthropol.* 6:74–87.

Krieger, A. D. 1947. Artifacts from the Plainview bison bed. In Sellards et al. Fossil bison and associated artifacts from Plainview, Texas. *Bull. Geol. Soc. Amer.* 58:938–54.

———. 1957. Notes and news:early man. *Amer. Antiquity* 22:321–23.

———. 1964. Early man in the New World. In Jennings and Norbeck, eds., pp. 23–81.

Kroeber, A. L. 1917. The tribes of the north Pacific coast of North America. In *Proc. 19th Int. Cong. of Americanists,* pp. 385–401.

———. 1925. *Handbook of the Indians of California.* Bur. of Amer. Ethnol. Bull. 78.

Krzywicki, L. 1934. *Primitive society and its vital statistics.* London:MacMillan.

Kunstadter, P. 1972. Demography, ecology, social structure and settlement patterns. In Harrison and Boyce, eds., pp. 313–51.

Kurtén, B. 1965. The Pleistocene Felidae of Florida. *Florida State Bull., Biol. Sci.* 9:215–73.

———. 1968. *Pleistocene mammals of Europe.* Chicago:Aldine.

Laming, A. 1960. Novas perspectivas sôbre a pré-história do sul do Brazil. *Anhembi* 38:228–35.

Laming, A. and Emperaire, J. 1959a. A jazida José Vieira:um sitio guarani e pre-cerâmico do interior do Paraná. *Arqueol.* 1–142.

———. 1959b. Bilan de trois compagnes de fouilles archéologiques au Brésil Méridional. *J. Soc. des Américanistes* 47:199–212.

Lanning, E. P. 1963. A pre-agricultural occupation on the central coast of Peru. *Amer. Antiquity* 28:360–71.

———. 1967a. *Peru before the Incas.* Englewood Cliffs: Prentice-Hall.

———. 1967b. Preceramic archaeology of the Ancón-Chillón region, central coast of Peru. Report to NSF on research carried out under grant GS869, 1965–66.

———. 1970. Pleistocene man in South America. *World Archaeol.* 2:90–111.

———. 1974a. Western South America. In Gorenstein et al., eds., pp. 65–86.

———. 1974b. Eastern South America. In Gorenstein et al., eds., pp. 87–109.

Lanning, E. P. and Hammel, E. A. 1961. Early lithic industries of western South America. *Amer. Antiquity* 27:139–54.

Lanning, E. P. and Patterson, T. C. 1967. Early man in South America. *Sci. Amer.* 217:44–50.

Lathrap, D. W. 1968a. The"hunting" economies of the tropical forest zone of South America: an attempt at historical perspective. In Lee and Devore, eds., pp. 23–29.

———. 1968b. Aboriginal occupation and changes in river channel on the Central Ucayali, Peru. *Amer. Antiquity* 33:62–79.

———. 1970. *The Upper Amazon.* London:Thames and Hudson.

———. 1973. Summary or model building: how does one achieve a meaningful overview of a continent's prehistory.(Review of Willey, *Introduction to American archaeology,* vol. 2). *Amer. Anthropol.* 75:1755-67.

——. In press. Our father the Cayman, our mother the gourd:Spinden revisited, or a unitary model for the emergence of agriculture in the New World. In Reed, ed., in press.

Laughlin, W. S. 1968. Hunting:an integrating biobehavior system and its evolutionary importance. In Lee and DeVore, eds., pp. 304-20.

Leakey, L. S. B. 1951. *Olduvai Gorge.* London:Cambridge Univ. Press.

——. 1957. Preliminary report on a Chellean I living site at BKii, Olduvai Gorge, Tanganyika Territory. In *Proc. 3rd Pan-African Congr. on Prehist., Livingstone 1955,* pp. 217-18. London:Chatto and Windus.

——. 1965. *Olduvai Gorge 1951-1961.* London:Cambridge Univ. Press.

Leakey, L. S. B. et al. 1972. *Pleistocene man at Calico.* San Bernardino:San Bernardino County Mus. Ass.

Leakey, M. D. 1967. Preliminary survey of the cultural material from beds I and II, Olduvai Gorge, Tanzania. In Bishop and Clark, eds., *Background to evolution in Africa,* pp. 417-46. Chicago:Univ. Chicago Press.

Lee, R. B. 1963. Population ecology of man in the early upper Pleistocene of South Africa. *Proc. Prehist. Soc.* 29:235-57.

——. 1968. What hunters do for a living or how to make out on scarce resources. In Lee and DeVore, eds., pp. 30-43.

——. 1969. !Kung Bushman subsistence: an input-output analysis. In Vayda, ed. *Ecological studies in cultural anthropology,* pp. 47-79. New York: Nat. Hist. Press.

——. 1972a. Population growth and the beginnings of sedentary life among the !Kung Bushmen. In Spooner, ed., pp. 329-42.

——. 1972b. The intensification of social life among the !Kung Bushmen. In Spooner, ed., pp. 343-50.

——. 1972c. Work effort, group structure, and land use in contemporary hunter-gatherers. In Ucko et al. eds., pp. 177-85.

Lee, R. B. and DeVore, I. 1968. Problems in the study of hunters and gatherers. In Lee and DeVore, eds., pp. 3-20.

Lee, R. B. and DeVore, I., eds. 1968. *Man the Hunter.* Chicago: Aldine.

Lee, T. E. 1957. The antiquity of the Shequiandah site. *Canadian Field-Naturalist* 71:117-37.

Leonhardy, F. C. and Rice, D. G. 1970. A proposed culture typology for the lower Snake River region, southeastern Washington. *Northwest Anthropol. Res. Notes* 4:1-29.

LePaige, G. 1959. Antiguas culturas atacameñas en la cordillera chilena:época Paleolítica. *Rev. Universitaria, Univ. Católica de Chile* 43:139-65.

——. 1960. Antiguas culturas atacameñas en la cordillera chilena: época Paleolítica. *Rev. Universitaria, Univ. Católica de Chile* 44-45:191-206.

——. 1964. *El precerámico en la cordillera atacameña y los cementerios de periodo agro-alfarero de San Pedro de Atacama.* Anal. 3. Univ. Norte, Antofagasta.

Leroi-Gourhan, A. 1969. Pollen grains of Gramineae and Cerealia from Shanidar and Zawi Chemi. In Ucko and Dimbleby, eds., pp. 143–48.

Lewis, H. T. 1972. The role of fire in the domestication of plants and animals in southwest Asia:a hypothesis. *Man* 7:195–222.

Lewis, T. M. and Kneburg, M. 1959. The archaic culture in the middle south. *Amer. Antiquity* 25:161–83.

Lewis, T. M. and Lewis, M. 1961. *Eva, an archaic site.* Knoxville:Univ. of Tennessee Press.

Li, H. 1966. Tung-nan-ya Tsai-p'ei Chih-wu chih Ch'i-yuan. Hong Kong: Inaugural address.

Logan, W. D. 1952. *Graham Cave:An archaic site in Montgomery County, Missouri.* Missouri Archaeol. Soc. Mem. 2.

Long, A. and Martin, P. S. 1974. The Death of American ground sloths. *Science* 186:638–40.

Lorenzo, J. L. 1955. Los concheros de la costa de Chiapas. *Anal., Inst. Nacional de Antropol. e Hist.* 7:41–50. Mexico D.F.

———. 1961. *La revolución neolítica en Mesoamérica.* Mexico D. F. pub. II, Dept. Prehist. Inst. Nacional de Antropol. e Hist.

———. 1967. *Sobre metodo arqueológico.* Mexico D. F. Bol., Inst. Nacional de Antropol. e Hist. 28.

Lorenzo, J. L. and Gonzales Quintero, L. 1970. El más antiguo teosinte. Mexico D.F. *Bol. Inst. Nacional de Antropol. e Hist.* 42:41–43.

Lothrop, S. K. 1928. *The Indians of Tierra del Fuego.* New York: Mus. Amer. Indian,Contrib. 10.

———. 1932. Indians of the Paraná Delta, Argentina. *Ann. New York Acad. Sci.* 33:77–232.

———. 1946. Indians of the Paraná Delta and La Plata littoral. In Steward, ed., 1946–59. vol. 1, pp. 177–90.

———. 1961. Early migrations to Central and South America:an anthropological problem in the light of other sciences. *J. Roy. Anthropol. Inst.* 91:97–123.

Loud, L. L. and Harrington, M. R. 1929. Lovelock Cave. *Univ. California Pub. in Amer. Archeol. and Ethnol.* 25:1–183.

Louw, J. T. 1960. *Prehistory of the Matjes River Rock Shelter.* Nat. Mus., Bloemfontein, mem. 1.

Louw, A. W. 1969. Bushman Rock Shelter, Ohrigstad, eastern Transvaal:a preliminary investigation, 1965. *South African Archaeol. Bull.* 24:39–51.

Lund, P. W. 1950. *Memórias sobre a paleontología Brasiliera.* Instituto Nacional do Liuro, Rio de Janeiro.

Lundelius, E. L. 1967. Late Pleistocene and Holocene faunal history of central Texas. In Martin and Wright, eds., pp. 287–319.

Lynch, T. F. 1967. *The nature of the central Andean preceramic.* Occasional papers of the Idaho State Univ. Mus. 21.

———. 1971. Preceramic transhumance in the Callejón de Huaylas, Peru. *Amer. Antiquity* 36:139–48.

———. 1972. Current research. *Amer. Antiquity* 37:274–78.

———. 1973. Harvest timing, transhumance and the process of domestication. *Amer. Anthropol.* 75:1254–59.

———. 1974. The antiquity of man in South America. *Quaternary Res.* 4:356–77.

———. 1976. The entry and postglacial adaptation of man in Andean South America. Paper prepared for the 9th International Congress of Prehistoric and Protohistoric Sciences.

Lynch, T. F. and Kennedy, K. A. R. 1970. Early human cultural and skeletal remains from Guitarrero Cave, northern Peru. *Science* 169:1307–10.

McArthur, M. 1960. Food consumption and dietary levels in groups of Aborigines living on naturally occurring foods. In Mountford, ed., *Records of the Australian-American scientific expedition to Arnhem Land 2.* Melbourne:Melbourne Univ. Press.

McBurney, C. B. M. 1967. *The Haua Fteah (Cyrenaica) and the stone age of the southeast Mediterranean.* London:Cambridge Univ. Press.

McCarthy, F. and McArthur, M. 1960. The food quest and the time factor in Aboriginal economic life. In Mountford, ed., *Records of the Australian-American Scientific Expedition to Arnhem Land 2,* pp. 145–94. Melbourne:Melbourne Univ. Press.

MacDonald, G. F. 1968. *Debert, a Paleo-Indian site in central Nova Scotia.* Nat. Mus. Canada Anthropol. Papers 16.

McGregor, J. C. 1965. *Southwestern Archaeology.* Urbana: Univ. Illinois Press.

MacNeish, R. S. 1958. Preliminary archaeological investigations of the Sierra de Tamaulipas, Mexico. *Trans. Amer. Phil. Soc.* (n.s.) 48:6.

———. 1964. Ancient Mesoamerican civilization. *Science* 143:531–37.

———. 1967. A summary of subsistence. In Byers, ed., pp. 290–310.

———. 1969. *First annual report of the Ayacucho archaeological-botanical project.* Andover: Peabody Found.

———. 1970. Social implications of changes in population and settlement patterns of the 12,000 years of prehistory of the Tehuacan Valley of Mexico. In Deprez, ed., *Population and Economics,* pp. 215–49. Manitoba:Univ. Manitoba Press.

———. 1971. Early man in the Andes. *Sci. Amer.* 224:36–46.

———. 1972. The evolution of community patterns in the Tehuacan Valley of Mexico and speculations about the cultural processes. In Ucko et al., eds., pp. 67–93.

MacNeish, R. S. et al. 1970. *Second annual report of the Ayacucho archaeological-botanical project.* Andover:Peabody Found.

MacNeish, R. S. et al. 1975. *The central Peruvian prehistoric interaction sphere.* Andover: Peabody Found.

Maingard, J. F. 1937. *Some notes on health and disease among the Bushmen of the southern Kalahari.* Johannesburg:Univ. Witwatersrand Press.

Maldonado-Koerdell, M. and Aveleyra Arroyo de Anda, L. 1949. Nota preliminar sobre dos artefactos del Pleistocene Superior hallados en la region de Tequixquiac, Mexico. In *El Mexico antiguo,* 7:154–61. Mexico D.F.

Mangelsdorf, P. 1947. The origin and evolution of maize. *Advances in Genet.* 1: 161–207.

Mangelsdorf, P. C. and Lister, R. H. 1956. Archaeological evidence on the evolution of maize in northwestern Mexico. *Harvard Bot. Mus. Leaflets* 17:6.

Mangelsdorf, P. C. et al. 1964. Origins of agriculture in Middle America. In West, ed., *Handbook of Middle American Indians I. Natural Environments and Early Cultures,* pp. 427–45. Austin: Univ. Texas Press.

Mangelsdorf, P. C. et al. 1967. Prehistoric wild and cultivated maize. In Byers, ed., pp. 178–200.

Mangelsdorf, P. C. and Reeves, R. G. 1939. *The origin of Indian corn and its relatives.* Texas Agr. Exp. Sta. Bull. 574.

Márquez Miranda, F. 1942. Hallazgos arqueológicos Chaqueños. *Relaciones, Soc. Argentina de Antropol.* 3:7–27.

Marshall, L. K. 1961. Sharing, talking and giving:relief of social tensions among !Kung bushmen. *Africa* 30:231–49.

——. 1962. !Kung bushmen religious beliefs. *Africa* 31:325–55.

Martin, J. F. 1973. On the estimation of the sizes of local groups in a hunting-gathering environment. *Amer. Anthropol.* 75:1448–68.

Martin, P. Sch. 1958. Pleistocene ecology and biogeography of North America. In Hubbs, ed., *Zoogeography,* pp. 375–420. Amer. Ass. Advance. Sci. Pub. 51.

——. 1966. African and Pleistocene overkill. *Nature* 212:339–42.

——. 1967a. Overkill at Olduvai Gorge. *Nature* 215:212–13.

——. 1967b. Prehistoric Overkill. In Martin and Wright, eds., pp. 75–120.

——. 1973. The discovery of America. *Science* 179:969–74.

Martin, P. Sch. and Guilday, J. E. 1967. Bestiary for Pleistocene biologists. In Martin and Wright, eds., pp. 1–62.

Martin, P. Sch. and Wright, H. E., eds. 1967. *Pleistocene extinctions:the search for a cause.* New Haven: Yale Univ. Press.

Martin, P. S. and Plog, F. 1973. *The archaeology of Arizona.* Garden City:Nat. Hist. Press.

Martin, P. S. et al. 1952. Mogollon cultural continuity and change. *Fieldiana: Anthropology* 40.

Mason, R. 1962. *The prehistory of the Transvaal.* Johannesburg:Witwatersrand Univ. Press.

Mason, R. J. 1962. The Paleo Indian tradition in eastern North America. *Current Anthropol.* 3:227–79.

Massey, W. C. 1961. The cultural distinction of aboriginal Baja California. In *Homenaje a Pablo Martínez del Rio,* pp. 411–22. Mexico D.F.

Mayer-Oakes, W. J. 1963. Early man in the Andes. *Sci. Amer.* 208:116–28.

——. 1966. El Inga projectile points—surface collections. *Amer. Antiquity* 31: 644–61.

Mayer-Oakes, W. J. and Bell, R. E. 1960. Early man site found in highland Ecuador. *Science* 131:1805–06.

Meggers, B. J. 1966. *Ecuador.* New York:Praeger.

Meggers, B. J. and Evans, C. 1957. *Archaeological investigations at the mouth of the Amazon.* Washington: Bur. Amer. Ethnol., Bull. 167.

——. 1962. The Machalilla culture; an early formative complex on the Ecuadorian Coast. *Amer. Antiquity* 28:186–92.

——. (eds.) 1963. *Aboriginal cultural development in Latin America: an interpretive review.* Washington: Smithsonian Miscellaneous Collections 146:1.

Meggers, B. J. et al. 1965. *Early formative period of coastal Ecuador.* Washington: Smithsonian Institution Contrib. to Anthropol. 1.

Meggitt, M. 1962. *Desert People.* Chicago: Univ. Chicago Press.

Mehringer, P. J. 1967. The environment of the late Pleistocene megafauna in the arid southwestern United States. In Martin and Wright, eds., pp. 247–66.

Meighan, C. W. 1959a. California cultures and the concept of an archaic stage. *Amer. Antiquity* 24:289–305.

——. 1959b. Varieties of prehistoric cultures in the Great Basin region. *Masterkey* 33:46–59.

Meighan, C. W. and Haynes, C. V. 1970. The Borax Lake site revisited. *Science* 167:1213–21.

Meldgaard, J. et al. 1963. Excavations at Tepe Guran, Luristan. *Acta Archaeol.* 34.

Mellaart, J. 1970. *Excavations at Hacilar.* Edinburgh: Edinburgh Univ. Press.

Mellars, P. A. 1973. The character of the middle-upper Paleolithic transition in southwest France. In C. Renfrew, ed., pp. 255–276.

Menghin, O. F. A. 1952a. Fundamentos cronológicos de la prehistoria de Patagonia. *RUNA* 5:23–43.

——. 1952b. Las pinturas rupestres de la Patagonia. *RUNA* 5:5–22.

——. 1955. Culturas precerámicas en Bolivia. *RUNA* 6:125–32.

——. 1956. El Altoparanaense. *Ampurias* 17–18:171–200.

——. 1957a. Das Protolithikum in Amerika. Buenos Aires: Centro Argentino de Estudios Prehistóricos, *Acta Praehist.* 1:5–40.

——. 1957b. El poblamiento prehistórico de Misiones. *Anal. Arqueol. y Etnol.* 12:19–40.

——. 1959–60. Estudios de prehistoria Araucana. Buenos Aires: Centro Argentino de Estudios Prehistóricos, *Acta Praehist.* 3–4:49–120.

——. 1962. Los sambaquís de la costa atlántica del Brasil meridional. *Amerindia* 1:53–81.

Menghin, O. F. A. and Bórmida, M. 1950. Investigaciones prehistóricas en cuevas de Tandilia, Provincia de Buenos Aires. *RUNA* 3:5–36.

Menghin, O. F. A. and González, A. R. 1954. Excavaciones arqueológicas en el yacimiento de Ongamira, Córdoba (Rep. Arg.). *Notas del Mus. de La Plata* 17:215–74.

Merrick, H. and Pastron, A. 1969. Contributions to a colloquium on the history of diet and subsistence in Africa. 1969 meeting of the Bay Area Africanists, University of California, Berkeley.

Métraux, A. 1946. Ethnography of the Chaco. In Steward, ed., 1946–59, vol. 1, pp. 197–370.

Meyers, J. T. 1971. The origins of agriculture: an evaluation of hypotheses. In Struever, ed., pp. 101–21.

Miller, E. T. 1967. Pesquisas arqueológicas efetuadas no nordeste do Rio Grande
do Sul. In *Programa Nacional de Pesquisas Arqueologicas:Resultados Prelimi-
nares do Primeiro Ano 1965-1966*, pp. 15-38. Museu Paraense Emilio Goeldi,
Publicações Avulsas 6.
Miller, T. O. 1969. Prehistória da região de Río Claro, S. P. : Tradições em Diver-
genia. *Cadernos Rioclarenses de Ciencias Humanas* 1:22-52.
Mirambell Silva, L. 1967. Excavaciones en un sitio Pleistocénico de Tlapacoya,
México. *Inst. Nacional de Antropol. e Hist., Bol.* 29:37-41.
Møhl, V. 1970. Fangstdyrene ved de Danske Strande. *Kuml,* pp. 297-329.
Monod, T. 1963. The late Tertiary and Pleistocene in the Sahara. In Howell and
Bourlière, eds., pp. 117-229.
Montané, J. C. 1968. Paleo-Indian remains from Laguna de Tagua Tagua, central
Chile. *Science* 161:1137-38.
———. 1972. Las evidencias del poblamiento temprano del Chile. *Pumapunku* 5:
40-53.
Morgan, L. H. 1877. *Ancient Society.* New York:Holt.
Moseley, M. E. 1972. Subsistence and demography:an example of interaction
from prehistoric Peru. *Southwestern J. Anthropol.* 28:25-49.
———. 1975. *The maritime foundations of Andean civilization.* Menlo Park:Cum-
mings.
Moseley, M. E. and Willey, G. R. 1973. Aspero, Peru: a reexamination of the site
and its implications. *Amer. Antiquity* 38:452-68.
Mostny, G. 1968. Association of human industries with Pleistocene fauna in central Chile. *Current Anthropol.* 9:214-15.
Mountjoy, J. B. et al. 1972. Mantanchén complex:new radiocarbon dates on early
coastal adaptations in west Mexico. *Science* 175:1242-43.
Movius, H. L. 1953. The Mousterian cave of Teshik Tash, southeastern Uzbekistan,
central Asia. *Amer. School Prehist. Res. Bull.* 17:11-71.
Mulloy, W. B. 1952. The northern plains. In Griffen, ed., *The archaeology of the
eastern United States*, pp. 124-138. Chicago:Univ. Chicago Press.
———. 1954. The McKean site. *Southwestern J. Anthropol.* 10:432-60.
———. 1958. A preliminary historical outline for the northwestern plains. *Univ.
Wyoming Pub.* 22:1-235.
Munson, P. J. 1968. Recent archaeological research in the Dhar Tichitt region of
south-central Mauretania. *West African Archaeol. Newsletter* 10.
———. 1970. Corrections and additional comments concerning the "Tichitt tradition." *West African Arch. Newsletter* 12:47-48.
———. In press. Archaeological data on the origins of cultivation in the southwestern Sahara and their implications for West Africa. In Harlan et al., eds., in press.
Murdock, G. P. 1959. *Africa.* New York: McGraw Hill.
———. 1968. The current status of the world's hunting and gathering people. In
Lee and DeVore, eds., pp. 13-22.
Murray, J. 1970. *The first European agriculture.* Chicago:Aldine.
Narr, K. J. 1956. Early food producing populations. In Thomas et al., eds., pp.
134-51.

——. 1961. *Urgeschichte der Kultur.* Stuttgart: A. Kroner.

——. 1963. *Kultur, Umwelt and Leiblichkeit des Eiszeitmenschens.* Stuttgart:G. Fischer.

Neel, J. V. 1970. Lessons from a primitive people. *Science* 170:815-22.

Neel, J. V. et al. 1964. Studies on the Xavante Indians of the Brazilian Mato Grosso. *Amer. J. Human Genet.* 16:52-140.

Neill, W. T. 1958. A stratified early site at Silver Springs, Florida. *Florida Anthropologist* 11:33-52.

Newell, R. R. 1973. The post-glacial adaptations of the indigenous populations of the northwest European plain. In Koslowski, ed., pp. 399-440.

Newman, T. M. 1966. *Cascadia Cave.* Idaho State Univ. Mus. Occasional Papers 18.

Niemeyer, H. and Schiappacasse, V. 1963. Investigaciones arqueológicas en las terrazas de Conanoxa, Valle de Camarones (Prov. de Tarapacá). *Rev. Universitaria* 48:101-66. Univ. Católica de Chile.

Nishiyama, I. 1971. Evolution and domestication of the sweet potato. *Bot. Mag.* 84:377-87.

Nordenskiold, E von. 1902-03. Präcolumbische Wohn-und Begräbnisplätze an der Südwestgrenze von Chaco. *Kongliga Svenska Vetenskapsacademiens Handlingar* 36:7:1-21.

Nougier, L. R. 1954. Essai sur le peuplement préhistorique de la France. *Population* 9:241-74.

Noy, T. et al. 1973. Excavations at Nahal Oren, Israel. *Proc. Prehist. Soc.* 39:75-99.

Nuñez, L. 1965. Desarrollo cultural prehispánica del norte de Chile. *Estudios Arqueol.* 1:37-115. Univ. Chile, Antofagasta.

Oakley, K. P. 1961a. On man's use of fire with comments on tool making and hunting. In Washburn, ed., pp. 176-93.

——. 1961b. A bone harpoon from Gambles Cave, Kenya. *Antiquaries J.* 41:76-87.

——. 1964. *Frameworks for dating fossil man.* London:Weidenfeld and Nicolson.

O'Brien, P. J. 1972. The sweet potato:its origin and dispersal. *Amer. Anthropol.* 74:342-64.

Orr, P. C. 1956. *Radiocarbon dates from Santa Rosa Island I.* Santa Barbara Mus. Nat. Hist. Bull. 2.

——. 1960. *Radiocarbon dates from Santa Rosa Island II.* Santa Barbara Mus. Nat. Hist. Bull. 3.

——. 1962. The Arlington Sping site, Santa Rosa Island, California. *Amer. Antiquity* 27:417-19.

Ortiz Troncoso, O. R. 1964. Investigaciones en conchales de Reloca (Prov. Maule, Chile). In *Arqueología de Chile Central y Areas Vecinas.* pp. 59-62. Publicación de los Trabajos presentados al Tercer Congreso Internacional de Arqueología Chilena. Santiago:Imprenta "Los Andes."

Ossa, P. 1974. Fechamiento del complejo Paiján en el Valle de Santa Catalina (Moche): el abrigo de Quirihuac. Paper presented to the 2nd Congreso Peruano del Hombre y la cultura Andina. Trujillo

Ossa, P. and Moseley, M. E. 1972. La Cumbre: a preliminary report on research into the early lithic occupations of the Moche Valley, Peru. *Ñawpa Pacha* 9: 1-16.

Owen, R. C. 1965. The patrilocal band: a linguistically and culturally hybrid unit. *Amer. Anthropol.* 67:675-90.

Parmalee, P. W. and Klippel, W. E. 1974. Freshwater mussels as a prehistoric food resource. *Amer. Antiquity* 39:421-34.

Patterson, T. C. 1966. Early cultural remains on the central coast of Peru. *Ñawpa Pacha* 4:145-53.

———. 1968. Current research (highland South America). *Amer. Antiquity* 33: 133-35.

———. 1971a. Central Peru: its population and economy. *Archaeol.* 24:316-21.

———. 1971b. The emergence of food production in central Peru. In Struever, ed., pp. 181-207.

Patterson, T. C. and Heizer, R. F. 1965. A preceramic stone tool collection from Viscachani, Bolivia. *Ñawpa Pacha* 3:107-14.

Patterson, T. C. and Lanning, E. P. 1964. Changing settlement patterns on the central Peruvian Coast. *Ñawpa Pacha* 2:113-23.

Patterson, T. C. and Moseley, M. E. 1968. Late preceramic and early ceramic cultures of the central coast of Peru. *Ñawpa Pacha* 6:115-33.

Perkins, D. 1964. Prehistoric Fauna from Shanidar, Iraq. *Science* 144:1565-66.

Perrot, J. 1962. Palestine-Syria-Cilicia. In Braidwood and Willey, eds., pp. 147-64.

———. 1966. Le gisement Natufien de Mallaha ('Eynan) Israel. *L'Anthropol.* 70: 437-84.

———. 1968. La Préhistoire Palestinienne. In *Supplément au Dictionnaire de la Bible,* pp. 286-446.

Petersen, E. B. 1973. A survey of the late paleolithic and mesolithic of Denmark. In Koslowski, ed., pp. 77-128.

Peterson, W. 1975. A demographers view of prehistoric demography. *Current Anthropol.* 16:207-26.

Pickersgill, B. 1969. The archaeological record of chili peppers (*Capsicum* spp.) and the sequence of plant domestication in Peru. *Amer. Antiquity* 34:54-66.

———. 1972. Cultivated plants as evidence for cultural contacts. *Amer. Antiquity* 37:97-104.

Pickersgill, B. and Heiser, C. B. In press. Origins and distributions of plants in the New World tropics. In Reed, ed., in press.

Piggott, S. 1950. *Prehistoric India.* Baltimore:Penguin.

———. 1965. *Ancient Europe.* Edinburgh:Edinburgh Univ. Press.

Pilling, A. R. 1955. Relationships of prehistoric cultures of coastal Monterey county, California. *Kroeber Anthropol. Soc. Papers* 12:70-87.

Polgar, S. 1964. Evolution and the ills of mankind. In Tax(ed.) *Horizons of anthropology,* Chicago:Aldine, pp. 200-11.

———. 1972. Population history and population policies from an anthropological perspective. *Current Anthropol.* 13:203-09.

———. 1975. Population, evolution, and theoretical paradigms. In Polgar, ed., pp. 1-25.

———. (ed.) 1975. *Population, Ecology and Social Evolution.* The Hague:Mouton.

Pollard, G. C. 1971. Cultural change and adaptation in the central Atacama desert of northern Chile. Ñawpa Pacha 9:41-64.

Pollard, G. C. and Drew, I. M. 1975. Llama herding and settlement in Prehispanic northern Chile: Application of an analysis for determining domestication. Amer. Antiquity 40:296-305.

Portères, R. 1950. Vieilles agricultures de l'Afrique intertropicale:centres d'origine et de diversification variétale primaire et berceaux d'agriculture antérieurs au XVIe siècle. L'Agronomie Tropicale 5:489-507.

———. 1951. Géographie alimentaire, berceaux agricoles et migrations des plantes cultivées en Afrique intertropicale. Compte Rendu Sommaire des Séances de la société de biogéographie 239:16-21.

Posnansky, M. and Cole, G. H. 1963. Recent Excavations at Magosi, Uganda:a preliminary report. Man 63:104-06.

Prest, V. K. 1969. Retreat of Wisconsin and Recent ice in North America. Geological Survey of Canada Map 1257a.

Price, T. D. 1973. A proposed model for the procurement systems in the Mesolithic of northwestern Europe. In Koslowski, ed., pp. 455-76.

PRONAPA 1970. Brazilian archaeology in 1968, an interim report on the national program of archaeological research. Amer. Antiquity 35:1-23.

Purseglove, J. W. In press. The origins and migrations of crops in tropical Africa. In Harlan et al., eds., in press.

Quimby, G. I. 1962. A year with a Chippewa Family 1763-1764. Ethnohist. 9: 217-39.

Radmilli, A. M. 1960. Considerazioni sul mesolitico italiano. Ann. Univ. di Ferrara 1:29-48.

Ranere, A. J. 1972. Early human adaptations to New World tropical forests:the view from Panama. Ph.D. dissertation, Univ. of California, Davis.

Rauth, J. W. and Hurt, W. R. 1960. The shellmound of Saquarema, Paraná, Brazil. Mus. News 21:9:1-7. Vermillion.

Ravines, R. 1967. El abrigo de Carú y sus relaciones culturales con otros sitios tempranos del sur del Peru. Ñawpa Pacha 5:39-57.

Reed, C. A. 1960. A review of the archaeological evidence on animal domestication in the prehistoric Near East. In Braidwood and Howe, pp.119-46.

———. In press-a. A model for the origins of agriculture in the Near East. In Reed, ed., in press.

———. In press-b. Introduction. In Reed, ed., in press.

———. Ed., in press. The Origins of Agriculture. The Hague:Mouton.

Reed, E. K. 1964. The greater southwest. In Jennings and Norbeck, eds., pp. 175-91.

Reichel-Dolmatoff, G. 1957. Momíl: a formative sequence from the Sinú Valley, Colombia. Amer. Antiquity 22:226-34.

———. 1959. The formative stage:an appraisal from the Colombian perspective. In Actas del XXXIII congreso internacional de Americanistas, vol. 2, pp. 152-64.

———. 1965a. Colombia. New York: Praeger.

———. 1965b. Excavaciones arqueológicas en Puerta Hormiga, Departmento de Bolívar. Publicaciones de la Univ. Los Andes, Antropol. 2. Bogota.

———. 1971. Early pottery from Colombia. Archaeol. 24:338-45.

Reichel-Dolmatoff, G. and A. 1946–50. Investigaciones arqueológicas en el Departmento del Magdalena:Parte I. Arqueología del Río Ranchería. *Bol. Arqueol.* 3:1–6. Bogota 1951.

———. 1956. Momíl, excavaciones en el Sinú. *Rev. Colombiana Antropol.* 5:109–333.

Renaud, E. B. 1938. *The Black Forks culture of southwest Wyoming.* Univ. Denver Archaeol. Surv. of the Western High Plains 10.

———. 1940. *Further research in the Black Forks Basin in southwest Wyoming, 1938–1939.* Univ. Denver Archaeol. Surv. of the Western High Plains 12.

Renfrew, C. 1968. Greek neolithic:backward or precocious. *Current Archaeol.* 7:168–72.

———. (ed.) 1973. *The explanation of culture change:models in prehistory.* Pittsburgh:Univ. Pittsburgh Press.

Renfrew, J. 1969. The archaeological evidence for the domestication of plants: methods and problems. In Ucko and Dimbleby, eds., pp. 149–72.

———. 1973. *Paleoethnobotany:the prehistoric food plants of the Near East and Europe.* New York: Columbia Univ. Press.

Richardson, J. 1972. The Pre-Columbian distribution of the bottle gourd (*Lagenaria siceraria*): a re-evaluation. *Econ. Bot.* 26:265–73.

Ritchie, W. A. 1969. *The archaeology of New York state.* Garden City:Nat. Hist. Press.

Robbins, L. H. 1967. A recent archaeological discovery in the Turkana district of northern Kenya. *Azania* 2:69–73.

———. 1974. *The Lothagam Site, a late stone age fishing settlement in the Lake Rudolph Basin, Kenya.* Michigan State Univ. Mus. Pub. Anthropology Series 1:2.

Roberts, F. H. 1935. A Folsom complex: preliminary report on investigations at the Lindenmeier site in northern Colorado. *Smithsonian Misc. Coll.* 94:4.

———. 1936. Additional information on the Folsom complex:report on the second season's investigations at the Lindenmeier site in northern Colorado. *Smithsonian Misc. Coll.* 95:10.

———. 1938. The Lindenmeier site in northern Colorado contributes additional data on the Folsom complex. *Explorations and Field Work of the Smithsonian Institution in 1937,* pp. 115–18. Washington.

———. 1940. Developments in the problem of the North American Paleo-Indian. In Essays in Historical Anthropology of North America. *Smithsonian Misc. Coll.* 100:51–116.

Rodden, R. J. 1962. Excavations at the early neolithic site at Nea Nikomedeia, Greek Macedonia (1961 season). *Proc. Prehist. Soc.* 28:267–88.

Rogers, M. J. 1939. *Early lithic industries of the lower basin of the Colorado River and adjacent desert areas.* San Diego Mus. Papers 3.

———. 1958. San Dieguito implements from the terraces of the Rincon Pantano and Rillito drainage system. *Kiva* 24:1–23.

Romer, A. 1933. *Vertebrate paleontology.* Chicago: Univ. Chicago Press.

Rossignol, M. 1962. Analyse pollinique de sédiments marins quaternaires en Israel, II–sédiments. *Pleistocènes Pollen et Spores* 14:121–48.

———. 1963. Analyse pollinique de sédiments dans la plaine de Haifa, Israel. *Israel J. Earth Sci.* 12:207–14.

Rouse, I. 1948. The Arawak. In Steward, ed., 1946–59, vol. 4, pp. 507–39.

Rouse, I. and Cruxent, J. M. 1963. *Venezuelan archaeology.* New Haven: Yale Univ. Press.

Rowe, J. H. 1961. Stratigraphy and seriation. *Amer. Antiquity* 26:324–30.

Royo y Gómez, J. 1960a. Características paleontológicas y geológicas del yacimiento de vertebratas de Muaco, Estado Falcón, con industria lítica humana. In *Memoria del III Congreso Geológico Venezolano,* vol. 2. pp. 501–05. Bol. Geol., publicacion especial, Caracas. no. 3.

———. 1960b. Pleistocene vertebrates from the Muaco deposit. *Soc. Vertebrate Paleontol. News Bull.* 58:31–32.

Rydén, S. 1948. Cord impression decorations in Chaco ceramics. *Archivos Ethnos* 1:1–6.

Sadek-Kooros, H. 1966. Jaguar Cave: an early man site in the Beaverhead Mountains of Idaho. Ph.D. dissertation, Harvard Univ.

Sahlins, M. 1968. Notes on the original affluent society. In Lee and DeVore, eds., pp. 85–89.

———. 1972. *Stone age economics.* Chicago: Aldine.

Sampson, C. G. 1974. *The stone age archaeology of southern Africa.* New York: Academic Press.

Sanders, W. T. and Price, B. 1968. *Mesoamerica.* New York:Random House.

Sanger, D. 1967. Prehistory of the Pacific Northwest Plateau as seen from the interior of British Columbia. *Amer. Antiquity* 32:186–97.

Sankalia, H. D. 1962. India. In Braidwood and Willey, eds., pp. 60–83.

———. 1974. *The prehistory and protohistory of India and Pakistan.* Poona: Deccan College.

Sanoja, M. 1963. Cultural development in Venezuela. In Meggers and Evans, eds., pp. 67–76.

———. 1966. Venezuelan archaeology looking toward the West Indies. *Amer. Antiquity* 31:232–36.

Sarma, A. V. N. 1974. Holocene paleoecology of south coastal Ecuador. *Proc. Amer. Phil. Soc.* 18:93–134.

Sauer, C. O. 1944. A geographical sketch of early man in America. *Geogr. Rev.* 34:529–73.

———. 1952. *Agricultural origins and dispersals.* New York: Amer. Geogr. Soc.

———. 1958. Age and area of American cultivated plants. In *33rd. International Congress of Americanists* 1, pp. 215–29.

Sauer, J. 1969. Identity of archaeologic grain amaranths from the valley of Tehuacán, Puebla, Mexico. *Amer. Antiquity* 34:80–81.

Sauer, J. and Kaplan, L. 1969. *Canavalia* beans in American prehistory. *Amer. Antiquity* 34:417–24.

Savory, H. N. 1968. *Spain and Portugal.* New York: Praeger.

Sayles, E. B. and Antevs, E. 1941. *The Cochise culture.* Gila Pueblo,Globe, Arizona, Medallion Papers 29.

Schiappacasse F., V. and Niemeyer F., H. 1964. Excavaciones de un conchal en el pueblo de Guanaqueros (Prov. de Coquimbo). In *Arqueología de Chile Central y Areas Vecinas*. pp. 235–62. Pub. de los Trabajos Presentados al Tercer Congreso Internacional de Arqueología Chilena. Santiago:Imprenta "Los Andes."

Schobinger, J. 1962. Investigaciones arqueológicas en la provincia de San Juan, Rep. Argentina (Informe Preliminar). In *Actas y Memorias 35th International Congress of Americanists* 1, pp. 615–20.

——. 1972. Nuevos hallazgos de puntas"colas de pescado" y consideraciones en torno al origen y dispersión de la cultura de cazadores superiores Toldense (Fell I) en Sudamérica. In *40th International Congress of Americanists*, pp. 33–50.

——. 1974. Current Research:South America. *Amer. Antiquity* 39:508–12.

Schrire, C. 1962. Oakhurst : a reexamination and vindication. *South African Archaeol. Bull.* 67:181–95.

Schultz, C. B. and Eiseley, L. C. 1935. Paleontological evidence of the antiquity of the Scottsbluff basin quarry and its associated artifacts. *Amer. Anthropol.* 37:306–19.

Schwanitz, F. 1966 *The Origin of Cultivated Plants* (English translation). Cambridge:Harvard Univ. Press.

Sears, W. H. 1964. The southeastern United States. In Jennings and Norbeck, eds., pp. 259–87.

Seddon, J. D. 1968. The origins and development of agriculture in east and southern Africa. *Current Anthropol.* 9:487–94.

Sellards, E. H. 1941. Stone images from Henderson County, Texas. *Amer. Antiquity* 7:29–38.

——. 1952. *Early man in America.* Austin: Univ. Texas Press.

Semenov, S. A. 1964. *Prehistoric Technology.* New York: Barnes and Noble.

Sengal, R. A. 1973. On mechanisms of population growth. *Current Anthropol.* 14:540–42.

Serrano, A. 1946. The sambaquís of the Brazilian coast. In Steward, ed., 1946–59, vol. 1, pp. 401–07.

——. 1954. Contenido e interpretación de la arqueología Argentina:el área litoral. *Rev. Univ. Nacional del Litoral* 29.

Service, E. K. 1962. *Primitive social organization: an evolutionary perspective.* New York: Random House.

Sharrock, F. W. 1966. *Prehistoric occupation patterns in southwest Wyoming and cultural relationships with the Great Basin and Plains culture areas.* Univ. Utah Anthropol. Papers 77.

Shaw, T. In press. Early crops in Africa: a review of evidence. In Harlan et al., eds., in press.

Sheffer, C. 1971. Review of Boserup:*The conditions of agricultural growth. Amer. Antiquity* 36:377–79.

Shutler, R. 1968, Tule Springs, its implications to early man studies in North America. *Eastern New Mexico Univ. Contrib. Anthropol.* 1:19–26.

Shutler, R., ed. 1971. Papers from a symposium on early man in North America. New Developments 1960–1970. *Arctic Anthropol.* 8:2.

Silva, F. A. 1967. Informes preliminares sobre a arqueología de Río Claro. *Programa nacional de Pesquisas Arqueológicas: Resultados Preliminares do Primeiro*

Ano 1965-1966. Mus. Paraense Emílio Goeldi, Pub. Avulsas 6:79–88.

Silva, J. E. 1964. Investigaciones arqueológicas en la costa de la zona central, Chile. In *Arqueología de Chile Central y Areas Vecinas.* pp. 263–74. Publicación de los Trabajos Presentados al Tercer Congreso Internacional de Arqueología Chilena. Santiago:Imprenta Los Andes.

Simmons, I. G. 1969. Evidence for vegetation changes associated with mesolithic man in Britain. In Ucko and Dimbleby, eds., pp. 113–22.

Simoons, F. J. 1965. Some questions on the economic prehistory of Ethiopia. *J. Afr. Hist.* 6:1–12.

Simpson, R. D. 1958. The Manix Lake archaeological survey. *Masterkey* 32:4–10.

——. 1960. Archaeological survey of the eastern Calico Mountains. *Masterkey* 34:25–35.

——. 1961. *Coyote Gulch:archaeological investigations of an early lithic locality in the Mojave Desert of San Bernardino County.* Archaeol. Surv. Ass. Southern California 5.

Singh, G. 1971.The Indus valley culture seen in the context of post glacial climate and ecological studies in north-west India. *Archaeol. and Phys. Anthropol. Oceania* 6:177–89.

Singh, P. 1974. *Neolithic cultures of western Asia.* New York:Seminar Press.

Slaughter, B. H. 1967. Animal ranges as a clue to late Pleistocene extinctions. In Martin and Wright, eds., pp. 155–67.

Smith, C. E. 1967. Plant remains. In Byers, ed., pp. 220–55.

Smith, P. E. L. 1964. The Solutrean culture. *Sci. Amer.* 211:86–94.

——. 1966. The late paleolithic of northeast Africa in light of recent research. *Amer. Anthropol.* 68 (2 pt. 2):326–55.

——. 1968. Ganj Dareh Tepe. *Iran* 6:158–60.

——. 1972a. Changes in population pressure in archaeological explanation. *World Archaeol.* 4:5–18.

——. 1972b. The consequences of food production. Reading: Addison-Wesley Module 31.

——. In press. Early food production in northern Africa as seen from southwestern Asia. In Harlan et al., eds., in press.

Smith, P. E. L. and Young, T. C. 1972. The evolution of early agriculture and culture in greater Mesopotamia:a trial model. In Spooner, ed., pp. 1–59.

Smolla, G. 1960. Neolithische Kulturerscheinungen:Studien zur Frage ihrer Herausbildungen. *Antiquitas*(series2)3:1–180.

Soergel, W. 1922. *Die Jagd der Vorzeit.* Jena:G. Fischer.

Solecki, Ralph. 1957. Shanidar Cave. *Sci. Amer.* 197:58–64.

——. 1964a. Zawi Chemi Shanidar, a post-Pleistocene village site in northern Iraq. In *Report of the VIth International Congress on the Quaternary,* 4:405–12.

——. 1964b. Shanidar Cave, a late Pleistocene site in northern Iraq. In *Report of the VIth International Congress on the Quaternary* 4:413–23.

Solecki, Rose, 1969. Milling tools and the epi-paleolithic in the Near East. *Etudes sur le quarternaire dans le monde.* Paris: VIIIe Congres INQUA, pp. 989–94.

Sonneville-Bordes, D. de 1960. *Le paléolithique supérieur en Périgord.*Bordeaux: Delmas.

——. 1963. Upper paleolithic cultures in western Europe. *Science* 142:347–55.

———. 1965. *L'age de la pierre.* Paris:Presses Univ. France.

Spath, C. D. 1973. Plant domestication: the case of *Manihot esculenta. J. Steward Anthropol. Soc.* 5(1):46–67.

Spencer, B. and Gillen, F. J. 1899. *The native tribes of central Australia.* London: Macmillan.

Spencer, R. F. and Jennings, J. D. 1965. *The native Americans; prehistory and ethnology of the North American Indians.* New York:Harper and Row.

Spooner, B., ed. 1972. *Population growth:anthropological implications.* Cambridge: MIT Press.

Stekelis, M. and Bar Josef, B. 1965. Un habitat du paléolithique supérieur à Ein Guev (Israel). Note Préliminaire. *L'Anthropologie* 69:176–83.

Stekelis M. et al. 1969. *Archaeological excavations at Ubeidiya 1964–1966.* Jerusalem: Israel Acad Sci. and Humanities.

Stephens, C. 1975. On mechanisms of population growth. *Current Anthropol.* 16 288–89.

Steward, J. H. 1929. Irrigation without agriculture. *Michigan Acad. Sci. Arts and Letters, Papers* 12:149–56.

———. 1937. *Ancient caves of the Great Salt Lake region.* Bur. Amer. Ethnol. Bull. 116.

———. 1938. *Basin Plateau aboriginal sociopolitical groups.* Bur. Amer. Ethnol. Bull. 120.

———. 1948. The circum-Caribbean tribes: an introduction. In Steward, ed., 1946–59, vol. 4, pp. 1–42.

———. (ed.) 1946–59. *Handbook of South American Indians.* Bur. Amer. Ethnol. Bull. 143.

Stewart, O. C. 1956. Fire as the first great force employed by man. In Thomas et al. eds. pp. 115–33.

Stini, W. A. 1971. Evolutionary implications of changing nutritional patterns in human populations. *Amer. Anthropol.* 73:1019–30.

Stott, D. H. 1962. Cultural and natural checks on population growth. In Montague, ed., *Culture and the Evolution of Man,* pp. 355–76. New York:Oxford Univ. Press.

Street, J. 1969. An evaluation of the concept of carrying capacity. *Prof. Geogr.* 21:1–4.

Strong, W. D. 1935. *An introduction to Nebraska archaeology.* Smithsonian Miscellaneous Collections 93:10.

———. 1957. *Paracas, Nazca and Tiahuanacoid cultural relationships in south coastal Peru.* Soc. Amer. Archaeol. Mem. 13.

Strong, W. D. and Evans, C. 1952. *Cultural stratigraphy in the Viru Valley, northern Peru.* Columbia Univ. Stud. Archaeol. and Ethnol. 4.

Struever, S. 1968. Woodland subsistence-settlement systems in the lower Illinois Valley. In Binford and Binford, eds., pp. 285–312.

———. (ed.) 1971. *Prehistoric Agriculture.* Garden City:Natural Hist. Press.

Struever, S. and Vickery, K. D. 1973. The beginning of cultivation in the Mid-West Riverine area of the United States. *Amer. Athropol.* 75:1197–1220.

Stuckenrath, R. 1963. University of Pennsylvania Radiocarbon dates VI. *Radiocarbon* 3:82–103.

Sturdy, D. A. 1975. Some reindeer economies in prehistoric Europe. In Higgs, ed.

Sturtevant, W. C. 1969. History and ethnography of some west Indian starches. In Ucko and Dimbleby, eds., pp. 177–199.

Suhm, D. et al. 1954. *An introductory handbook of Texas archaeology.* Bull. Texas Archaeol. Soc. 25.

Sulimirski, T. 1970. *Prehistoric Russia.* London:John Baker.

Sullivan, L. R. and Hellman, M. 1925. *The Punín calvarium.* Amer. Mus. Nat. Hist. Anthropol. Papers 23.

Sussman, R. M. 1972. Child transport, family size and increase in human population during the Neolithic. *Current Anthropol.* 13:258–59.

Swanson, E. H. 1962. *The emergence of Plateau culture.* Occasional Papers of the Idaho State College Museum 8.

———. 1966. The geographic foundations of Desert Culture. In *Great Basin anthropological conference, Reno 1964,* pp. 137–46. Univ. Nevada Desert Res. Inst. Technical Reports, Social Science and Humanities Publication 1.

Swedlund, A. ed. 1975. *Population studies in archaeology and biological anthropology: a symposium.* Soc. Amer. Archaeol. Mem. 30.

Tamers, M. A. 1966. IVIC natural radiocarbon measurements 2. *Radiocarbon* 8: 204–12.

———. 1969. IVIC natural radiocarbon measurements 4. *Radiocarbon* 11:396–422.

———. 1970. IVIC natural radiocarbon measurements 5. *Radiocarbon* 12:509–25.

———. 1971. IVIC natural radiocarbon measurements 6. *Radiocarbon* 13:32–44.

Taschini, M. 1964. Il livello mesolitico del Riparo Blanc al Monte Circeo. *Bull. Paleontologia Italiana* 15:65–88.

Taylor, D. W. 1965. The study of Pleistocene non-marine molluscs in North America. In Wright and Frey, eds., pp. 297–311.

Thomas, E. M. 1959. *The harmless people.* London:Secker and Warburg.

Thomas, H. A. 1971. Population dynamics of primitive societies. In Singer, ed., *Is there an optimum level of population.* New York: McGraw Hill.

Thomas, W. L. et al., eds. 1956. *Man's role in changing the face of the earth.* Chicago: Univ. Chicago Press.

Tixier, J. 1963. *Typologie de l'epipaléolithique du Maghreb.* Mem. centre de recherches anthropol., prehist. et ethnogr. Alger.

Tode, A. et al. 1953. Die Untersuchung der Paläolithischen Freilandstation von Salzgitter-Lebenstedt. *Eiszeitalter und Gegenwart* 3:144–220.

Torres, L. M. 1907. Arqueología de la cuenca del Río Paraná. *Rev. Mus. La Plata* 14:53–122.

Towle, M. A. 1961. *The ethnobotany of pre-Columbian Peru.* Chicago:Aldine.

Treganza, A. E. 1952. *Archaeological investigations in the Farmington Reservoir area, Stainislaus County, California.* Rep. Univ. California Anthropol. Rec. 2:2.

Treganza, A. E. and R. F. Heizer 1953. *Additional data on the Farmington complex: a stone implement assemblage of probable early postglacial date from central California.* Rep. Univ. California Archaeol. Surv. 22.

Treistman, J. 1972. *The prehistory of China.* Garden City:Nat. Hist. Press.

———. 1975. The Far East. In Stigler, ed., *Varieties of culture in the Old World,* pp. 106-28. New York:St. Martin's Press.

Tringham, R. 1971. *Hunters, fishers and farmers of eastern Europe 6000-3000 B.C.* London:Hutchinson.

———. 1973. The mesolithic of southeastern Europe. In Koslowski, ed., pp. 551–82.

True, D. L. et al. 1970. Archaeological investigations in northern Chile: Project Tarapacá–preceramic resources. *Amer. Antiquity* 35:170–84.

Tuohy, D. R. 1968. Some early lithic sites in western Nevada. *Eastern New Mexico Univ. Contrib. Anthropol.* 1(4):27–38.

Turnbull, C. 1965. *Wayward Servants.* Garden City: Natural History Press.

———. 1968. The importance of flux in two hunting societies. In Lee and DeVore, eds., pp. 133–37.

Ucko, P. J. and Dimbleby, G. W., eds., 1969. *The domestication and exploitation of plants and animals.* London:Duckworth.

Ucko, P. J. et al., eds. 1972. *Man, Settlement and Urbanism.* London:Duckworth.

Vallois, H. V. 1961. The social life of early man: the evidence from the skeletons. In Washburn, ed., pp. 214–235.

Valoch, K. 1968. Evolution of the Paleolithic in central and eastern Europe. *Current Anthropol.* 9:351–90.

van Campo, M. and Bouchud, J. 1962. Flore accompagnant le squellette d'enfant mousterien découvert au Roc de Marsal, commune du Bugue (Dordogne) et première étude de la Faune du Gisement. *Compt. Rendus Hebdomadaires de l'Académie des Sciences* 254:897–99.

van der Hammen, T. 1957a. The stratigraphy of the late Glacial. *Geol. en Mijinbouw* 19:250–54.

———. 1957b. The age of the Usselo Culture. *Geol. en Mijinbouw* 19:396–97.

———. 1958. Las terrazas del Río Magdalena y la posición estratigráfica de los hallazgos de Garzón. *Rev. Colombiana Antropol.* 6:261–70.

van Loon, M. 1966. Mureybat:an early village in inland Syria. *Archaeol.* 19:215–16.

———. 1968. The Oriental Institute excavations at Mureybat, Syria; preliminary report on the 1965 campaign. *Near Eastern Stud.* 27:265–82.

van Zeist, W. 1967. Late Quaternary vegetation history of western Iran. *Rev. Paleobotany and Palynology* 2:301–11.

———. 1969. Reflections on prehistoric environments in the Near East. In Ucko and Dimbleby, eds., pp. 35–46.

van Zeist, W. and Bottema, S. 1966. Paleobotanical investigations at Ramad. *Ann. Archéol. Arabes Syriennes* 16:179–80.

van Zeist, W. and Casparie, W. A. 1968. Wild einkorn wheat and barley from Tell Mureybat in northern Syria. *Acta Bot. Neerland* 17:44–53.

van Zeist, W. and Wright, H. E. 1963. Preliminary pollen studies at Lake Zeribar, Zagros mountains, southwestern Iran. *Science* 140:65–67.

Vaufrey, R. 1955. *Préhistoire de l'Afrique.* Vol. 1. Pub. Inst. Hautes Études de Tunis.

Vereschagin, N. K. 1967. Primitive hunters and pleistocene extinction in the Soviet Union. In Martin and Wrights, eds., pp. 365–398.

Vermeersch, P. 1970. L'Elkabien. *Chronique d'Egypte* 45:45-68.

Vértes, L. 1966. Des vestiges humains et des outils du Paléolithique inférieur (450,000 av. J.C.) découverts en Hongrie. *Archaeol.* 12:66-71.

Vishnu-Mittre. In press. Changing economy in ancient India. In Reed, ed., in press.

Vita-Finzi, C. and Higgs, E. S. 1970. Prehistoric economy in the Mount Carmel area of Palestine: site catchment analysis. *Proc. Prehist. Soc.* n.s. 36:1-37.

Wallace, W. J. 1954. The Little Sycamore site and the early milling stone cultures of southern California. *Amer. Antiquity* 20:112-23.

———. 1955. A suggested chronology for southern California coastal archaeology. *Southwestern J. Anthropol.* 11:214-30.

———. 1962. Prehistoric cultural development in the southern California deserts. *Amer. Antiquity* 28:172-80.

Walter, H. V. 1948. *The prehistory of the Lagoa Santa region (Minas Gerais.)* Belo Horizonte, Brazil: Oficina Gráficas de Papelería e Tipografía Brazil de Vellosa and Cia., Ltd.

Warren, C. N. 1967. The San Dieguito complex: a review and hypothesis. *Amer. Antiquity* 32:168-85.

Warren, C. N. and Ranere, A. J. 1968. Outside Danger Cave:a review of early man in the Great Basin. *Eastern New Mexico Univ. Contrib. Anthropol.* 1 (4):6-18.

Warren, C. N. and True, D. L. 1961. The San Dieguito complex and its place in California prehistory. In *UCLA Archaeological Survey Annual Report 1960-1961* pp. 246-337.

Washburn, S. L. ed. 1961. *Social Life of Early Man.* Chicago: Aldine.

Washburn, S. L. and Lancaster, C. S. 1968. The evolution of hunting. In Lee and DeVore, eds., pp. 293-303.

Watanabe, H. 1968. Subsistence and ecology of northern food gatherers with special reference to the Ainu. In Lee and DeVore, eds., pp. 69-79.

Waterbolk, H. T. 1962. The lower Rhine basin. In Braidwood and Willey, eds., pp. 227-53.

———. 1968. Food production in prehistoric Europe. *Science* 162:1093-1102.

Watson, P. J. 1966. Prehistoric miners of Salts Cave, Kentucky. *Archaeol.* 19: 237-43.

Watson, R. A. and Watson, P. J. 1969. *Man and Nature.* New York:Harcourt Brace.

Weaver, M. P. 1972. *The Aztecs, Maya and their Predecessors.* New York:Seminar Press.

Webb, W. S. 1946. Indian Knoll Site Oh2, Ohio County, Kentucky. *Univ. Kentucky Rep. Anthropol. and Archaeol.*, vol. 4, no. 3, pt. 1, pp. 111-365.

Webb, W. S. and DeJarnette, D. L. 1942. *An archaeological survey of Pickwick Basin and the adjacent portions of the states of Alabama, Mississippi, and Tennessee.* Bur. Amer. Ethnol. Bull. 129.

———. 1948. *The Flint River site, Mao-48.* Geological survey of Alabama, Mus. Paper 23, Univ. Alabama.

Weberbauer, A. 1936. Phytogeography of the Peruvian Andes. In Macbride, ed.,

1936-60, *Flora of Peru*, pt. 1, no. 1, pp. 13-80. Chicago: Field Museum of Natural History.

Wedel, W. R. 1961. *Prehistoric man on the Great Plains*. Norman:Univ. Oklahoma Press.

———. 1964. The Great Plains. In Jennings and Norbeck, eds., pp. 193-220.

Wedel, W. R. et al. 1968. Mummy Cave:prehistoric record from the Rocky Mountains of Wyoming. *Science* 160:184-85.

Weiss, K. M. 1973. *Demographic models for Anthropology*. Soc. Amer. Archaeol. Mem. 27.

———. 1975. The application of demographic models to anthropological data. *Human Ecol.* 3:87-104.

Wendorf, D. F., ed. 1968. *The prehistory of Nubia*. Dallas: Southern Methodist Univ. Press.

Wendorf, D. F. and Hester, J. J. 1962. Early man's utilization of the Great Plains environment. *Amer. Antiquity* 28:159-71.

Wendorf, D. F. and Said, R. 1967. Paleolithic Remains in Upper Egypt. *Nature* 215: 244-47.

Wendorf, D. F. and Schild, R. In press. The use of ground grain during the late paleolithic of the lower Nile Valley, Egypt. In Harlan et al., eds., in press.

Wendorf, D. F. et al. 1961. *Paleoecology of the Llano Estacado*. Fort Burgwin Res. Center Pub. 1.

Wendorf, D. F. et al. 1970. Late Paleolithic sites in Upper Egypt. *Archaeologia Polona* 12:19-42.

Wendt, W. E. 1964. Die Prakeramische Seidlung am Rio Seco, Peru. *Baessler Archiv* 11:225-75.

———. 1966. Two prehistoric sites in Egyptian Nubia. *Postilla* 102:1-46.

West, R. G. and McBurney, C. B. M. 1954. The Quaternary deposits at Hoxne, Suffolk and their archaeology. *Proc. Prehist. Soc.* 20:131-54.

Wettlaufer, B. and Mayer-Oakes, W. J. 1960. *The Long Creek site*. Saskatchewan Dept. of Nat. Resources, Anthropol Ser. 2.

Whalen, N. M. 1973. Agriculture and the Cochise. *Kiva* 39:89-96.

Wheat, J. B. 1972. *The Olsen-Chubbuck Site: a Paleo Indian bison kill*. Soc. Amer. Archaeol. Mem. 26.

Wheeler, Sir M. 1968. *Early India and Pakistan*. New York:Praeger.

Whitaker, T. W. et al. 1957. Cucurbit materials from three caves near Ocampo, Tamaulipas. *Amer. Antiquity* 22:352-58.

White, L. 1959. *The evolution of culture*. New York:McGraw Hill.

Whitehouse, R. D. 1968. Settlement and economy in southern Italy in the neothermal period. *Proc. Prehist. Soc.* 34:332-67.

———. 1971. The last hunter-gatherers in southern Italy. *World Archaeol.* 2:239-54.

Wilke, P. J. et al. 1972. Harvest selection and domestication in seed plants. *Antiquity* 46:203-09.

Wilkinson, P. F. 1972. Ecosystem models and demographic hypotheses:predation and prehistory in North America. In Clarke, ed., pp. 543-76.

————. 1975. The relevance of musk ox exploitation to the study of prehistoric animal economies. In Higgs, ed., pp. 9–54.

Willey, G. R. 1953. *Prehistoric settlement patterns in the Viru Valley, Peru.* Bur. Amer. Ethnol. Bull. 155.

————. 1966. *An introduction to American archaeology,* vol.1. Englewood Cliffs: Prentice-Hall.

————. 1971. *An introduction to American archaeology,* vol.2. Englewood Cliffs: Prentice-Hall.

Willey, G. R. and Corbett, J. M. 1954. *Early Ancon and early Supe culture:Chavin horizon sites of the central Peruvian coast.* Columbia Univ. Stud. Archaeol. and Ethnol. 3.

Williams, B. J. 1968. The Birhor of India and some comments on band organization. In Lee and DeVore, eds., pp. 126–31.

————. 1974. *A model of band society.* Soc. Amer. Archaeol. Mem. 29.

Williams, S. and Stoltman, J. E. 1965. An outline of southeastern United States prehistory with particular emphasis on the Paleo Indian era. In Wright and Frey, eds., pp. 669–84.

Wilmsen, E. N. 1968. Lithic analysis in paleoanthropology. *Science* 161:982–87.

————. 1970. *Lithic analysis and cultural inference; a Paleo Indian case.* Arizona Univ. Anthropol. Papers 16.

Wing, E. S. In press. Animal domestication in the Andes. In Reed, ed., in press.

Winters, H. D. 1969. *The Riverton culture.* Illinois Archaeol. Surv. Monogr. 1.

Witthoft, J. 1954. A note on fluted point relationships. *Amer. Antiquity* 19: 271–73.

————. 1956. Paleo-Indian cultures in eastern and southeastern North America. In Johnson, ed., Chronology and Development of early cultures in North America. Andover:mimeographed.

Wobst, H. M. 1974. Boundary conditions for Paleolithic social systems:a simulation approach. *Amer. Antiquity* 39:147–77.

————. 1975. The demography of finite populations and the origin of the incest taboo. *Amer. Antiquity* 40:75–81.

————. 1976. Locational relationships in paleolithic society. *J. Human Evolution* 5:49–58.

Woodburn, J. 1968a. An introduction to Hadza ecology. In Lee and DeVore, eds., pp. 49–55.

————. 1968b. Stability and flexibility in Hadza residential groupings. In Lee and DeVore, eds., pp. 103–10.

————. 1972. Ecology, nomadic movement and the composition of the local group among hunters and gatherers: an East African example and its implications. In Ucko et al., eds., pp. 193–206.

Wormington, H. M. 1957. *Ancient man in North America* (4th ed.). Denver Mus. Nat. Hist. Popular Series 4.

————. 1971. Comments on early man in North America. In Shutler, ed., pp. 83–91.

Wright, G. A. 1969. *Obsidian analyses and prehistoric Near Eastern trade:7500– 3500 B.C.* Univ. Michigan Anthropol. Papers 37.

——. 1971. Origins of food production in southwestern Asia:a survey of ideas. *Current Anthropol.* 12:447–77.

Wright, H. E. 1968. Natural environment of early food production north of Mesopotamia. *Science* 161:334–39.

——. 1970. Environmental changes and the origin of agriculture in the Near East. *Bioscience* 20:210–13.

——. In press. Environmental change and the origin of agriculture. In Reed, ed., in press.

Wright, H. E. and Frey, D. G., eds. 1965. *The Quaternary of the United States.* Princeton:Princeton Univ. Press.

Wynne-Edwards, V. C. 1962. *Animal dispersion in relation to social behavior.* Edinburgh:Oliver and Boyd.

Yarnell, R. A. In press. Native plant husbandry north of Mexico. In Reed, ed., in press.

Yellen, J. and Harpending, H. 1972. Hunter-gatherer populations and archaeological inference. *World Archaeol.* 4:244–53.

Yen, D. E. 1971. Construction of the hypothesis for distribution of the sweet potato. In Riley et al., eds., *Man across the sea,* pp. 328–42. Austin: Univ. Texas Press.

Yesner, D. R. 1975. Nutrition and population dynamics of hunter-gatherers. Paper read at the annual meeting of the American Anthropological Association.

Yudkin, J. 1969. Archaeology and the nutritionist. In Ucko and Dimbleby, eds., pp. 547–54.

Zevallos Menendez, C. 1971. *La agricultura en el formativo temprano del Ecuador (cultura Valdivia).* Guayaquil:Casa de la cultura.

Zohary, D. 1969. The progenitors of wheat and barley in relation to domestication and agricultural dispersal in the Old World. In Ucko and Dimbleby, eds., pp. 47–66.

Zohary, D. and Hopf, M. 1973. The domestication of pulses in the Old World. *Science* 182:887–94.

Zubrow, E. 1971. Carrying capacity and dynamic equilibrium in the prehistoric southwest. *Amer. Antiquity* 36:127–38.

——. 1975. *Prehistoric Carrying Capacity:a model.* Menlo Park:Cummings.

Zucchi, A. 1973. Prehistoric human occupation of the western Venezuelan llanos. *Amer. Antiquity* 38:182–89.

INDEX

Abortion, 43–45, 51, 54
Acheulian, 66, 68–69, 90; aquatic resources of, 94–95; gathering during, 91–92, 95, 96; hunting during, 90–91, 93–94, 96–97, 170; population expansion during, 98, 100–02
Achona complex, 240
Aegilops, 139
Africa: plant domestication in, 32, 106–11; population pressure in, 86, 102–06; tool kits in, 66, 92–94, 96, 97–104
Agave, 219
Agricultural revolution, 4. *See also* Domestication, plant
Ahrensburgian culture, 115
Alaka phase, 270
Allan, William, 20
Alleröd, 121
Amahuaca Indians, 33
Amaranthus, 216
Amazon Basin, 223, 225, 250–51, 274–78
Americas. *See* Mexico; North America; South America
Ammerman, A. J., 51
Ampajango complex, 246
Ananatuba ceramic complex, 275
Anderson, E., 23
Andes, 225–27, 250, 260–61
Animals: big game, 90–97, 100, 133, 152, 157, 168–75, 184–86, 209, 248; domestication of, 8, 60, 130–31, 140–41, 259; in Europe, 116–20, 123; migratory, 93, 100, 112–13, 118–20; in North America, 157, 159–60, 165, 168–75, 184, 200–01, 209; in South America, 248, 259, 273. *See also* Extinction, animal; Hunting; *individual species*
Antelian assemblages, 133

Aquatic resources: in Africa, 94–95, 100, 103, 104–05, 109; in Asia, 21, 152–53, 154–55, 156; in Europe, 94, 111, 114–15, 123–25; in Middle East, 135–36, 140; in North America, 21, 174, 180, 189, 201, 206–07, 208, 210–11, 219; in South America, 248–49, 251, 256–60, 263, 265–69, 271, 273–74. *See also* Shellfish
Arachis, 225. *See also* Peanuts
Archaeological cultures. *See* Tool complexes
Archaic economies, 25, 188–99; and Desert Culture, 202–06; evolution of, 157–60, 173–74, 176–81; tool kits of, 190, 196–98, 204–05
Argentina, 261–62, 273
Arnhem Land, 31
Artemisia, 143
Asch, N., 193
Asia, 151–56; aquatic resources in, 21, 152–53, 154–55, 156; and expansion, 86–87, 160–64, 167–68; hunting in, 96–97, 152–53; plant domestication in, 24, 25, 153, 155, 156; tool kits in, 90
Assemblages. *See* Tool complexes; Tool kits
Austral, A. G., 245–46
Australian aborigines, 20, 62–63
Australopithicine populations, 86
Avocados, in Mexico, 216
Axes, in Middle East, 134
Ayacucho, 236, 237, 245, 253–54

Baboons, in Africa, 93
Balcones phase tools, 197
Baradostian assemblage, 133
Barfield, L., 243
Barker, G. W. W., 119, 120

331

PRAISE FOR *RISKY GOSPEL*

"Some Christians are paralyzed by fear or indecision or constant second-guessing or by the blur of activity in their lives. In this book, Owen Strachan points out what's immobilizing you right now, and to give you the gospel coaching to run the race again. Read this book to stir up courage in yourself, or to equip you to do so for a brother or sister in need."

—RUSSELL D. MOORE, PhD, PRESIDENT, ETHICS AND RELIGIOUS LIBERTY COMMISSION, SOUTHERN BAPTIST CONVENTION

"Owen Strachan takes us through a journey of the struggles of the ordinary Christian life. Refreshingly honest, he shares about his own current and past temptation to fear and shrink back all while challenging the reader to look to God's Word and the gospel for the strength to take risks. In his words, 'Jesus came to embolden us not to anesthetize us.' With humor and wisdom, Owen casts a vision for a risky life in our faith, marriages, work, churches, and culture. In the end we see that risking it all for the gospel isn't risking at all."

—TRILLIA NEWBELL, AUTHOR OF *UNITED: CAPTURED BY GOD'S VISION FOR DIVERSITY*

"My favorite books are those that lead me to want to thank the authors for writing them. This is one such work. Owen Strachan clarifies a compelling vision for fruitfulness and faithfulness in light of the work of Christ, but he does not stop there. He exhorts us to reach our full potential in Him by embracing the call to risk and shows that we can boldly live out this gospel calling no matter our walk of life. Thank you, Owen, for such an encouraging book."

—JOE CRISPIN, FORMER PROFESSIONAL BASKETBALL PLAYER; WRITER AND SPEAKER

"Could it be that the safest place in the world is in risking everything for the gospel? 'Risky Christianity is Christianity,' argues Strachan. And so he calls us to give what we cannot keep to gain what we cannot lose. Many popular books and leaders will lead you down a risky and wrong path. But Strachan offers a reliable and readable account of what it means to risk everything for the gospel in life's major arenas. Please read, Christian, and learn to think and live well."

—JONATHON LEEMAN, AUTHOR OF *REVERBERATION* AND *THE CHURCH AND THE SURPRISING OFFENSE OF GOD'S LOVE*

"Here's a book that would have done me a lot of good in my teens and twenties. But even a little later in life I'm grateful for how Owen impressed upon me the privilege of risking a life that's secure in God. Read this book and embark on a life of 'empowered dependence' and 'winsome courage.'"

—COLLIN HANSEN, EDITORIAL DIRECTOR
FOR THE GOSPEL COALITION

"This is a different kind of book from those around it. Whether you're a tired parent or a wide-eyed millennial (or some combination of both), Owen is writing to you. The vision of God in these pages is so glorious and the grasp of the practical so relevant, that you will be compelled to live bold for the fame of Jesus. There is no shortcut to Christian maturity, but *Risky Gospel* touches all the bases. Read it. Trust God. Build something."

—JONATHAN PARNELL,
CONTENT STRATEGIST, DESIRINGGOD.ORG

"*Risky Gospel* isn't a book for perfect Christians who have it all together. Nor is it for those seeking merely to numb the pain. With humility and humor, as a friend, Strachan shows us the awesomeness of God, reminding us that Jesus came to embolden—not anesthetize—his followers. *Risky Gospel* is a rock inviting readers to take a flying leap of faith."

—ERIC TEETSEL, EXECUTIVE DIRECTOR,
MANHATTAN DECLARATION

"There's a rumbling, thunderous joy rolling through the pages of *Risky Gospel*. Owen candidly describes how the Christian life is to be one of boldness, conviction, and intentionality. The kind of zeal he's talking about isn't haphazard and reactionary, but it flows from living in light of the joy-inciting gospel of Jesus Christ. What a timely message for the Church around the world!"

—GLORIA FURMAN, CROSS-CULTURAL WORKER,
AUTHOR OF *GLIMPSES OF GRACE*

"I love this book. I love this book because it calls you to risk and do things that matter. I love this book because it calls you to risk with the strength available in Christ. And I love this book because it not only tells you what to accomplish and how, but it actually gets you excited about doing it. I'm excited about sharing this book with those in my ministry. You should be too."

—HEATH LAMBERT, EXECUTIVE DIRECTOR OF NANC,
ASSOCIATE PROFESSOR OF BIBLICAL COUNSELING AT THE
SOUTHERN BAPTIST THEOLOGICAL SEMINARY/BOYCE COLLEGE